Graduate Texts in Mathematics **98**

Springer
New York
Berlin
Heidelberg
Barcelona
Budapest
Hong Kong
London
Milan
Paris
Santa Clara
Singapore
Tokyo

Graduate Texts in Mathematics

continued after index

Theodor Bröcker
Tammo tom Dieck

Representations of Compact Lie Groups

With 24 Illustrations

Springer

Theodor Bröcker
Universität Regensburg
Fachbereich Mathematik
Universitätsstrasse 31
8400 Regensburg
Federal Republic of Germany

Tammo tom Dieck
Mathematisches Institut
Universität Göttingen
Bunsenstrasse 3–5
3400 Göttingen
Federal Republic of Germany

ISBN 978-3-642-05725-0

Mathematics Subject Classification (1991): 22E47

Library of Congress Cataloging in Publication Data
Bröcker, Theodor.
Representations of compact lie groups.
(Graduate texts in mathematics; 98)
Bibliography: p.
Includes indexes.
1. Lie groups. 2. Representations of groups.
I. Dieck, Tammo tom. II. Title. III. Series.
QA387.B68 1985 512′.55 84-20282

Springer-Verlag Berlin Heidelberg New York
a part of Springer Science+Business Media

Auch ging es mir, wie jedem, der reisend oder lebend mit Ernst gehandelt, daß ich in dem Augenblicke des Scheidens erst einigermaßen mich wert fühlte, hereinzutreten. Mich trösteten die mannigfaltigen und unentwickelten Schätze, die ich mir gesammlet.

G.

Preface

This book is based on several courses given by the authors since 1966. It introduces the reader to the representation theory of compact Lie groups.

We have chosen a geometrical and analytical approach since we feel that this is the easiest way to motivate and establish the theory and to indicate relations to other branches of mathematics. Lie algebras, though mentioned occasionally, are not used in an essential way. The material as well as its presentation are classical; one might say that the foundations were known to Hermann Weyl at least 50 years ago.

Prerequisites to the book are standard linear algebra and analysis, including Stokes' theorem for manifolds. The book can be read by German students in their third year, or by first-year graduate students in the United States.

Generally speaking the book should be useful for mathematicians with geometric interests and, we hope, for physicists.

At the end of each section the reader will find a set of exercises. These vary in character: Some ask the reader to verify statements used in the text, some contain additional information, and some present examples and counter-examples. We advise the reader at least to read through the exercises.

The book is organized as follows. There are six chapters, each containing several sections. A reference of the form III, (6.2) refers to Theorem (Definition, etc.) (6.2) in Section 6 of Chapter III. The roman numeral is omitted whenever the reference concerns the chapter where it appears. References to the Bibliography at the end of the book have the usual form, e.g. Weyl [1].

Naturally, we would have liked to write in our mother tongue. But we hope that our English will be acceptable to a larger mathematical community, although any personal manner may have been lost and we do not feel competent judges on matters of English style.

Arunas Liulevicius, Wolfgang Lück, and Klaus Wirthmüller have read the manuscript and suggested many improvements. We thank them for their generous help. We are most grateful to Robert Robson who translated part of the German manuscript and revised the whole English text.

Contents

CHAPTER I

Lie Groups and Lie Algebras

In this chapter we explain what a Lie group is and quickly review the basic concepts of the theory of differentiable manifolds. The first section illustrates the notion of a Lie group with classical examples of matrix groups from linear algebra. The spinor groups are treated in a separate section, §6, but the presentation of the general theory of representations in this book presupposes no knowledge of spinor groups. They only appear as examples which, although important, may be skipped. In §§2, 3, and 4 we construct the exponential map and exploit it to obtain elementary information about the structure of subgroups and quotients, and in §5 we explain how to construct an invariant integral using differential forms. We quote Stokes' theorem to get a result about mapping degrees which we shall use in Chapter IV.

1. The Concept of a Lie Group and the Classical Examples

The concept of a Lie group arises naturally by merging the algebraic notion of a group with the geometric notion of a differentiable manifold. However, the classical examples, as well as the methods of investigation, show the theory of Lie groups to be a significant geometric extension of linear algebra and analytic geometry.

(1.1) Definition. A ***Lie group*** is a differentiable manifold G which is also a group such that the group multiplication

$$\mu: G \times G \to G$$

(and the map sending g to g^{-1}) is a differentiable map. A ***homomorphism of Lie groups*** is a differentiable group homomorphism between Lie groups.

For us the word ***differentiable*** means infinitely often differentiable. Throughout this book we use the words differentiable, ***smooth***, and C^∞ as synonymous.

The identity map on a Lie group is a homomorphism, and composing homomorphisms yields a homomorphism—Lie groups and homomorphisms form a category. One may define the usual categorical notions: in particular, an ***isomorphism*** (denoted by $\cong$) is an invertible homomorphism.

We will use e or 1 to denote the identity element of G, although we will sometimes use E when considering a matrix group and 0 when considering an additive abelian group.

The reader should know what a group is, and the concept of a differentiable manifold should not be new. Nonetheless, we review a few facts about manifolds.

(1.2) Definition. An ***n-dimensional*** (differentiable) ***manifold*** M^n is a Hausdorff topological space with a countable (topological) basis, together with a maximal ***differentiable atlas***. This atlas consists of a family of ***charts*** $h_\lambda\colon U_\lambda \to U'_\lambda \subset \mathbb{R}^n$, where the ***domains of the charts***, $\{U_\lambda\}$, form an open cover of M^n, the U'_λ are open in $\mathbb{R}^n$, the charts (***local coordinates***) h_λ are homeomorphisms, and every ***change of coordinates*** $h_{\lambda\mu} = h_\mu \circ h_\lambda^{-1}$ is differentiable on its domain of definition $h_\lambda(U_\lambda \cap U_\mu)$.

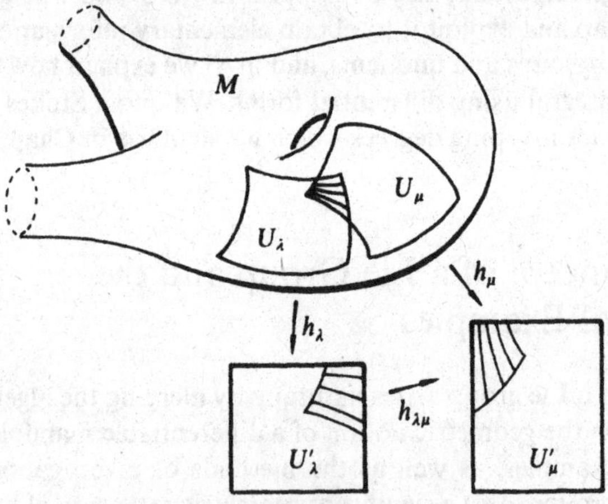

Figure 1

The atlas is maximal in the sense that it cannot be enlarged to another differentiable atlas by adding more charts, so any chart which could be added to the atlas in a consistent fashion is already in the atlas.

A continuous map $f\colon M \to N$ of differentiable manifolds is called ***differentiable*** if, after locally composing with the charts of M and N, it induces a differentiable map of open subsets of Euclidean spaces.

The reader may find an elementary introduction to the basic concepts of differentiable manifolds in the books by Bröcker and Jänich [1] or Guillemin and Pollak [1], but we will assume little in the way of background. We now turn to the examples which, as previously mentioned, one more or less knows from linear algebra.

(1.3) Every finite-dimensional vector space with its additive group structure is a Lie group in a canonical way. Thus, up to isomorphism, we get the groups $\mathbb{R}^n$, $n \in \mathbb{N}_0$.

(1.4) The ***torus*** $\mathbb{R}^n/\mathbb{Z}^n = (\mathbb{R}/\mathbb{Z})^n \cong (S^1)^n$ is a Lie group. Here $S^1 = \{z \in \mathbb{C} \mid |z| = 1\}$ is the unit circle viewed as a multiplicative subgroup of $\mathbb{C}$, and the isomorphism $\mathbb{R}/\mathbb{Z} \to S^1$ is induced by $t \mapsto e^{2\pi i t}$. The n-fold product of the circle with itself has the structure of an abelian Lie group due to the following general remark:

(1.5) If G and H are Lie groups, so is $G \times H$ with the direct product of the group and manifold structures on G and H.

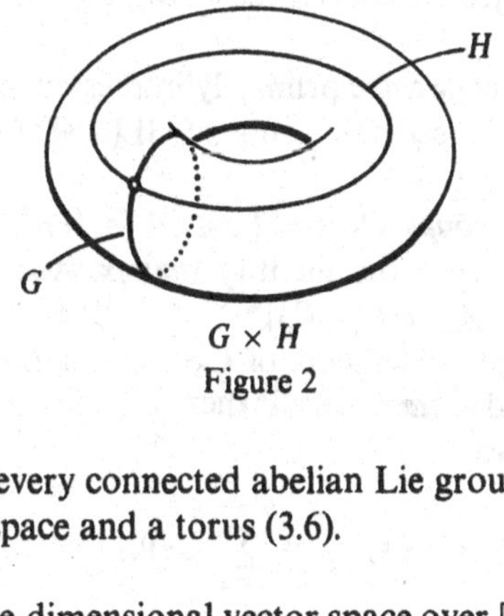

$G \times H$
Figure 2

It will turn out that every connected abelian Lie group is isomorphic to the product of a vector space and a torus (3.6).

(1.6) Let V be a finite-dimensional vector space over $\mathbb{R}$ or $\mathbb{C}$. The set $\mathrm{Aut}(V)$ of linear automorphisms of V is an open subset of the finite-dimensional vector space $\mathrm{End}(V)$ of linear maps $V \to V$, because $\mathrm{Aut}(V) = \{A \in \mathrm{End}(V) \mid \det(A) \neq 0\}$ and the determinant is a continuous function. Thus $\mathrm{Aut}(V)$ has the structure of a differentiable manifold. After the introduction of coordinates, the group operation of $\mathrm{Aut}(V)$ is matrix multiplication, which is algebraic and hence differentiable. Therefore $\mathrm{Aut}(V)$ has a canonical structure as a Lie group, and we get the groups

$$GL(n, \mathbb{R}) = \mathrm{Aut}_{\mathbb{R}}(\mathbb{R}^n) \quad \text{and} \quad GL(n, \mathbb{C}) = \mathrm{Aut}_{\mathbb{C}}(\mathbb{C}^n).$$

Linear maps $\mathbb{R}^n \to \mathbb{R}^k$ may be described by $(k \times n)$-matrices, and, in particular, $GL(n, \mathbb{R})$ is canonically isomorphic to the group of invertible $(n \times n)$-matrices. Thus we will think of $GL(n, \mathbb{R})$, its classical subgroups $SL(n, \mathbb{R})$, $O(n)$, $SO(n)$, ..., and $GL(n, \mathbb{C})$ as matrix groups.

The group $GL(n, \mathbb{R})$ has two connected components on which the sign of the determinant is constant. Automorphisms with positive determinant form an open and closed subgroup $GL^+(n, \mathbb{R})$. It is connected because performing elementary row and column operations which do not involve multiplication by a negative scalar does not change components.

These linear groups yield many others once one knows, as we will show in (3.11) and (4.5), that a closed subgroup of a Lie group and the quotient of a Lie group by a closed normal subgroup inherit Lie group structures.

(1.7) As a result we get the groups

$$SL(n, \mathbb{R}) = \{A \in GL(n, \mathbb{R}) \mid \det(A) = 1\}, \quad \text{and}$$
$$SL(n, \mathbb{C}) = \{A \in GL(n, \mathbb{C}) \mid \det(A) = 1\},$$

the ***special linear groups*** over $\mathbb{R}$ and $\mathbb{C}$. We also get the ***projective groups***

$$PGL(n, \mathbb{R}) = GL(n, \mathbb{R})/\mathbb{R}^* \quad \text{and} \quad PGL(n, \mathbb{C}) = GL(n, \mathbb{C})/\mathbb{C}^*,$$

where $\mathbb{R}^* = \mathbb{R}\backslash\{0\}$ and $\mathbb{C}^* = \mathbb{C}\backslash\{0\}$ are embedded as the subgroups of scalar multiples of the identity matrix. The projective groups are groups of transformations of projective spaces, see (1.16), Ex. 11.

In this book, however, we are primarily interested in compact groups, so we recall the following closed subgroups of $GL(n, \mathbb{R})$ from linear algebra:

(1.8) The ***orthogonal groups*** $O(n) = \{A \in GL(n, \mathbb{R}) \mid {}^tA \cdot A = E\}$, where ${}^tA$ denotes transpose and E is the identity matrix. Analogously there is the ***unitary group*** $U(n) = \{A \in GL(n, \mathbb{C}) \mid {}^*A \cdot A = E\}$, where ${}^*A = {}^t\bar{A}$ is the conjugate transpose of A. Elements of $O(n)$ are called ***orthogonal*** and elements of $U(n)$ are called ***unitary***. On $\mathbb{R}^n$ there is an ***inner product***, the ***standard Euclidean scalar product***

$$\langle x, y\rangle = \sum_{\nu=1}^{n} x_\nu \cdot y_\nu,$$

and on $\mathbb{C}^n$ one has the ***standard Hermitian product***

$$\langle x, y\rangle = \sum_{\nu=1}^{n} x_\nu \cdot \bar{y}_\nu.$$

$O(n)$ (resp. $U(n)$) consists of those automorphisms which preserve the inner product on $\mathbb{R}^n$ (resp. $\mathbb{C}^n$), i.e., those automorphisms A for which

$$\langle Ax, Ay\rangle = \langle x, y\rangle.$$

$O(n)$ is also split into two connected components by the values ± 1 of the determinant, and one of these is the ***special orthogonal group***

$$SO(n) = \{A \in O(n) \mid \det(A) = 1\}.$$

The connectedness of SO(n) follows from (4.7), but one may also, for example, join $A \in \mathrm{SO}(n)$ to E by an arc in $\mathrm{GL}^+(n, \mathbb{R})$ and apply Gram–Schmidt orthogonalization to this arc (see Lang [2], VI, §2).

The ***special unitary group*** is defined analogously:

$$\mathrm{SU}(n) = \{A \in \mathrm{U}(n) \mid \det(A) = 1\}.$$

These groups are compact, being closed and bounded in the finite-dimensional vector space End(V).

(1.9) ***Quaternions.*** There is up to isomorphism only one proper finite field extension of $\mathbb{R}$, namely the field $\mathbb{C}$ of complex numbers. There is, however, a skew field containing $\mathbb{C}$ of complex dimension 2 and real dimension 4, called the ***quaternion algebra*** $\mathbb{H}$, which may be described as follows: The $\mathbb{R}$-algebra $\mathbb{H}$ is the algebra of (2×2) complex matrices of the form

$$\begin{bmatrix} a & b \\ -\bar{b} & \bar{a} \end{bmatrix},$$

with matrix addition and multiplication.

If such a matrix is nonzero, its determinant, $|a|^2 + |b|^2$, is nonzero, and its inverse is another matrix of the same form. Thus every nonzero $h \in \mathbb{H}$ has a multiplicative inverse, so $\mathbb{H}$ is a ***division algebra*** (also called skew field). We consider $\mathbb{C}$ as a subfield of $\mathbb{H}$ via the canonical embedding $\mathbb{C} \to \mathbb{H}$ given by

$$c \mapsto \begin{bmatrix} c & 0 \\ 0 & \bar{c} \end{bmatrix},$$

so we may think of $\mathbb{C}$, and therefore also $\mathbb{R}$, as subfields of $\mathbb{H}$.

The field $\mathbb{R}$ is the center of $\mathbb{H}$. For the center, $Z = \{z \in \mathbb{H} \mid zh = hz \text{ for all } h \in \mathbb{H}\}$, certainly contains $\mathbb{R}$, and, were Z larger than $\mathbb{R}$, then Z as a proper finite field extension of $\mathbb{R}$, would be isomorphic to $\mathbb{C}$. But $Z \neq \mathbb{H}$, so choosing $x \in \mathbb{H}$ with $x \notin Z$ we get a proper finite (commutative!) field extension $Z(x) \cong \mathbb{C}(x)$, which is impossible; see also (1.16), Ex. 14.

The algebra $\mathbb{H}$ is a complex vector space, $\mathbb{C}$ acting by left multiplication. As such it has a ***standard basis*** comprised of two elements

$$1 = \begin{bmatrix} 1 & 0 \\ 0 & 1 \end{bmatrix} \quad \text{and} \quad j = \begin{bmatrix} 0 & 1 \\ -1 & 0 \end{bmatrix},$$

with the rules for multiplication

$$zj = j\bar{z} \quad \text{for } z \in \mathbb{C} \text{ and } j^2 = -1.$$

This basis gives the ***standard isomorphism*** of complex vector spaces

$$\mathbb{C}^2 \to \mathbb{H}, \qquad (a, b) \mapsto a + bj = \begin{bmatrix} a & b \\ -\bar{b} & \bar{a} \end{bmatrix}.$$

The quaternion algebra $\mathbb{H}$ has a ***conjugation anti-automorphism***

$$\imath\colon \mathbb{H} \to \mathbb{H}, \qquad h = a + bj \mapsto \imath(h) = \bar{h} = \bar{a} - bj, \qquad a, b \in \mathbb{C}.$$

Viewing h as a complex matrix, $\imath(h) = {}^*h$, where ${}^*h$ is the adjoint matrix. Conjugation is $\mathbb{R}$-linear, coincides with complex conjugation on $\mathbb{C}$, and obeys the laws

$$\imath(h \cdot k) = \imath(k) \cdot \imath(h) \quad \text{and} \quad \imath^2 = \text{id}.$$

The ***norm*** on $\mathbb{H}$ is defined analogously to the complex norm by

$$\mathrm{N}(h) = h \cdot \bar{h} = \bar{h} \cdot h.$$

Note that $\mathrm{N}(a + bj) = |a|^2 + |b|^2$ is real and nonnegative, and that $\mathrm{N}(h) = 0$ precisely if $h = 0$. As with the complex numbers, the multiplicative inverse of $h \in \mathbb{H}$ is $\bar{h} \cdot \mathrm{N}(h)^{-1}$, and if $h \in \mathbb{C}$, $\mathrm{N}(h) = |h|^2$. If one views h as a (2×2) complex matrix, $\mathrm{N}(h) = \det(h)$.

As a real vector space $\mathbb{H}$ has a ***standard basis*** consisting of the four elements

$$1 = \begin{bmatrix} 1 & 0 \\ 0 & 1 \end{bmatrix}, \quad i = \begin{bmatrix} i & 0 \\ 0 & -i \end{bmatrix}, \quad j = \begin{bmatrix} 0 & 1 \\ -1 & 0 \end{bmatrix}, \quad \text{and} \quad k = \begin{bmatrix} 0 & i \\ i & 0 \end{bmatrix},$$

with rules for multiplication

$$i^2 = j^2 = k^2 = -1,$$
$$ij = -ji = k, \qquad jk = -kj = i, \quad \text{and} \quad ki = -ik = j.$$

The quaternions $ai + bj + ck$, $a, b, c \in \mathbb{R}$, are called ***pure*** quaternions, and, as a real vector space, $\mathbb{H}$ splits into $\mathbb{R}$ and the space of pure quaternions isomorphic to $\mathbb{R}^3$. Each $h \in \mathbb{H}$ has unique expression as $h = r + q$ with $r \in \mathbb{R}$ and $q \in \mathbb{R}^3$ (pure). Conjugation may be expressed in this notation as

$$\imath(r + q) = r - q,$$

and therefore $\mathrm{N}(r + q) = r^2 - q^2$. Thus on the subspace $\mathbb{R}^3$ of pure quaternions, $\mathrm{N}(q) = -q^2$, so q^2 is a nonpositive real number. The pure quaternions may be characterized by this property using only the ring structure of $\mathbb{H}$. If $h = r + q$, $r \in \mathbb{R}$, q pure, then $h^2 = r^2 + q^2 + 2rq$ is real if and only if $r = 0$ or $q = 0$, and is nonpositive real if and only if $r = 0$.

With the ***standard isomorphism*** of real vector spaces $\mathbb{R}^4 \to \mathbb{H}$ sending (a, b, c, d) to $a + bi + cj + dk$, the norm on $\mathbb{H}$ corresponds to the Euclidean norm, the square of the Euclidean absolute value on $\mathbb{R}^4$. With the standard isomorphism $\mathbb{C}^2 \cong \mathbb{H}$, the quaternionic norm corresponds to the standard Hermitian norm on $\mathbb{C}^2$. The group

$$\mathrm{Sp}(1) = \{h \in \mathbb{H} \mid N(h) = 1\}$$

is called the ***quaternion group***, or group of unit quaternions. In matrix notation Sp(1) consists of the matrices

$$\begin{bmatrix} a & b \\ -\bar{b} & \bar{a} \end{bmatrix}, \qquad a, b \in \mathbb{C}, \qquad |a|^2 + |b|^2 = 1,$$

and thus is the same as SU(2). The standard isomorphism $\mathbb{H} \cong \mathbb{R}^4$ identifies Sp(1) with the unit sphere, S^3. This group is the universal covering of the rotation group SO(3), see (6.17), (6.18), and plays an important role in theoretical physics. We will meet the quaternion algebra again in §6 in the guise of the Clifford algebra C_2.

(1.10) The $\mathbb{H}$-Linear Groups. The basic statements of linear algebra may also be formulated for skew fields. An endomorphism $\varphi\colon \mathbb{H}^n \to \mathbb{H}^n$, which is linear with respect to multiplication on the left by scalars from $\mathbb{H}$, may be described by an $(n \times n)$-matrix $(\varphi_{\lambda\nu})$ with coefficients in $\mathbb{H}$ as follows: If $e_\nu \in \mathbb{H}^n$ is the νth unit vector, then $\varphi_{\lambda\nu}$ is defined by $\varphi(e_\nu) = \sum_\lambda \varphi_{\lambda\nu} e_\lambda$. Thus if $h = (h_1, \ldots, h_n) \in \mathbb{H}^n$, we have

$$\varphi(h) = \varphi\left(\sum_\nu h_\nu e_\nu\right) = \sum h_\nu \varphi(e_\nu) = \sum_{\nu,\lambda} h_\nu \varphi_{\lambda\nu} e_\lambda,$$

and

$$\varphi(h)_\lambda = \sum_\nu h_\nu \varphi_{\lambda\nu}.$$

Consequently we may canonically identify the $\mathbb{H}$-linear group

$$\mathrm{GL}(n, \mathbb{H}) = \mathrm{Aut}_{\mathbb{H}}(\mathbb{H}^n)$$

with the group of invertible $(n \times n)$-matrices with coefficients in $\mathbb{H}$, as we did with linear groups earlier. In this case matrices are multiplied as follows:

$$(\psi_{\mu\lambda}) \cdot (\varphi_{\lambda\nu}) = \left(\sum_\lambda \varphi_{\lambda\nu} \cdot \psi_{\mu\lambda}\right).$$

An $\mathbb{H}$-endomorphism of $\mathbb{H}^n$ is invertible precisely if it is invertible as an $\mathbb{R}$-linear map, so, as before, $\mathrm{Aut}_{\mathbb{H}}(\mathbb{H}^n)$ is open in the $\mathbb{H}$-vector space $\mathrm{End}_{\mathbb{H}}(\mathbb{H}^n)$ and $\mathrm{GL}(n, \mathbb{H})$ is a $4n^2$-dimensional Lie group.

The standard isomorphism $\mathbb{H} = \mathbb{C} + \mathbb{C}j = \mathbb{C}^2$ induces a standard isomorphism of complex vector spaces

$$\mathbb{H}^n = \mathbb{C}^n + \mathbb{C}^n \cdot j = \mathbb{C}^n \oplus \mathbb{C}^n = \mathbb{C}^{2n},$$

and, accordingly, an $\mathbb{H}$-linear endomorphism φ of $\mathbb{H}^n$ may be thought of as a special kind of $\mathbb{C}$-linear endomorphism of $\mathbb{C}^{2n}$:

$$\mathbb{C}^n \oplus \mathbb{C}^n = \mathbb{C}^n + \mathbb{C}^n \cdot j \xrightarrow{\varphi} \mathbb{C}^n + \mathbb{C}^n \cdot j = \mathbb{C}^n \oplus \mathbb{C}^n,$$

namely, one which commutes with the $\mathbb{R}$-linear (but not $\mathbb{C}$-linear!) map

$$j\colon \mathbb{C}^n \oplus \mathbb{C}^n \to \mathbb{C}^n \oplus \mathbb{C}^n,$$
$$(u, v) = u + vj \mapsto j(u + vj) = -\bar{v} + \bar{u}j = (-\bar{v}, \bar{u})$$

coming from left multiplication by j. The condition that φ commute with left multiplication by j is equivalent to the condition that, as an endomorphism of $\mathbb{C}^n \oplus \mathbb{C}^n$, the map φ is given by a matrix of the form

$$\begin{bmatrix} A & -\bar{B} \\ B & \bar{A} \end{bmatrix}, \qquad A, B \in \operatorname{End}_{\mathbb{C}}(\mathbb{C}^n).$$

Note that an $\mathbb{H}$-linear endomorphism may be represented uniquely in the form $A + Bj$, where A and B are complex $(n \times n)$-matrices.

(1.11) There is an inner product on $\mathbb{H}^n$, the ***standard symplectic scalar product***: If $h = (h_1, \ldots, h_n)$ and $k = (k_1, \ldots, k_n)$, then

$$\langle h, k \rangle = \sum_{v=1}^{n} h_v \bar{k}_v.$$

The corresponding norm is given by $\langle h, h \rangle = \sum_v h_v \bar{h}_v = \sum_v N(h_v) \geq 0$. The ***symplectic group***, $\mathrm{Sp}(n)$, is the group of norm-preserving automorphisms of $\mathbb{H}^n$:

$$\mathrm{Sp}(n) = \{\varphi \in \mathrm{GL}(n, \mathbb{H}) \mid N(\varphi(h)) = N(h) \text{ for all } h \in \mathbb{H}^n\}.$$

A norm-preserving automorphism leaves the inner product invariant ((1.16), Ex. 10). If we identify $\mathbb{H}^n$ with $\mathbb{C}^{2n}$ as above, the standard norms on $\mathbb{H}^n$ and $\mathbb{C}^{2n}$ correspond, so $\mathrm{Sp}(n)$ is identified with the subgroup of $\mathrm{U}(2n)$ of matrices of the form

$$\begin{bmatrix} A & -\bar{B} \\ B & \bar{A} \end{bmatrix} \in \mathrm{U}(2n), \qquad A, B \in \operatorname{End}(\mathbb{C}^n).$$

Thus we will view $\mathrm{Sp}(n)$ as a group of complex matrices. A complex $(2n \times 2n)$-matrix in $\mathrm{Sp}(n)$ is called a ***symplectic matrix***.

(1.12) The map $\mathbb{C}^{2n} = \mathbb{H}^n \xrightarrow{j} \mathbb{H}^n = \mathbb{C}^{2n}$ from (1.10), which sends $(u, v) = u + vj$ to $(-\bar{v}, \bar{u}) = j(u + vj)$ is not $\mathbb{C}$-linear. It is composed of the $\mathbb{C}$-linear map induced by right multiplication by j followed by complex conjugation $c\colon \mathbb{C}^{2n} \to \mathbb{C}^{2n}$, where $c(w) = \bar{w}$. Right multiplication by j may be written as

$$J\colon \mathbb{C}^{2n} \to \mathbb{C}^{2n}, \qquad (u, v) \mapsto (-v, u)$$

and expressed by the matrix

$$J = \begin{bmatrix} 0 & -E \\ E & 0 \end{bmatrix}, \qquad E = \text{identity matrix in } \mathrm{GL}(n, \mathbb{C}).$$

Hence a unitary matrix $A \in \mathrm{U}(2n)$ is symplectic if and only if $AcJ = cJA$. Since $Ac = c\bar{A}$, this means $c\bar{A}J = cJA$, and therefore $\bar{A}J = JA$. And because $A \in \mathrm{U}(2n)$, ${}^tA = \bar{A}^{-1}$, so we end up with

$${}^tAJA = J.$$

This equation expresses the fact that the linear transformation A fixes the bilinear form

$$(u, v) \mapsto {}^tuJv,$$

defined by the matrix J.

Dropping the condition that A be unitary gives the ***complex symplectic group***

$$\mathrm{Sp}(n, \mathbb{C}) = \{A \in \mathrm{GL}(2n, \mathbb{C}) \mid {}^tAJA = J\}.$$

(1.13) As a matter of principle, one should always consider the three cases $\mathbb{R}$, $\mathbb{C}$, and $\mathbb{H}$, and these are the only three finite-dimensional real division algebras. This is the content of the Frobenius theorem. For a proof see Jacobson [2], 7.7, p. 430. Further information and historical remarks on quaternions may be found in Chapters 6 and 7 by Koecher and Remmert in Ebbinghaus *et al.* [1].

We have defined subgroups

$$\mathrm{GL}(n, \mathbb{H}) \supset \mathrm{Sp}(n), \text{ symplectic scalar product},$$

$$\mathrm{GL}(n, \mathbb{C}) \supset \mathrm{U}(n), \text{ Hermitian scalar product},$$

$$\mathrm{GL}(n, \mathbb{R}) \supset \mathrm{O}(n), \text{ Euclidean scalar product},$$

in a completely analogous fashion. We refer to each of the scalar products involved simply as ***inner product***.

More generally, to every bilinear map of a finite-dimensional real vector space V into a real vector space H

$$V \times V \to H, \qquad (v, w) \mapsto \langle v, w \rangle,$$

there belongs a Lie group $G = \{A \in \mathrm{Aut}(V) \mid \langle Av, Aw \rangle = \langle v, w \rangle$ for all $v, w \in V\}$. Many important Lie groups with a geometric flavor arise in this way, for example the ***Lorentz group***, which comes from the scalar product on $\mathbb{R}^4$

$$\langle x, y \rangle = x_1y_1 + x_2y_2 + x_3y_3 - x_4y_4.$$

Some of the linear groups with which we shall be concerned are depicted, together with some of their inclusions, in the following diagram.

$$
\textbf{(1.14)}\qquad
\begin{array}{ccccccccc}
 & & & & \mathrm{GL}^+(2n, \mathbb{R}) & & & & \\
 & & & & \uparrow & & & & \\
\mathrm{GL}^+(n, \mathbb{R}) & \to & \mathrm{GL}(n, \mathbb{R}) & \to & \mathrm{GL}(n, \mathbb{C}) & \to & \mathrm{GL}(n, \mathbb{H}) & \to & \mathrm{GL}(2n, \mathbb{C}) \\
\uparrow & & \uparrow & & \uparrow & & \uparrow & & \uparrow \\
\mathrm{SO}(n) & \to & \mathrm{O}(n) & \to & \mathrm{U}(n) & \to & \mathrm{Sp}(n) & \to & \mathrm{Sp}(n, \mathbb{C}) \\
 & & & & \downarrow & & \downarrow & & \\
 & & & & \mathrm{SO}(2n) & & \mathrm{U}(2n) & &
\end{array}
$$

(1.15) Finally, we should point out that every finite group is a zero-dimensional compact Lie group. Many things we will say about representations in general are of interest in the special case of finite groups. We will encounter the following important finite groups:

The ***symmetric groups***

$$\mathrm{S}(n) = \text{the group of all permutations of } \{1, \ldots, n\}.$$

The ***alternating groups***

$$\mathrm{A}(n) = \text{the group of all even permutations of } \{1, \ldots, n\}.$$

The ***cyclic groups***

$$\mathbb{Z}/n = \mathbb{Z}/n\mathbb{Z} = \text{the cyclic group of order } n.$$

(1.16) Exercises

1. Let G be a Lie group. Use the fact that $\mu\colon G \times G \to G$ is differentiable to show that the map $G \to G$, $g \mapsto g^{-1}$, is differentiable. *Hint*: Use the implicit function theorem in a neighborhood of the unit element.

2. Show that $\mathrm{O}(n)$ is a Lie group as follows: Let S be the space of symmetric matrices and consider the map $f\colon \mathrm{End}(\mathbb{R}^n) \to S$ defined by $f(A) = {}^tAA$. Then $\mathrm{O}(n) = f^{-1}(E)$, and E is a regular value of f (i.e., $\mathrm{rank}(df_A) = \dim(S)$ for all $A \in f^{-1}(E)$). Use the same method to show that $\mathrm{U}(n)$ is a Lie group.

3. Show that G_0, the connected component of the unit element, is a normal subgroup of the Lie group G.

4. Show that a connected Lie group is generated by every neighborhood of the unit element.

5. Show that a discrete normal subgroup of a connected Lie group must be contained in the center of the group.

6. For the inclusions in diagram (1.14): Show that $\mathrm{U}(n) \subset \mathrm{SO}(2n)$ and $\mathrm{GL}(n, \mathbb{C}) \subset \mathrm{GL}^+(2n, \mathbb{R})$ by viewing $\mathbb{C}^n$ as a real vector space. Describe complex and unitary matrices as real $(2n \times 2n)$-matrices of a special form. Show $\mathrm{GL}(n, \mathbb{C}) \cap \mathrm{SO}(2n) = \mathrm{U}(n)$ and $\mathrm{GL}(n, \mathbb{R}) \cap \mathrm{U}(n) = \mathrm{O}(n)$.

7. Explicitly describe an injective homomorphism $\mathrm{O}(n) \to \mathrm{SO}(n + 1)$.

8. Let $D \subset \mathrm{SL}(n, \mathbb{R})$ be the group of upper triangular matrices with positive elements on the diagonal. Show that the map

$$D \times \mathrm{O}(n) \to \mathrm{GL}(n, \mathbb{R}), \qquad (A, C) \mapsto A \cdot C$$

is a diffeomorphism. (*Hint*: This is the content of the Gram–Schmidt orthogonalization process, see Lang [2], VI, §2.) Thus $\mathrm{GL}(n, \mathbb{R}) \cong \mathrm{O}(n) \times \mathbb{R}^{(1/2)n(n+1)}$ as a differentiable manifold.

Show in the same way that $B \times \mathrm{U}(n) \to \mathrm{GL}(n, \mathbb{C})$, $(A, C) \mapsto A \cdot C$, is a diffeomorphism, where B is the group of triangular complex matrices with positive real diagonals. Thus $\mathrm{GL}(n, \mathbb{C}) \cong \mathrm{U}(n) \times \mathbb{R}^{n \cdot n}$ as a manifold. Also show that $\mathrm{SL}(n, \mathbb{R}) \cong \mathrm{SO}(n) \times \mathbb{R}^{(1/2)n \cdot (n+1) - 1}$ as manifolds, and in particular $\mathrm{SL}(2, \mathbb{R}) \cong S^1 \times \mathbb{R}^2$.

9. Let $P \subset \mathrm{GL}(n, \mathbb{R})$ be the set of positive-definite symmetric matrices. Show that multiplication induces a bijection $P \times \mathrm{O}(n) \to \mathrm{GL}(n, \mathbb{R})$. (*Hint*: If $A \in \mathrm{GL}(n, \mathbb{R})$, then $A \cdot {}^tA \in P$, so $A \cdot {}^tA = B^2$ for some $B \in P$, and $B^{-1}A \in \mathrm{O}(n)$.) Let $H \subset \mathrm{GL}(n, \mathbb{C})$ be the set of positive-definite Hermitian matrices. Show that multiplication induces a bijection $H \times \mathrm{U}(n) \to \mathrm{GL}(n, \mathbb{C})$.

10. Show that symplectic maps $A\colon \mathbb{H}^n \to \mathbb{H}^n$ leave the symplectic scalar product invariant: If $A \in \mathrm{Sp}(n)$, $h, k \in \mathbb{H}^n$, and $\langle h, k\rangle = \sum_\nu h_\nu \cdot \bar{k}_\nu$ by definition, then $\langle Ah, Ak\rangle = \langle h, k\rangle$.

11. The real projective space $\mathbb{R}P^n$ of lines through the origin in $\mathbb{R}^{n+1}$ may be given the structure of an n-dimensional manifold, and $\mathrm{PGL}(n+1, \mathbb{R})$ is a group of transformations (diffeomorphisms) of this manifold. Give the necessary definitions, and then repeat for the complex projective space $\mathbb{C}P^n$ and $\mathrm{PGL}(n+1, \mathbb{C})$.

12. Show:
(i) $\mathrm{O}(2n+1) \cong \mathrm{SO}(2n+1) \times \mathbb{Z}/2$ as groups; and
(ii) $\mathrm{O}(2n) \cong \mathrm{SO}(2n) \times \mathbb{Z}/2$ and $\mathrm{U}(n) \cong \mathrm{SU}(n) \times S^1$ as manifolds.
In case (ii) describe the multiplication $\mathrm{SO}(2n) \times \mathbb{Z}/2$ inherits from the group $\mathrm{O}(2n)$ (semidirect product).

There is a surjective homomorphism

$$S^1 \times \mathrm{SU}(n) \to \mathrm{U}(n), \qquad (\zeta, A) \mapsto \zeta \cdot A.$$

Show that the kernel is cyclic of order n.

13. Show that if one identifies $\mathbb{R}^3$ with the subspace of pure quaternions in $\mathbb{H}$, the vector product in $\mathbb{R}^3$ is given by $p \times q =$ pure part of $p \cdot q$.

14. Verify that $\mathbb{R}$ is the center of $\mathbb{H}$ by direct calculation.

2. Left-Invariant Vector Fields and One-Parameter Groups

For our next topic we discuss tangent spaces of manifolds and see what they look like for Lie groups. Intuitively, the tangent space at a point p of a submanifold $M \subset \mathbb{R}^n$ is the space of velocity vectors $\dot{\alpha}(0)$ of arcs $\alpha\colon \mathbb{R} \to M$ with $\alpha(0) = p$.

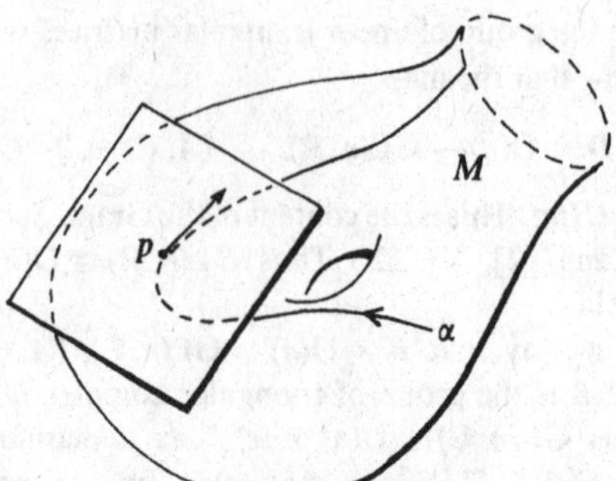

Figure 3

There is an invariant description of this space, which may be given as follows: First we restrict our attention to the local situation. Let M be an n-dimensional manifold with $p \in M$. Two differentiable maps f, g defined locally at p with values in N have ***equal germs at*** p if $f \mid U = g \mid U$ for some neighborhood U of p in M. This is an equivalence relation: an equivalence class is called a ***germ*** and denoted $f\colon (M, p) \to (N, q)$ where $f(p) = q$. Thus such a germ is represented by a map $f\colon U \to N$, where U is a neighborhood of p, and $g\colon V \to N$ represents the same germ if f and g agree on a smaller neighborhood $W \subset U \cap V$. The set $\mathscr{E}_p$ of all germs of real-valued functions $(M, p) \to \mathbb{R}$ is an $\mathbb{R}$-algebra in a natural way, addition and multiplication being done on representatives.

(2.1) Definition. A ***tangent vector*** at $p \in M^n$ is a linear map $X\colon \mathscr{E}_p \to \mathbb{R}$ satisfying the following product rule (a ***derivation*** of the $\mathbb{R}$-algebra $\mathscr{E}_p$):

$$X(\varphi \cdot \psi) = X(\varphi) \cdot \psi(p) + \varphi(p) \cdot X(\psi).$$

One should think of $X(\varphi)$ as the directional derivative of φ in the direction X. The set $\mathrm{T}_p M$ of all tangent vectors at p is a real vector space in a natural way and is called the ***tangent space*** of M at the point p. The germ of a differentiable map $f\colon (M, p) \to (N, q)$ induces a homomorphism of $\mathbb{R}$-algebras

$$f^*\colon \mathscr{E}_q \to \mathscr{E}_p, \qquad \varphi \mapsto \varphi \circ f$$

and hence the ***tangent map*** (the ***differential***)

$$\mathrm{T}_p f\colon \mathrm{T}_p M \to \mathrm{T}_q N, \qquad X \mapsto X \circ f^*.$$

Thus $\mathrm{T}_p f(X)\varphi = X(\varphi \circ f)$. The map $f \mapsto \mathrm{T}_p f$ is ***functorial***, which means $\mathrm{T}_p f(\mathrm{id}) = \mathrm{id}$, and the maps coming from a composition

$$(M, p) \xrightarrow{f} (N, q) \xrightarrow{g} (L, r)$$

obey $(\mathrm{T}_q g) \circ (\mathrm{T}_p f) = \mathrm{T}_p(g \circ f)\colon \mathrm{T}_p M \to \mathrm{T}_r L$.

It follows from functoriality that an invertible germ has an invertible differential, and therefore a chart $h: U \to U' \subset \mathbb{R}^n$, $p \in U$, induces an isomorphism $\mathrm{T}_p h: \mathrm{T}_p M = \mathrm{T}_p U \to \mathrm{T}_{h(p)} U' = \mathrm{T}_{h(p)} \mathbb{R}^n$. The right-hand side is easily understood because one has:

(2.2) Proposition. *If V is a finite-dimensional real vector space, then $\mathrm{T}_p V$ is canonically isomorphic to V for all $p \in V$.*

Proof. We define a homomorphism $V \to \mathrm{T}_p V$ by sending the vector v to the derivation $X_v: \mathscr{E}_p \to \mathbb{R}$ given by differentiation in the direction v:

$$X_v(\varphi) = \left.\frac{\partial}{\partial t}\right|_{t=0} \varphi(p + tv).$$

The map $V \to \mathrm{T}_p V$ is clearly injective (choose φ linear), so we must show it to be surjective. For this we may assume $(V, p) = (\mathbb{R}^n, 0)$. In particular, the derivations $\partial/\partial x_i$, in the directions of the canonical basis vectors of $\mathbb{R}^n$, lie in the image of our map. Hence if $X \in \mathrm{T}_0 \mathbb{R}^n$ with $X(x_i) = a_i$, where x_i is the ith coordinate function, the derivation $Y = \sum a_i(\partial/\partial x_i)$ is also in the image of our map. Now for any derivation Z, the product rule implies that $Z(1) = Z(1) + Z(1)$, so $Z(1) = 0$ and $Z(c) = 0$ for any constant c. Thus $X - Y$ vanishes on constants, and also on each x_i by construction. But this is enough to show that $X = Y$. For given any φ with $\varphi(0) = 0$,

$$\varphi(x) = \sum \varphi_i(x) \cdot x_i, \qquad \varphi_i(x) = \int_0^1 D_i \varphi(tx)\, dt,$$

where D_i is differentiation with respect to the ith variable. Thus any tangent vector in $\mathrm{T}_0 \mathbb{R}^n$ vanishing on each x_i vanishes on φ by linearity and the product rule again. □

Note by the way that a derivation is completely determined by its values on linear functions.

After the introduction of suitable charts around p and q, a differentiable germ $f: (M, p) \to (N, q)$ may be described by a germ $(\mathbb{R}^m, 0) \to (\mathbb{R}^n, 0)$, which we will also call f.

$$\begin{array}{ccc} (M, p) & \xrightarrow{\;f\;} & (N, q) \\ {\scriptstyle \text{chart}}\downarrow & & \downarrow{\scriptstyle \text{chart}} \\ (\mathbb{R}^m, 0) & \xrightarrow[f]{} & (\mathbb{R}^n, 0) \end{array}$$

The tangent map $\mathrm{T}_0 f$ is calculated as follows:

$$\mathrm{T}_0 f\left(\frac{\partial}{\partial x_j}\right)(\varphi) = \left.\frac{\partial}{\partial x_j}\right|_0 (\varphi \circ f) = \sum_{i=1}^{n} \frac{\partial f_i}{\partial x_j}(0) \cdot \frac{\partial \varphi}{\partial y_i}(0),$$

so

$$\mathrm{T}_0 f\left(\frac{\partial}{\partial x_j}\right) = \sum_{i=1}^{n} \frac{\partial f_i}{\partial x_j}(0) \cdot \frac{\partial}{\partial y_i}.$$

That means that, with respect to the bases $(\partial/\partial x_j)$ and $(\partial/\partial y_i)$ for $\mathrm{T}_0 \mathbb{R}^m$ and $\mathrm{T}_0 \mathbb{R}^n$, the tangent map $\mathrm{T}_0 f$ is described by the ***Jacobian matrix***

$$Df = \left(\frac{\partial f_i}{\partial x_j}\right).$$

The family of all tangent spaces $(\mathrm{T}_p M \mid p \in M)$ fits together into a global object

$$\mathrm{T}M = \bigcup_{p \in M} \mathrm{T}_p M \quad \text{(disjoint union)},$$

the ***tangent bundle.***

The tangent bundle is comprised of the ***total space*** $\mathrm{T}M$, the ***base space*** M, the ***fibers*** $\mathrm{T}_p M$, and the ***projection*** $\pi\colon \mathrm{T}M \to M$, defined by sending $v \in \mathrm{T}_p M$ to p.

Each chart $h\colon U \to U'$ of M gives rise to a ***bundle chart*** $\mathrm{T}(h)$. This bundle chart is linear on fibers and is given by

$$\mathrm{T}M \supset \mathrm{T}U \xrightarrow{Th} \mathrm{T}U' = U' \times \mathbb{R}^n, \qquad Th \mid \mathrm{T}_p M = \mathrm{T}_p h,$$

where we use the previously mentioned fact that there is a ***canonical*** isomorphism $\mathrm{T}_p U' \cong \mathbb{R}^n$. These charts form an atlas, which makes $\mathrm{T}M$ into a $2n$-dimensional manifold. The projection π is ***locally trivial***, namely the diagram

$$\begin{array}{ccc} \mathrm{T}M \supset \mathrm{T}U & \xrightarrow{Th} & U' \times \mathbb{R}^n \\ \downarrow{\scriptstyle \pi} & & \downarrow{\scriptstyle \mathrm{pr}_1} \\ M \supset U & \xrightarrow{h} & U' \end{array}$$

commutes, and Th is a linear isomorphism on fibers.

For a Lie group the situation is simple, insofar as the tangent bundle is trivial, i.e., the tangent bundle is globally isomorphic to the product of the base space and a fiber. Such an isomorphism is obtained as follows: Every group element $x \in G$ defines a ***left translation***

$$l_x\colon G \to G, g \mapsto xg$$

with inverse $l_x^{-1} = l_{x^{-1}}$. Let e be the unit of G and let $\mathrm{L}G = \mathrm{T}_e G$. Then there is an ***isomorphism of vector bundles***

$$\begin{array}{ccc} G \times \mathrm{L}G & \cong & \mathrm{T}G, \qquad (g, X) \mapsto \mathrm{T}_e l_g(X). \\ & {\scriptstyle \mathrm{pr}_1} \searrow \quad \swarrow {\scriptstyle \pi} & \\ & G & \end{array} \tag{2.3}$$

That is to say, the diagram commutes, and the restrictions to the fibers $LG \cong \{g\} \times LG \to T_g G$ are linear isomorphisms.

(2.4) Definitions. The vector space $LG := T_e G$ is called the ***Lie algebra*** of G. The word "algebra" is not yet justified, but we will explain the algebra structure soon. A homomorphism of Lie groups $f: G \to H$ induces a homomorphism $Lf = T_e f: LG \to LH$ of Lie algebras in a functorial fashion.

A differentiable ***vector field*** on a manifold M is a differentiable section of the tangent bundle, which is to say a differentiable map $X: M \to TM$ such that

$$\pi \circ X = \mathrm{id}_M,$$

or, equivalently, $X(p) \in T_p M$. Saying that X is ***differentiable*** is the same as saying that if $f: M \to \mathbb{R}$ is differentiable, then so is the map $Xf: M \to \mathbb{R}$, $p \mapsto X(p)(f)$. In local coordinates $(x_1, \ldots, x_n)$—or on an open set in $\mathbb{R}^n$—a differentiable vector field may be written in the form

$$x \mapsto \sum_{i=1}^{n} a_i(x) \frac{\partial}{\partial x_i},$$

with smooth functions a_i.

A vector field X on a Lie group is called ***left-invariant*** if the diagram

$$\begin{array}{ccc} TG & \xrightarrow{Tl_x} & TG \\ {\scriptstyle X}\uparrow & & \uparrow{\scriptstyle X} \\ G & \xrightarrow[l_x]{} & G \end{array}$$

commutes for every $x \in G$.

(2.5) Remarks. Given $v \in LG$, there is a constant section $x \mapsto (x, v)$ of $G \times LG$, and the trivialization (2.3) transforms this section into the vector field $X_v: G \to TG$, $x \mapsto T_e l_x(v)$. The map $v \mapsto X_v$ defines a canonical isomorphism between LG and the vector space of left-invariant vector fields on G. From now on we will identify LG with this space, and we will denote a left-invariant vector field on G by $X \in LG$.

A vector field $X: M \to TM$ asks to be integrated. A germ of a curve $\alpha: (\mathbb{R}, \tau) \to (M, p)$ defines a tangent vector

$$\left.\frac{\partial}{\partial t}\right|_\tau \alpha = \dot{\alpha}(\tau) \in T_p M$$

mapping $\mathscr{E}_p \to \mathbb{R}$ by sending φ to $\partial/\partial t|_\tau \varphi(\alpha(t))$. Using the canonical isomorphism $\mathbb{R} = T_\tau \mathbb{R}$, (2.2), in which $1 \in \mathbb{R}$ corresponds to the basis vector $\partial/\partial t \in T_\tau \mathbb{R}$, this can be expressed as $\dot{\alpha}(\tau) = T_\tau \alpha(1)$. In other words, a parametrized curve defines a tangent vector which, as a derivation, is differentiation with respect to the parameter. We will frequently describe and calculate tangent vectors as velocity vectors of curves.

An ***integral curve*** $\alpha\colon]a, b[\to M$ of a vector field X on M is a differentiable curve with the property $\dot{\alpha}(t) = X(\alpha(t))$ for $a < t < b$.

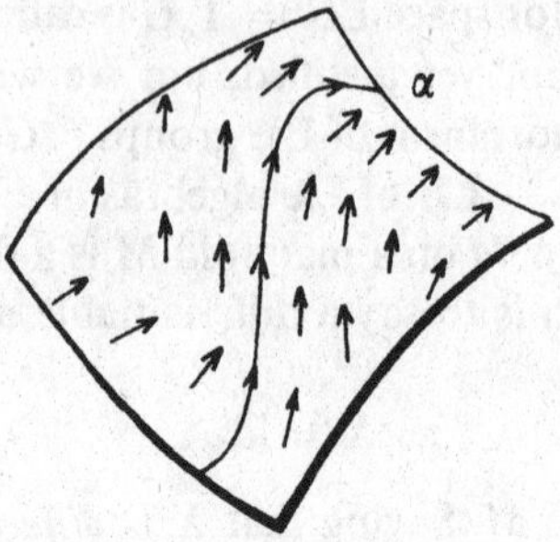

Figure 4

In local coordinates,

$$X(x) = \sum_{i=1}^{n} a_i(x) \frac{\partial}{\partial x_i} \quad \text{and} \quad \alpha(t) = (\alpha_1(t), \ldots, \alpha_n(t)).$$

Thus, with respect to the basis $(\partial/\partial x_i)$ we have $\dot{\alpha} = (\dot{\alpha}_1, \ldots, \dot{\alpha}_n)$, and the condition that α be an integral curve is $\dot{\alpha}_i(t) = a_i(\alpha_1(t), \ldots, \alpha_n(t))$ for $i = 1, \ldots, n$. The theory of differential equations (Bröcker [1], III; Lang [1], VI) tells us that exactly one maximal integral curve passes through each point of M. More precisely: Given the vector field X on M, there is an open set $A \subset \mathbb{R} \times M$ such that $A \cap (\mathbb{R} \times p)$ is an open interval containing the origin for each $p \in M$, together with a differentiable map, the (local) ***flow*** of the vector field,

$$\Phi\colon A \to M, \qquad (t, p) \mapsto \alpha_p(t).$$

The curve $t \mapsto \alpha_p(t)$ is the unique maximal integral curve of X with ***initial value*** $\alpha_p(0) = p$, and $]a_p, b_p[$ is its interval of definition. The flow Φ satisfies the following *flow equations*

$$\Phi(0, p) = p \quad \text{and} \quad \Phi(s, \Phi(t, p)) = \Phi(s + t, p),$$

wherever the left-hand side is defined. This follows because the curves

$$s \mapsto \Phi(s, \Phi(t, p)) \quad \text{and} \quad s \mapsto \Phi(s + t, p)$$

are both integral curves of X starting at $\Phi(t, p)$. Setting $\Phi_t(p) = \Phi(t, p)$ makes these equations more appealing to the eye and the intuition:

(2.6) $$\Phi_0 = \mathrm{id}_M, \qquad \Phi_s \circ \Phi_t = \Phi_{s+t}.$$

The vector field can be recovered from the flow:

$$X(p) = \frac{\partial}{\partial t}\bigg|_0 \Phi(t, p) \in \mathrm{T}_p M.$$

Returning to the case of a Lie group G, if X is a left-invariant vector field on G, and α is an integral curve of X, then $l_x\alpha$ is also an integral curve of X for every $x \in G$. In other words, $\alpha_x = l_x\alpha_e$, and, as a consequence, if α_e is defined on the interval $]-\varepsilon, \varepsilon[$, then α may be extended beyond any time t by at least $\pm\varepsilon$. This means that all the intervals of definition $]a_x, b_x[$ of maximal integral curves are equal to all of $\mathbb{R}$, and the flow associated to X is global. So in this case the existence of the flow Φ can be seen quite easily: $\Phi(t, g) = g\alpha^X(t)$, where α^X is the integral curve for the field $X \in LG$ starting at $\alpha^X(0) = e$.

(2.7) Remark. A ***one-parameter group*** of a Lie group G is a homomorphism of Lie groups

$$\alpha\colon \mathbb{R} \to G$$

(the homomorphism, not just its image!). The correspondence

$$\alpha \mapsto \dot{\alpha}(0) \in LG$$

defines a canonical bijection between the set of one-parameter groups of G and the Lie algebra of G.

PROOF (of last statement). Interpreting LG as the space of left-invariant vector fields on G, the inverse map is given by

$$X \mapsto \alpha^X = \text{integral curve of } X \text{ starting at } e.$$

This is, in fact, a one-parameter group. For if Φ is the flow associated to X, $\alpha^X(s + t) = \Phi_{t+s}(e) = \Phi_t(\Phi_s(e))$ by (2.6), and $\Phi_t(ge) = g\Phi_t(e)$ because of left invariance of X and hence Φ. Setting $g = \Phi_s(e)$, we have $\Phi_t(\Phi_s(e)) = \Phi_t(\Phi_s(e)\cdot e) = \Phi_s(e)\Phi_t(e) = \alpha^X(s)\cdot\alpha^X(t)$, so α^X is a homomorphism. The composition $X \mapsto \alpha^X \mapsto \dot{\alpha}^X(0)$ is the identity, and to see that $\alpha \mapsto \dot{\alpha}(0) = X \mapsto \alpha^X$ is also the identity, note that the one-parameter group α defines a flow $\Phi\colon \mathbb{R} \times G \to G$, $(t, g) \mapsto g\cdot\alpha(t)$, with $\partial/\partial t|_0 \Phi(g, t) = Tl_g(\dot{\alpha}(0))$. This is the same flow as the one corresponding to the left-invariant vector field X, so their integral curves starting at e coincide: $\alpha = \alpha^X$. □

(2.8) Examples. A finite-dimensional real vector space V, interpreted as a Lie group, coincides with LV, and $\alpha^v(t) = tv$ is the one-parameter group corresponding to v.

Similarly, the torus $\mathbb{R}^n/\mathbb{Z}^n$ has $\mathbb{R}^n$ as its Lie algebra, and the one-parameter group corresponding to $v \in \mathbb{R}^n$ is $\alpha^v\colon t \mapsto tv \bmod \mathbb{Z}^n$.

The group of linear automorphisms $\mathrm{Aut}(V)$ has as its Lie algebra $\mathrm{End}(V)$, the vector space of all linear endomorphisms of V, since $\mathrm{Aut}(V)$ is an open submanifold of $\mathrm{End}(V)$. The one-parameter subgroup corresponding to $A \in \mathrm{End}(V)$ is

$$\alpha^A\colon \mathbb{R} \to \mathrm{Aut}(V), \qquad t \mapsto \exp(tA) = \sum_{\nu=0}^{\infty} \frac{1}{\nu!}(tA)^\nu.$$

This instructive example will be of considerable concern to us.

So far we have examined what the basic constructions from the theory of manifolds mean for Lie groups. Before turning to more detailed study of one-parameter groups, we should say a few words about the algebra structure of the Lie algebra of a Lie group. Although we hardly use it in this book, which takes a geometric point of view, this structure is an important fundamental concept. The algebra structure of a Lie algebra may be described as follows: Taking $X, Y \in \mathrm{L}G$ to be left-invariant vector fields, for each point $g \in G$ the fields X and Y yield derivations on function germs defined about g. Now, if φ is a germ of a function, then $X\varphi$ may also be viewed as a germ of a function, to which Y may be applied. The same is true with X and Y switched, and the ***Lie product*** $[X, Y]$ of X and Y is given as a derivation by

(2.9) $$[X, Y]\varphi = X(Y\varphi) - Y(X\varphi).$$

An easy calculation shows that $[X, Y](\varphi\psi) = [X, Y](\varphi)\cdot\psi + \varphi\cdot[X, Y](\psi)$ and hence, in contrast to the individual summands, (2.9) really does satisfy the product rule.

Referring to the group structure of G, we have an alternative description of the Lie product of left-invariant vector fields. Each element $g \in G$ gives rise to an inner automorphism

$$\mathrm{c}(g)\colon G \to G, \qquad x \mapsto gxg^{-1}.$$

This, in turn, gives rise to a homomorphism which will later be of considerable concern to us, namely the

(2.10) ***Adjoint representation***

$$\mathrm{Ad}\colon G \to \mathrm{Aut}(\mathrm{L}G), \qquad g \mapsto \mathrm{Lc}(g).$$

Here L denotes the differential at the unit—see (2.4). The homomorphism Ad induces a homomorphism of Lie algebras

$$\mathrm{ad} = \mathrm{LAd}\colon \mathrm{L}G \to \mathrm{LAut}(\mathrm{L}G) = \mathrm{End}(\mathrm{L}G)$$

sending X to the homomorphism $Y \mapsto [X, Y]$, so

(2.11) $$[X, Y] = \mathrm{ad}(X)Y.$$

We will compute the right-hand side more explicitly to show that it coincides with the earlier definition (2.9) of $[X, Y]$. If $X, Y \in \mathrm{L}G$, then $\mathrm{c}(\alpha^X(s))\alpha^Y(t) = \alpha^X(s)\cdot\alpha^Y(t)\cdot\alpha^X(-s)$, since $\alpha^X(s)^{-1} = \alpha^X(-s)$. Setting $a(s, t) = \mathrm{c}(\alpha^X(s))\alpha^Y(t)$, we have $\mathrm{Ad}(\alpha^X(s))Y = \partial/\partial t\,|_0\, a(s, t) \in \mathrm{L}G$, and

$$\mathrm{ad}(X)Y = \frac{\partial}{\partial s}\bigg|_0 \frac{\partial}{\partial t}\bigg|_0 a(s, t),$$

where we have used the identification of the vector space LG with its tangent space. If we view tangent vectors as derivations acting on germs of functions φ at the unit, this last equation means that

$$(\mathrm{ad}(X)Y)\varphi = \frac{\partial^2}{\partial s\, \partial t}\bigg|_0 \varphi(a(s, t)),$$

where $\varphi\colon (G, e) \to (\mathbb{R}, 0)$. There is something to check here, see (2.22), Ex. 9. To calculate the derivative with respect to s, apply the chain rule to the composition

$$\begin{aligned} &(s, t) \mapsto (s, t, -s), \\ &\mathbb{R}^2 \longrightarrow \mathbb{R}^3 \longrightarrow \mathbb{R} \\ &\qquad (\sigma, t, \tau) \mapsto \varphi(\alpha^X(\sigma) \cdot \alpha^Y(t) \cdot \alpha^X(\tau)). \end{aligned}$$

The result is that

$$\frac{\partial^2}{\partial s\, \partial t}\bigg|_0 \varphi(a(s, t)) = \frac{\partial^2}{\partial \sigma\, \partial t}\bigg|_0 \varphi(\alpha^X(\sigma) \cdot \alpha^Y(t)) - \frac{\partial^2}{\partial \tau\, \partial t}\bigg|_0 \varphi(\alpha^Y(t) \cdot \alpha^X(\tau)).$$

Thus if φ is a differentiable germ at the unit,

$$(\mathrm{ad}(X)Y)\varphi = \frac{\partial^2}{\partial s\, \partial t}\bigg|_0 \varphi(\alpha^X(s) \cdot \alpha^Y(t)) - \frac{\partial^2}{\partial s\, \partial t}\bigg|_0 \varphi(\alpha^Y(t) \cdot \alpha^X(s)).$$

But $(t, g) \mapsto g \cdot \alpha^Y(t)$ is the flow corresponding to Y, so

$$\frac{\partial}{\partial t}\bigg|_0 \varphi(g \cdot \alpha^X(s) \cdot \alpha^Y(t)) = Y\varphi(g \cdot \alpha^X(s)).$$

Repeating this for X,

$$\frac{\partial^2}{\partial s\, \partial t}\bigg|_0 \varphi(g \cdot \alpha^X(s) \cdot \alpha^Y(t)) = (XY\varphi)(g),$$

and doing the same thing to the other summand, we see that (2.11) and (2.9) describe the same left-invariant vector fields.

(2.12) Properties of the Lie Product. The Lie product $LG \times LG \to LG$, $(X, Y) \mapsto [X, Y]$, is bilinear and hence provides LG with the structure of a real algebra. The Lie product also satisfies

(i) $[X, X] = 0$, hence $[X, Y] = -[Y, X]$.
(ii) $[[X, Y], Z] + [[Y, Z], X] + [[Z, X], Y] = 0$, the ***Jacobi identity***.

These identities are easily verified from (2.9) (see (2.22), Ex. 3). An algebra over a field which satisfies the properties (2.12) is called a ***Lie algebra***.

We want to determine the Lie algebras of the linear groups we have previously introduced. The one-parameter groups for $X, Y \in \mathrm{LAut}(V) = \mathrm{End}(V)$ are functions $(\mathbb{R}, 0) \to (\mathrm{Aut}(V), E)$ such that

$$\alpha^X(s) = E + sX \mod s^2,$$
$$\alpha^Y(t) = E + tY \mod t^2,$$

so

$$a(s, t) = \alpha^X(s)\alpha^Y(t)\alpha^X(-s) = (E + sX)(E + tY)(E - sX)$$
$$= E + tY + st(XY - YX) \mod(s^2, t^2),$$

and taking the derivative $\partial^2/\partial s\, \partial t$ at zero yields

(2.13) $$[X, Y] = XY - YX.$$

Any associative algebra becomes a Lie algebra with this Lie product. This also gives the Lie product for the subspaces and quotients of $\mathrm{End}(V)$ which are the Lie algebras of subgroups and quotient groups of $\mathrm{Aut}(V)$. We will examine our classical linear groups one by one.

(2.14) The Lie algebra of an abelian group has trivial Lie product $[X, Y] = 0$, as is evident from (2.11). In particular, $LT^n = L\mathbb{R}^n = \mathbb{R}^n$, with the Lie product $[v, w] = 0$.

(2.15) $\mathrm{so}(n) = \mathrm{LSO}(n) \subset \mathrm{End}(\mathbb{R}^n)$ is the Lie algebra of skew-symmetric matrices, and consequently $\dim \mathrm{SO}(n) = \frac{1}{2}n(n-1)$. The Lie algebra $\mathrm{LSO}(n)$ may be computed as follows: If α^X is a one-parameter group in $\mathrm{SO}(n)$, then, modulo s^2,

$$E = {}^t\alpha^X(s) \cdot \alpha^X(s) = {}^t(E + sX)(E + sX) = E + s({}^tX + X),$$

so ${}^tX + X = 0$ and X is skew-symmetric. And if X is skew-symmetric, then ${}^t\alpha^X(s)\alpha^X(s) = \exp(s \cdot {}^tX)\exp(s \cdot X) = \exp(-sX) \cdot \exp(sX) = \exp(0) = E$, so α^X lies entirely in $\mathrm{SO}(n)$. We still have not proved that $\mathrm{SO}(n)$ is a manifold.

Similar calculations allow us to achieve our goal in other linear groups:

(2.16) $\mathrm{u}(n) = \mathrm{LU}(n) \subset \mathrm{End}(\mathbb{C}^n)$ is the Lie algebra of skew-Hermitian matrices. This shows $\dim \mathrm{U}(n) = n^2$.

(2.17) $\mathrm{sl}(n) = \mathrm{LSL}(n)$ is the Lie algebra of matrices in $\mathrm{End}(\mathbb{R}^n)$ with zero trace. In this case the calculation with a one-parameter group in $\mathrm{SL}(n)$ gives (mod s^2):

$$1 = \det(\alpha^X(s)) = \det(E + sX) = 1 + s\,\mathrm{Tr}(X),$$

which may be interpreted as saying $(\det \alpha^X)^{\cdot}(0) = \mathrm{Tr}(X) = 0$.

Conversely, we have

$$\left.\frac{d}{ds}\right|_0 \det \exp((t+s)X) = \det \exp(tX) \cdot \left.\frac{d}{ds}\right|_0 \det \exp(sX),$$

and $d/ds|_0 \det \exp(sX) = \mathrm{Tr}(X)$ as we have just seen. This yields the differential equation

$$(\det \exp(tX))^{\cdot} = \mathrm{Tr}(X) \cdot \det \exp(tX),$$

so if $\mathrm{Tr}(X) = 0$, $\det \exp(tX)$ is constant and, since $\det \exp(0) = 1$, we have $\exp(tX) \in \mathrm{SL}(n)$ for all t.

By combining (2.16) and (2.17), one gets

(2.18) $\mathrm{su}(n) = \mathrm{LSU}(n) \subset \mathrm{End}(\mathbb{C}^n)$ is the Lie algebra of skew-Hermitian matrices with trace zero.

(2.19) The Lie algebra $\mathrm{sp}(n) = \mathrm{LSp}(n)$ consists of the skew-Hermitian matrices in $\mathrm{End}(\mathbb{H}^n) \subset \mathrm{End}(\mathbb{C}^{2n})$, and these are obviously the complex $(2n \times 2n)$-matrices of the form

$$\begin{bmatrix} A & -\bar{B} \\ B & \bar{A} \end{bmatrix}, \qquad {}^t\bar{A} = -A, \qquad {}^tB = B.$$

The dimension of $\mathrm{Sp}(n)$ is $2n^2 + n$.

The adjoint representation (2.10) of the linear groups is given by

(2.20) $$\mathrm{Ad}(A)\colon \mathrm{End}(V) \to \mathrm{End}(V), \qquad X \mapsto AXA^{-1}.$$

To see this, note that

$$c(A)\alpha^X(t) = A\exp(tX)A^{-1} = \exp(tAXA^{-1}) = \alpha^{AXA^{-1}}(t)$$

and differentiate at $t = 0$, showing that the groups operate by conjugation as claimed.

(2.21) Physicists' Convention. By using a scaling factor on $LS^1 = \mathbb{R}$, we may assume that the exponential map of S^1 is given by

$$\exp\colon \mathbb{R} \to S^1, \qquad x \mapsto e^{ix}.$$

If we describe LS^1 as the space of skew-Hermitian (1×1)-matrices, the element $x \in \mathbb{R}$ corresponds to the matrix $(ix) \in \mathrm{End}_{\mathbb{C}} \mathbb{C}^1$, and multiplication by $(-i)$ gives the inverse correspondence. Physicists are accustomed to multiplying the Lie algebra of any linear Lie group by $(-i)$, and hence, denoting what we have called $A \in \mathrm{End}(V)$ by $B = -iA$, the exponential map is given by

$$\exp\colon B \mapsto e^{iB}.$$

If A is skew-Hermitian, then iA is Hermitian, so, in physicists' notation, the Lie algebra of $\mathrm{U}(n)$ is the space of Hermitian (self-adjoint) operators. One should always use the above method of translation when passing from our formulas to those of the physicists.

(2.22) Exercises

1. Show that a connected one-dimensional Lie group is isomorphic to $\mathbb{R}$ or S^1.

2. Show that a bijective homomorphism of Lie groups is an isomorphism.

3. Check the properties (2.12) of the Lie product.

4. Calculate the Lie algebras of the projective groups $PGL(n, \mathbb{R})$ and $PGL(n, \mathbb{C})$.

5. Check that the Lie algebras of $SO(n)$, $U(n)$, $SL(n, \mathbb{R})$, $SL(n, \mathbb{C})$, and $Sp(n)$ (which we have given in the text) are, in fact, closed under the Lie product of matrices and invariant under conjugation by elements of their corresponding groups. For example, if A, B are skew-symmetric $(n \times n)$-matrices, and $C \in SO(n)$, you must show that $AB - BA$ and CAC^{-1} are skew-symmetric.

6. Show that the image of the one-parameter group $t \mapsto (t, \sqrt{2}t) \bmod \mathbb{Z}^2$ is dense in T^2.

7. Specify $n - 1$ injective homomorphisms $\varphi_\nu \colon S^1 \to SO(n)$, $\nu = 1, \ldots, n - 1$, such that $SO(n)$ is generated by the elements $\{\varphi_\nu(z) \mid z \in S^1, \nu = 1, \ldots, n - 1\}$.

8. Show that there is a linear isomorphism $\varphi\colon so(3) \to \mathbb{R}^3$ such that $\varphi[X, Y] = \varphi(X) \times \varphi(Y)$ (vector product in $\mathbb{R}^3$) and $\varphi(AXA^{-1}) = A\varphi(X)$ for $A \in SO(3)$. This isomorphism sends a one-parameter group to the vector pointing in the direction of its rotation axis and scaled by its angular velocity. In components

$$\varphi\colon \begin{bmatrix} 0 & -z & y \\ z & 0 & -x \\ -y & x & 0 \end{bmatrix} \mapsto (x, y, z).$$

9. Let $a\colon (\mathbb{R}^2, 0) \to (M, p)$ be the germ of a differentiable map with $a(s, 0) = p$ for all s. Then $s \mapsto \partial/\partial t|_0 a(s, t) \in T_pM$ is the germ of a differentiable curve, and so $\partial/\partial s|_0 \partial/\partial t|_0 a(s, t) \in T_pM$ is a well-defined tangent vector. Show that this vector acts as a derivation on germs of functions at p by $\varphi \mapsto \partial^2/\partial s\, \partial t|_0 \varphi a(s, t)$. *Hint*: You may suppose that $(M, p) = (\mathbb{R}^n, 0)$ and that φ is linear.

3. The Exponential Map

Given a tangent vector X at the unit of a Lie group G, which determines a left-invariant vector field of G, there is the one-parameter group $\alpha^X\colon \mathbb{R} \to G$ with $\dot{\alpha}^X(0) = X$.

(3.1) Proposition. *The map*

$$\exp\colon LG \to G, \qquad X \mapsto \alpha^X(1),$$

which is called the ***exponential map****, is differentiable. Furthermore, its differential at the origin is the identity.*

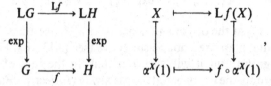

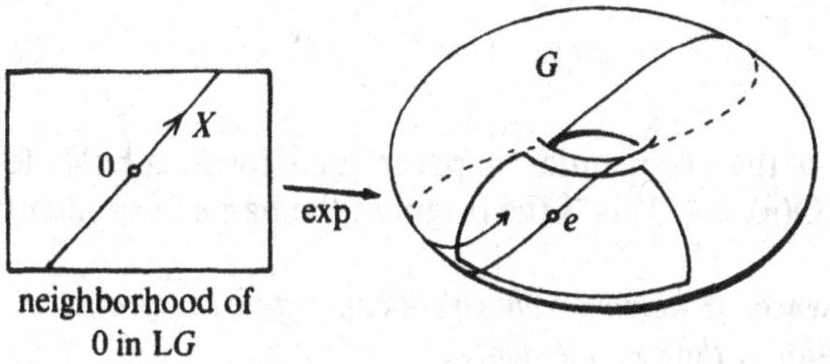

Figure 5

PROOF. The one-parameter groups $s \mapsto \alpha^{tX}(s)$ and $s \mapsto \alpha^{X}(ts)$ belong to the same vector $tX \in \mathrm{L}G$, and are therefore equal. In particular,

$$\exp(tX) = \alpha^{tX}(1) = \alpha^{X}(t),$$

and so $\partial/\partial t|_0 \exp(tX) = X$. Thus if exp is differentiable, $T_0 \exp = \mathrm{id}_{\mathrm{L}G}$. We show differentiability by considering the map

$$\mathbb{R} \times G \times \mathrm{L}G \to G \times \mathrm{L}G, \qquad (t, g, X) \mapsto (g \cdot \alpha^{X}(t), X).$$

This is simply the flow on $G \times \mathrm{L}G$ corresponding to the vector field $(g, X) \mapsto (X(g), 0)$, and hence is a differentiable map. Hence the restriction $1 \times e \times \mathrm{L}G \to G$, $(1, e, X) \mapsto \alpha^{X}(1)$, is also differentiable. □

Note that, from now on, we may describe the one-parameter group α^X by $t \mapsto \exp(tX)$.

The exponential map plays a central role—it describes how the Lie algebra determines the structure of the corresponding Lie group. Thus, as in a first course in analysis, it is the most important map in the theory of Lie groups.

(3.2) Naturality. A homomorphism of Lie groups $f: G \to H$ induces a commutative diagram

$$\begin{array}{ccc} \mathrm{L}G & \xrightarrow{\mathrm{L}f} & \mathrm{L}H \\ {\scriptstyle\exp}\downarrow & & \downarrow{\scriptstyle\exp} \\ G & \xrightarrow[f]{} & H \end{array} \qquad \begin{array}{ccc} X & \longmapsto & \mathrm{L}f(X) \\ \downarrow & & \downarrow \\ \alpha^X(1) & \longmapsto & f \circ \alpha^X(1) \end{array},$$

since $f \circ \alpha^X$ is a one-parameter group with initial vector $T_e f(\dot{\alpha}^X(0)) = \mathrm{L}f(X)$.

The exponential map is locally invertible (i.e., a local diffeomorphism) at the origin $0 \in \mathrm{L}G$ because its differential at 0 is the identity, by the inverse function theorem.

(3.3) Example. If $G = \mathrm{Aut}(V)$ is the group of automorphisms of a finite-dimensional vector space, then $\mathrm{L}G = \mathrm{End}(V)$, and (see (2.8))

$$\exp(A) = \sum_{\nu=0}^{\infty} \frac{1}{\nu!} A^{\nu}.$$

By naturality of the exponential map, the same formula holds for all linear groups, $\mathrm{U}(n)$, $\mathrm{O}(n)$, etc. This is the origin of the name "exponential map."

(3.4) Consequence. *A homomorphism of connected Lie groups is determined by its differential at the unit element.*

PROOF. Since the exponential map is natural and is a local diffeomorphism, the differential $\mathrm{L}f$ of a homomorphism f determines the homomorphism on a neighborhood of the unit. Thus $\mathrm{L}f$ determines f on the entire Lie group, since it is a general fact that a connected topological group G is generated by every neighborhood U of the identity. In fact, after replacing U by $U \cap U^{-1}$, if necessary, where $U^{-1} = \{u^{-1} | u \in U\}$, we may assume $U = U^{-1}$. Letting $U^n = \{u_1 \cdot \ldots \cdot u_n | u_i \in U\}$, we see that $\bigcup_{n=1}^{\infty} U^n$ is an open subgroup of G. As such it has open cosets. Since these are disjoint and G is connected, there is just one coset. □

Occasionally we will need to consider other differentiable maps $(\mathrm{L}G, 0) \to (G, e)$ whose differentials at the origin are the identity. The multiplication $\mu: G \times G \to G$ has differential at the point (e, e)

(3.5) $$\mathrm{T}_{(e,e)}\mu: \mathrm{L}G \oplus \mathrm{L}G \to \mathrm{L}G, \qquad (X, Y) \mapsto X + Y.$$

This is clear from the observation that the stated differential is linear and restricts to the identity on each summand of the left. Thus when the Lie algebra $\mathrm{L}G$ is in some way split as a direct sum of vector spaces

$$\mathrm{L}G = V_1 \oplus V_2 \oplus \cdots \oplus V_t,$$

the map

$$\mathrm{L}G \to G, \qquad (X_1, \ldots, X_t) \mapsto \exp(X_1) \ldots \exp(X_t), \qquad X_\nu \in V_\nu$$

has differential $\mathrm{id}_{\mathrm{L}G}$ at the origin and hence is locally invertible.

The exponential map need not be surjective (see (3.13), Ex. 1), even on a connected Lie group. But it will turn out that, in the case of a compact connected Lie group, the exponential map is always surjective. Furthermore, the exponential map of a connected Lie group G is, in general, a homomorphism only on lines through the origin. In fact, it is a homomorphism on all of $\mathrm{L}G$ precisely if G is abelian.

Indeed, the exponential map is a bijection on a neighborhood of the origin which, as we have seen, contains a set of generators of G. Since the

Lie algebra is commutative as an additive group, we conclude that, if the exponential map is a homomorphism, G is abelian. Conversely, if G is abelian, multiplication is a homomorphism $G \times G \to G$ which induces the map $\mathrm{L}G \times \mathrm{L}G \to \mathrm{L}G$, $(X, Y) \mapsto X + Y$ by (3.5). The statement then follows by naturality of the exponential map (3.2).

(3.6) Theorem. *A connected abelian Lie group is the product of a torus and a vector space*: $G \cong T^k \times \mathbb{R}^s$.

PROOF. Since the image of the exponential map contains a set of generators,

$$\exp\colon \mathrm{L}G \to G$$

is a surjective homomorphism. Its kernel $K \subset \mathrm{L}G$ is a discrete subgroup of $\mathrm{L}G$, because the exponential map is a local bijection at the origin. Therefore—as we will show in (3.8)—K is generated by linearly independent vectors $g_1, \ldots, g_k \in \mathrm{L}G$. We complete this system using $g_{k+1}, \ldots, g_n$ so that $g_1, \ldots, g_n$ form a basis of $\mathrm{L}G$. This basis determines an isomorphism $\mathrm{L}G \cong \mathbb{R}^n$ such that

$$K \cong \mathbb{Z}^k \times 0 \subset \mathbb{R}^k \times \mathbb{R}^{n-k} = \mathbb{R}^n.$$

Hence $\mathrm{L}G/K \cong \mathbb{R}^n/(\mathbb{Z}^k \times 0) = T^k \times \mathbb{R}^{n-k}$. The homomorphism

$$T_k \times \mathbb{R}^{n-k} \cong \mathrm{L}G/K \to G$$

is a bijective local diffeomorphism, and hence an isomorphism of Lie groups. □

(3.7) Corollary. *A compact abelian Lie group is isomorphic to the product of a torus and a finite abelian group.*

PROOF. For the purposes of this proof, we will write our Lie groups additively. By (3.6), the connected component of the unit of a compact abelian Lie group G is a torus T (where T might be $T^0 = \{e\}$). Thus there is a short exact sequence (i.e., the image of the inclusion i is the kernel of the projection p)

$$0 \to T \underset{i}{\overset{\subset}{\to}} G \underset{p}{\to} B \to 0.$$

The sequence splits, since T is divisible. In detail: the quotient group B is discrete, since T is open in G, and also compact and hence finite. We must find a section for p—i.e., a homomorphism $s\colon B \to G$ with $p \circ s = \mathrm{id}_B$. Once we have this section, we get the isomorphism

$$T \times B \to G, \qquad (t, b) \mapsto i(t) + s(b)$$

with inverse $g \mapsto (g - s(p(g)), p(g))$.

To construct s we use, for the sake of simplicity, that B is a product of cyclic groups $B = \mathbb{Z}/n_1 \times \cdots \times \mathbb{Z}/n_k$. Corresponding to generators $b_1, \ldots, b_k$

of these cyclic groups, we choose elements $c_1, \ldots, c_k \in G$ with $p(c_\nu) = b_\nu$. Then $n_\nu \cdot c_\nu \in T$, since $p(n_\nu c_\nu) = n_\nu \cdot p(c_\nu) = 0$ and T is the kernel of p. Since $T = \mathbb{R}^k/\mathbb{Z}^k$ is a torus, $n_\nu c_\nu = v_\nu \bmod \mathbb{Z}^k$ with $v_\nu \in \mathbb{R}^k$. Let $d_\nu = v_\nu/n_\nu \bmod \mathbb{Z}^k$, then $d_\nu \in T$ and $n_\nu d_\nu = n_\nu c_\nu$. Setting $g_\nu = c_\nu - d_\nu$, $p(g_\nu) = b_\nu$, and $n_\nu g_\nu = 0$. Thus the homomorphism induced by $b_\nu \mapsto g_\nu$ is a well-defined section $B \to G$. □

We still need to show the following:

(3.8) Lemma. *A discrete subgroup B of a finite-dimensional vector space V is generated by linearly independent vectors $g_1, \ldots, g_k$.*

Proof. We proceed by induction on $n = \dim V$. For $n = 1$ we may assume without loss of generality that $V = \mathbb{R}$, in which case either $B = 0$ or B is generated by its smallest positive element. Now let $n > 1$ and $B \neq 0$. Choose a Euclidean metric on V and an element $g_1 \in B$ of smallest positive norm. Then there is an orthogonal splitting $V = \mathbb{R} \cdot g_1 \oplus W$, where $W = (\mathbb{R} \cdot g_1)^\perp$. Consider the projection $p\colon B \subset \mathbb{R} \cdot g_1 \oplus W \to W$. We claim that the group $p(B) \subset W$ does not contain a nonzero element of norm smaller than $|g_1|/2$.

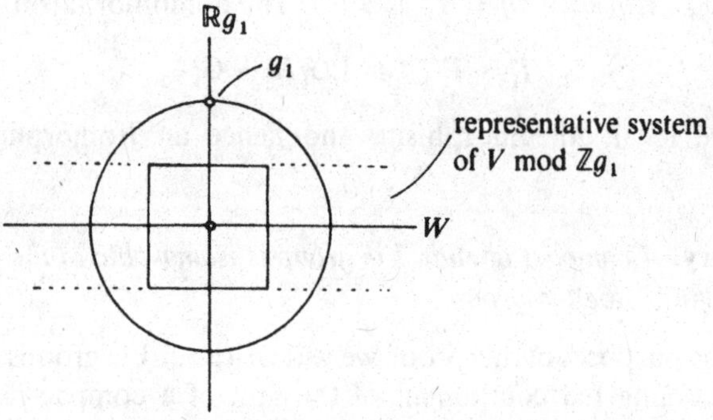

Figure 6

For given $0 < |p(g)| < |g_1|/2$, $g \in B$, there is an $m \in \mathbb{Z}$ such that the projection of $g + mg_1$ onto $\mathbb{R} \cdot g_1$ has norm at most $|g_1|/2$. Thus $g + mg_1 \in B$, but $0 < |g + mg_1| \leq |g_1|/\sqrt{2}$, contradicting the choice of g_1. This means that $p(B)$ is discrete, and by the induction hypothesis is generated in W by linearly independent vectors $h_2, \ldots, h_k$ with $k \leq n$. The kernel of p is generated by g_1, so we have a short exact sequence

$$0 \to \langle g_1 \rangle \to B \xrightarrow[p]{} \langle h_2, \ldots, h_k \rangle \to 0.$$

In this case it is easy to find a section for p—just choose any $g_2, \ldots, g_k \in B$ with $p(g_\nu) = h_\nu$. Therefore B is generated by $g_1, \ldots, g_k$ and these vectors are linearly independent because $p(g_2), \ldots, p(g_k)$ are linearly independent and $p(g_1) = 0$. □

(3.9) Definition. A ***(Lie-) subgroup*** of a Lie group G is an injective homomorphism of Lie groups

$$f: H \to G.$$

Thus every injective one-parameter group is a subgroup and the inclusions of the classical linear groups in (1.14) provide many examples of subgroups. Note, however, that a subgroup need not be an embedding of manifolds. Certainly a subgroup is always immersive, i.e., the tangent map $T_h f$ is injective for all $h \in H$. To see this, it is only necessary to ascertain that Lf is injective, and since the mappings exp are local diffeomorphisms at the origin, the injectivity of f implies the injectivity of Lf in the following commutative diagram:

$$\begin{array}{ccc} LH & \xrightarrow{Lf} & LG \\ {\scriptstyle \exp}\downarrow & & \downarrow{\scriptstyle \exp} \\ H & \xrightarrow[f]{} & G \end{array}$$

But the bijection $f: H \to f(H)$ need not be a homeomorphism! In fact, $f(H)$ may be a dense proper subset of G ((2.22), Ex. 6). The map

$$\mathbb{Z} \to S^1, \qquad n \mapsto e^{in}$$

is just such an example—it is a subgroup with dense image which is not an embedding. Maps like these will prove quite useful to us.

(3.10) Definition. A subset N of an m-dimensional manifold M is called an ***n-dimensional*** (or ***(m − n)-codimensional***) ***submanifold*** of M if every point $p \in N$ possesses a chart of M

$$h: (U, p) \to (U', 0) \subset \mathbb{R}^n \oplus \mathbb{R}^{m-n} = \mathbb{R}^m,$$

such that $h(N \cap U) = \mathbb{R}^n \cap U'$, where $\mathbb{R}^n = \mathbb{R}^n \oplus 0 \subset \mathbb{R}^m$.

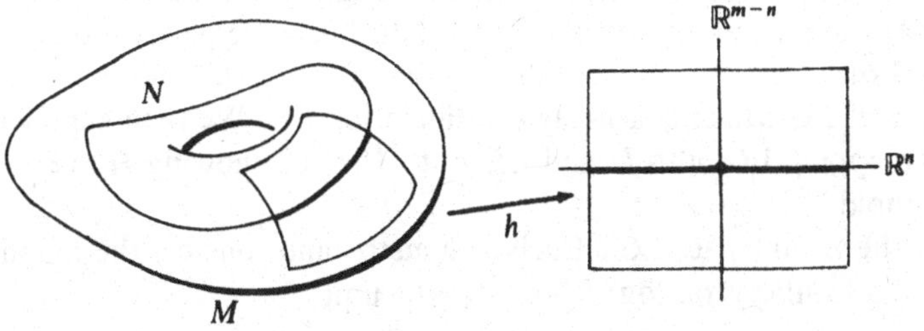

Figure 7

Restricting the charts in the definition to maps $N \cap U \to \mathbb{R}^n \cap U'$ provides us with an atlas and hence the structure of a differentiable manifold for N. Remember that, by convention, everything refers to the differentiable (smooth) category.

A map $f: N \to M$ is called an ***embedding*** if $f(N)$ is a submanifold of M and $f: N \to f(N)$ is a diffeomorphism. An easy point-set topology argument shows that an injective immersion $f: N \to M$ is an embedding if $f: N \to f(N)$ is a homeomorphism (Bröcker and Jänich [1], Prop. (5.7)).

A subset H of a Lie group G is called an ***abstract*** subgroup if, after forgetting the differentiable structure, it is a subgroup in the group-theoretic sense. Thus H is an abstract subgroup of G if $gh^{-1} \in H$ whenever $g, h \in H$. If H is also a submanifold of G, then H is also a Lie group, since the multiplication $H \times H \to H$ is just the restriction of the differentiable multiplication $G \times G \to G$.

(3.11) Theorem. *An abstract subgroup H of a Lie group G is a submanifold of G if and only if H is closed in G.*

Proof. If H is a submanifold of G, then H is locally closed in G. Thus there is a neighborhood U of the unit $e \in G$ such that $H \cap U$ is closed in U. Given $y \in \overline{H}$, let $x \in yU^{-1} \cap H$, so $x \in H$ and $y \in xU$.

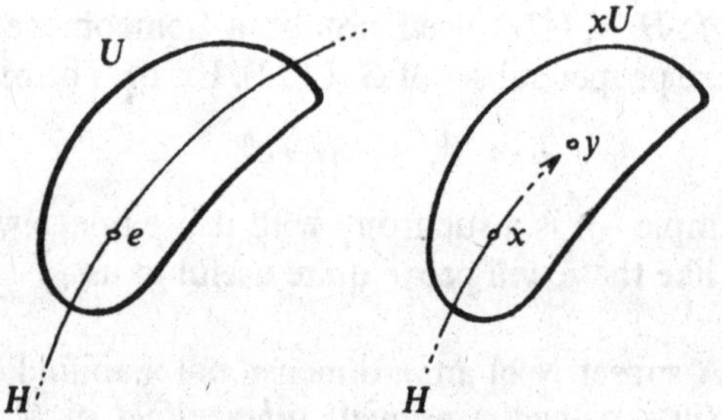

Figure 8

Then $y \in \overline{H} \cap xU$, which implies $x^{-1}y \in \overline{H} \cap U = H \cap U$. Hence $y \in H$ and H is closed.

Conversely, let H be a closed abstract subgroup of the Lie group G. By the usual translation argument it suffices to find a neighborhood U of the unit of G such that $H \cap U$ is a submanifold of U. To this end, we first find a subspace of LG which is a likely candidate for LH. We then exponentiate, locally mapping LG onto U and LH onto $H \cap U$, showing $H \cap U$ to be a submanifold.

Give the vector space LG a Euclidean metric and consider the exponential map and a local inverse log: $U \to U'$ at the unit:

$$\begin{array}{ccc} \mathrm{L}G & \xrightarrow{\exp} & G \supset H \\ \cup & & \cup \\ U' & \underset{\log}{\overset{}{\rightleftarrows}} & U \\ \ni & & \ni \\ 0 & & e \end{array}$$

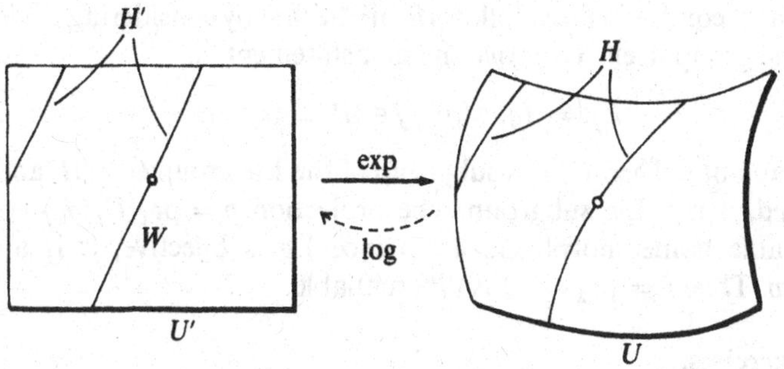

Figure 9

Let $H' = \log(H \cap U)$. The proof now proceeds in three steps, (i), (ii), and (iii).

(i) Let (h_n) be a sequence converging to zero in H' such that $(h_n/|h_n|) \to X \in LG$. Then $\exp(tX) \in H$ for all $t \in \mathbb{R}$.

Proof of (i). As $n \to \infty$, $(t/|h_n|) \cdot h_n \to tX$ and $|h_n| \to 0$. Since $|h_n| \to 0$, we can find $m_n \in \mathbb{Z}$ such that $(m_n \cdot |h_n|) \to t$, so $\exp(m_n \cdot h_n) = \exp(m_n \cdot |h_n| \cdot (h_n/|h_n|)) \to \exp(tX)$. But $\exp(m_n \cdot h_n) = \exp(h_n)^{m_n} \in H$, and H is closed, so $\exp(tX) \in H$.

(ii) The set $W = \{sX \mid X = \lim(h_n/|h_n|),\ h_n \in H',\ s \in \mathbb{R}\}$ is a linear subspace of LG.

Proof of (ii). Let $X, Y \in W$ and $h(t) = \log(\exp(tX) \cdot \exp(tY))$. Then as $t \to 0$ with $t > 0$, $h(t)/t \to X + Y$ by (3.5), and $h(t)/|h(t)| = h(t)/t \cdot t/|h(t)| \to (X + Y)/|X + Y|$, from which (ii) follows.

It remains to show:

(iii) $\exp(W)$ is a neighborhood of the unit in H.

Proof of (iii). Let D be the orthogonal complement of W in LG. The map

$$W \oplus D \to G, \qquad (X, Y) \mapsto \exp(X) \cdot \exp(Y)$$

is locally invertible at the origin. Suppose that (iii) is false. Choose $(X_n, Y_n) \in W \oplus D$ with $\exp(X_n) \cdot \exp(Y_n) \in H$, $Y_n \neq 0$, and $(X_n, Y_n) \to 0$ as $n \to \infty$. Since D is a (closed) subspace, we may find a $Y \in D$ such that, after passing to a subsequence, $Y_n/|Y_n| \to Y$. Note that $|Y| = 1$, so $Y \neq 0$. But since $\exp(X_n) \in H$, and H is a subgroup, $\exp(Y_n) \in H$. Thus $Y \in W$, a contradiction. □

(3.12) Proposition. *Let $f\colon G \to H$ be a group homomorphism between Lie groups which is continuous as a map between manifolds. Then f is differentiable, and hence is really a Lie group homomorphism. In particular, a topological group has at most one Lie group structure.*

PROOF. The second statement follows from the first by considering the identity map of the group itself. To prove the first statement, let

$$\Gamma_f = \{(g, f(g)) | g \in G\} \subset G \times H$$

be the graph of f. Then Γ_f is a subgroup of the Lie group $G \times H$, and, since it is closed, it is a Lie subgroup. The projection $p = \mathrm{pr}_1 | \Gamma_f : \Gamma_f \to G$ is a differentiable homeomorphism, and, since Lp is bijective, it is a diffeomorphism. Thus $f = \mathrm{pr}_2 \circ p^{-1}$ is differentiable. □

(3.13) Exercises.

1. Show that the exponential map of the group SL(2, ℝ) is not surjective. What values can the trace Tr exp(A) take, if $A \in \mathrm{sl}(2, \mathbb{R})$? Calculate the image of the exponential map.
2. Show that the exponential map is surjective for SO(n) and U(n). *Hint*: Each matrix in U(n) is conjugate to a diagonal matrix.
3. Show that an abelian Lie group is the product of a vector space, a torus, and a countable discrete abelian group.
4. Show that the adjoint representation defines a surjective homomorphism Sp(1) = SU(2) → SO(3) with kernel consisting of the two-element set $\{E, -E\}$. As a manifold, SO(3) is diffeomorphic to real projective space $\mathbb{R}P^3$, and this homomorphism is the universal covering $S^3 \to \mathbb{R}P^3$.
5. Show that in every Lie group there is a neighborhood of the unit not containing any subgroup other than $\{e\}$.
6. Show that a compact connected (complex-) holomorphic Lie group is abelian, i.e., it is isomorphic to $\mathbb{C}^n/B$, where B is a discrete subgroup of $\mathbb{C}^n$. *Hint*: The adjoint representation $G \to \mathrm{Aut}(LG)$ is trivial, since holomorphic functions on a compact manifold are locally constant.

4. Homogeneous Spaces and Quotient Groups

We wish to describe the geometry of both the right multiplication of a closed subgroup $H \subset G$ on the Lie group G and the left multiplication of G on G/H. But first we need to introduce some terminology.

(4.1) Definition. Let G be a Lie group. A (left) ***G-space*** is a topological space X together with a continuous ***left operation*** (also called ***left action***) of G on X

$$\Phi: G \times X \to X, \qquad (g, x) \mapsto \Phi(g, x) = g \cdot x$$

such that

$$e \cdot x = x \quad \text{and} \quad (gh) \cdot x = g \cdot (h \cdot x).$$

We will usually denote the G-space just by its underlying topological space X. A map $f: X \to Y$ of G-spaces is called ***equivariant***, or a G-map, if for all $x \in X$ and $g \in G$ we have

$$f(g \cdot x) = g \cdot f(x).$$

A subset A of a G-space X is called ***invariant*** if $g \cdot a \in A$ for all $a \in A$ and $g \in G$. An invariant point is called a ***fixed point***. The subgroup

$$G_x = \{g \in G \mid g \cdot x = x\}$$

of G (closed if X is Hausdorff) is called the ***isotropy group*** of the point $x \in X$, and the invariant subspace

$$G \cdot x = \{g \cdot x \mid g \in G\}$$

of X is called the ***orbit*** of x. The set of all orbits in X is denoted by X/G, and there is a canonical projection

$$\pi = \pi_G: X \to X/G, \qquad x \mapsto G \cdot x.$$

We provide the ***orbit space*** X/G with the quotient topology with respect to this projection. If X/G consists of a single point—in other words, if X contains only one orbit—the G-action is called ***transitive***.

A G-space $\Phi: G \times X \to X$ is said to be ***differentiable***, and is called a ***G-manifold***, if X is a differentiable manifold and Φ is differentiable. We will only be concerned with differentiable G-spaces.

The map $\Phi: G \times X \to X$ may be described by its adjoint

$$G \to \mathrm{Homeo}(X) = \text{group of homeomorphisms } X \to X,$$
$$g \mapsto [x \mapsto g \cdot x].$$

The equations $e \cdot x = x$ and $(gh) \cdot x = g \cdot (h \cdot x)$ just say that this map is well defined and a homomorphism.

If x has isotropy group H, then $g \cdot x$ has isotropy group gHg^{-1}. Thus a conjugacy class of isotropy groups is attached to each orbit. The G-space is divided into disjoint orbits, and invariant subsets are precisely the sets $\pi^{-1}(A)$, where $A \subset X/G$.

Just as there are left actions, there are ***right actions***

$$X \times G \to X, \qquad (x, g) \mapsto x \cdot g,$$

with $x \cdot e = x$ and $x \cdot (gh) = (x \cdot g) \cdot h$. Each right operation canonically gives rise to a left operation via the formula

$$g \cdot x = x \cdot g^{-1},$$

and vice versa.

There is no end to examples: Each finite-dimensional vector space V is an $\mathrm{Aut}(V)$-space, and any homomorphism $\varphi\colon G \to \mathrm{Aut}(V)$ defines a G-action on V by

$$g \cdot v = \varphi(g)v.$$

The G-spaces which interest us at the moment are homogeneous spaces, which we provisionally define as follows: Let $H \subset G$ be a closed subgroup of a Lie group. Then group multiplication defines a right operation $G \times H \to G$, $(g, h) \mapsto gh$, of the group H on the manifold G. The orbit space G/H of right H-cosets of G is called ***homogeneous space***. The group G acts on G/H on the left via

$$G \times G/H \to G/H, \qquad (g, xH) \mapsto gxH.$$

But in order to come to a fuller understanding, we must first have a better description of both the projection $\pi\colon G \to G/H$ and the right action of H on G. For this purpose we use a concept basic to the theory of transformation groups—that of a principal bundle.

(4.2) Definition. Let H be a Lie group. An ***H-principal bundle*** is a ***locally trivial*** right H-space

$$E \times H \to E.$$

The space E is called the ***total space***, H is called the ***structure group***, $B = E/H$ is called the ***base space***, and the canonical map $\pi\colon E \to B$ is called the ***projection*** of the bundle. The orbit $\pi^{-1}\{b\}$ is called the ***fiber*** over $b \in B$. The ***axiom of local triviality*** says: Each point of the base space possesses a neighborhood U together with an equivariant homeomorphism (called a ***bundle chart***)

$$\varphi\colon \pi^{-1}(U) \to U \times H,$$

where we view $U \times H$ as a right H-space, with H operating on the right via multiplication on the second factor: $(u, h) \cdot k = (u, hk)$.

A ***bundle map*** $f\colon E \to E'$ is an H-equivariant continuous map of total spaces.

Locally, that is to say on the open sets $\pi^{-1}(U)$, the total space with its projection looks like the product $U \times H$ with projection onto the first factor.

Two bundle charts $\varphi_j\colon \pi^{-1}U_j \to U_j \times H$, $j = 1, 2$, give a change of bundle charts over $U_1 \cap U_2 = U_{12}$ of the form

$$U_{12} \times H \xrightarrow{\varphi_1^{-1}} \pi^{-1}U_{12} \xrightarrow{\varphi_2} U_{12} \times H, \qquad (u, h) \to (u, \gamma_{12}(u) \cdot h),$$

where the continuous map $\gamma_{12}\colon U_{12} \to H$ is defined by $\varphi_2\varphi_1^{-1}(u, e) = (u, \gamma_{12}(u))$. Here it is important to distinguish right from left: The group H

operates on E on the right, whereas in order to change charts, one multiplies by $\gamma_{12}(u) \in H$ on the left. Note that left and right multiplications commute.

The perhaps simplest example of a nontrivial principal bundle, that is, one which is not globally the product of its base space and fiber, is the double covering of the circle by itself $\pi: S^1 \to S^1$, $z \mapsto z^2$ as a $\mathbb{Z}/2$-principal bundle. Here $S^1 = \{z \in \mathbb{C} \mid |z| = 1\}$.

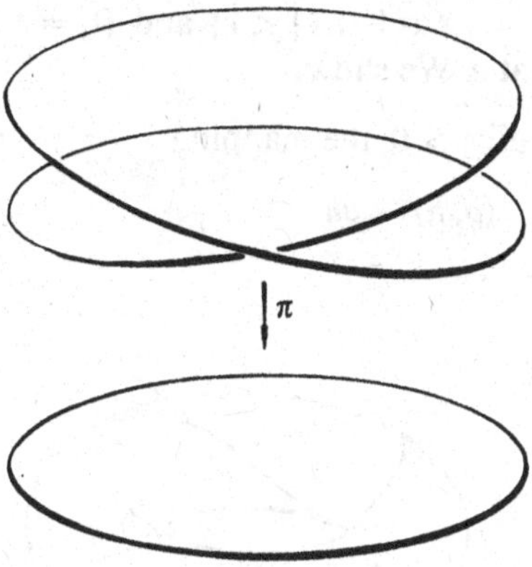

Figure 10

An H-principal bundle is called ***differentiable*** if the total and base spaces are differentiable manifolds, the action of H and the bundle projection are smooth maps, and the bundle charts in the definition may be chosen to be diffeomorphisms. If it exists, the differentiable structure of the base space is determined by that of the total space, for by looking at the bundle charts one sees that a map $f: B \to M$ of the base space is differentiable if and only if the composition $E \xrightarrow{\pi} B \xrightarrow{f} M$ is differentiable. In other words B is the differentiable quotient of E with respect to π.

As one would expect, a family $(\varphi_\lambda: \pi^{-1}U_\lambda \to U_\lambda \times H \mid \lambda \in \Lambda)$ is called a ***bundle atlas*** if the sets U_λ, $\lambda \in \Lambda$, cover the base space.

These concepts will serve here as a geometric description of the projection of a Lie group onto its homogeneous space, $G \to G/H$.

(4.3) Theorem. *Let G be a Lie group and H a closed subgroup. The operation of H on G by right multiplication defines an H-principal bundle with total space G, structure group H, base space G/H, and projection $\pi: G \to G/H$, $g \mapsto gH$. In particular, G/H is a differentiable manifold and π is a submersion, i.e., the rank of $T\pi$ is* dim G/H *at every point.*

PROOF. (i) First we must show that the quotient topology on G/H is Hausdorff. Examples in the exercises ((4.15), Ex. 1 and 2) point out that some care should be taken when proving this. So let $xH \neq yH$ be two points in G/H. Choose a compact neighborhood K of x disjoint from the closed coset yH. Then KH is a neighborhood of xH disjoint from yH, so it suffices to show that KH is closed in G. This is a special case of the general fact that if K is compact and A is closed, then KA is closed in G. To verify this, consider the map $f: K \times G \to G, (k, g) \mapsto k^{-1}g$. Now $x \in KA$ if and only if $k^{-1}x \in A$ for some $k \in K$, so $KA = \mathrm{pr}_2(f^{-1}A)$. But $f^{-1}A$ is closed, and, since K is compact, $\mathrm{pr}_2: K \times G \to G$ is a (proper) closed map. Thus KA is closed.

Next we need to describe the local product structure of G as an H-space. Choose a Euclidean metric on LG, and let L$G = V \oplus$ LH be an orthogonal linear splitting. Let $V_\varepsilon = \{X \in V \mid |X| < \varepsilon\}$ and $D_\varepsilon = \exp(V_\varepsilon)$. We call D_ε a (transverse) ***slice*** to H at e. We show:

(ii) For sufficiently small $\varepsilon > 0$, the mapping

$$\mu: D_\varepsilon \times H \to G, \qquad (g, h) \mapsto gh$$

is an open embedding.

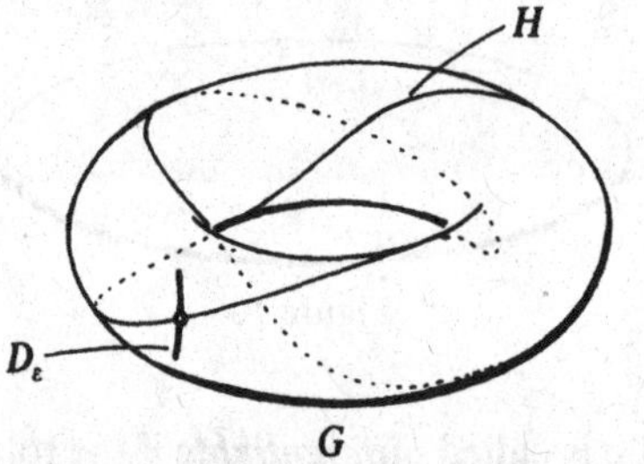

Figure 11

Proof of (ii). The differential of μ at the point (e, e) is the identity on both summands $\mathrm{T}_e D_\varepsilon = V$ and $\mathrm{T}_e H = \mathrm{L}H$ of $\mathrm{T}_{(e,e)}(D_\varepsilon \times H)$. Thus if we choose ε sufficiently small, μ gives a diffeomorphism $D_\varepsilon \times U \to D_\varepsilon \cdot U$ for some neighborhood U of the unit in H. This implies that μ is a local diffeomorphism everywhere, for $\mu | (D_\varepsilon \times Uh) = h \circ (\mu | D_\varepsilon \times U) \circ h^{-1}$.

It remains to show that we can make μ injective by choosing ε small enough. So let $d_1, d_2 \in D_\varepsilon$, $h_1, h_2 \in H$ with $d_1 h_1 = d_2 h_2$. Setting $h = h_1 h_2^{-1}$, we have $d_1 h = d_2$ and we may choose ε so that d_1 and d_2 are so near to the unit that $h = d_1^{-1} d_2$ must lie in U. Since μ is injective on $D_\varepsilon \times U$ and $\mu(d_1, h) = \mu(d_2, e)$, we then get $h_1 = h_2$ and $d_1 = d_2$, showing global injectivity.

(iii) It is more or less obvious how to proceed from here: The sets

$$U_g = gD_\varepsilon \cdot H, \qquad g \in G$$

are invariant under right H-operation, and the U_g/H, $g \in G$, constitute the open cover of the base space G/H needed for an atlas of the manifold G/H and the H-principal bundle G. Charts for G/H are given by the maps $h_g: U_g/H \to D_\varepsilon$ whose inverses are defined by the composition

$$h_g^{-1}: D_\varepsilon = D_\varepsilon \times e \subset D_\varepsilon \times H \underset{\mu}{\to} D_\varepsilon \cdot H \underset{g}{\to} gD_\varepsilon H = U_g \underset{\pi}{\to} U_g/H.$$

Charts for the bundle are given by diffeomorphisms φ_g with inverses

$$\varphi_g^{-1}\colon U_g/H \times H \xrightarrow[h_g \times \mathrm{id}]{} D_\varepsilon \times H \xrightarrow[\mu]{} D_\varepsilon \cdot H \xrightarrow[g]{} gD_\varepsilon H = U_g. \qquad \square$$

We frequently denote a bundle (imprecisely) by a sequence

fiber → total space → base space.

For example,

$$H \to G \to G/H.$$

(4.4) Corollary. *If $N \subset G$ is a closed normal subgroup of a Lie group, then G/N, with its canonical differentiable and group structures, is itself a Lie group. If N is contained in the kernel of a homomorphism $f\colon G \to H$, there is a unique factorization of homomorphisms of Lie groups:*

$$\begin{array}{ccc} G & \xrightarrow{f} & H \\ {\scriptstyle \pi}\downarrow & \nearrow & \\ G/N & & \end{array}$$

PROOF. The group multiplication $G/N \times G/N \to G/N$ is differentiable, as is clear from the commutative diagram

$$\begin{array}{ccc} G \times G & \longrightarrow & G \\ {\scriptstyle \pi \times \pi}\downarrow & & \downarrow{\scriptstyle \pi} \\ G/N \times G/N & \longrightarrow & G/N \end{array}$$

since $\pi \times \pi$ is locally (up to diffeomorphism) the projection of a product onto one of its factors. Similarly, the homomorphism $G/N \to H$ is differentiable. The statement about kernels is also easy. $\square$

Left translation induces a left action of G on its homogeneous space G/H. Since left multiplication and right multiplication commute, that is, $(gx)h = g(xh)$, the translations $l_g\colon G \to G$, $x \mapsto gx$, are bundle maps of the bundle $\pi\colon G \to G/H$. We may now give a more fitting definition of homogeneous space:

(4.5) Definition. A ***homogeneous space*** of a Lie group G is a manifold M with a differentiable transitive left operation of G on M.

(4.6) Proposition. *Let M be a homogeneous space of the Lie group G, let $p \in M$ and let $G_p \subset G$ be the isotropy group of p. Then the map*

$$f_p\colon G/G_p \to M, \qquad gG_p \mapsto gp$$

is well defined and is a diffeomorphism.

PROOF. The definition of G_p implies f_p is well defined and injective, and transitivity implies that f_p is surjective. Since the map $\alpha_p: G \to M$ defined by $\alpha_p(g) = g \cdot p$ is differentiable, so is f_p. It remains to show that f_p is immersive, i.e., that its differential at every point is injective (see (4.15), Ex. 4). Now f_p is G-equivariant and therefore it has constant rank. Since it is injective, it must be immersive by the rank theorem of calculus. □

On one hand, this proposition allows us to recognize as homogeneous spaces many spaces which are well known or, in any case, worth knowing. On the other hand, if we are given a set M on which a Lie group G acts transitively (as a group without differentiable structure) in such a way that the isotropy group G_p of some $p \in M$ is closed, then we can use the bijection $G/G_p \to M$ of the proposition to provide M with the structure of a homogeneous space with this bijection becoming a diffeomorphism. We give some classical examples.

(4.7) The group $SO(n)$ operates linearly on $\mathbb{R}^n$ and transitively when restricted to the sphere $S^{n-1} = \{x \in \mathbb{R}^n \,|\, |x| = 1\}$. The nth unit vector $e_n = (0, \ldots, 0, 1)$ has isotropy group $SO(n-1) \subset SO(n)$, and hence we have a diffeomorphism

$$SO(n)/SO(n-1) \to S^{n-1}, \qquad [A] \mapsto Ae_n$$

and the $SO(n-1)$-principal bundle

$$SO(n-1) \to SO(n) \to S^{n-1}.$$

Incidentally, this yields an inductive proof that $SO(n)$ is connected, for if a bundle has connected fiber and connected base space, then its total space is connected.

The action of $O(n)$ on S^{n-1} leads to a similar fibration

$$O(n-1) \to O(n) \to O(n)/O(n-1) = S^{n-1}.$$

(4.8) The unitary group acts linearly on $\mathbb{C}^n$ and transitively when restricted to $S^{2n-1} = \{x + iy \in \mathbb{C}^n \,|\, |x|^2 + |y|^2 = 1\}$. The isotropy group of the last unit vector in $\mathbb{C}^n$ is $U(n-1) \subset U(n)$. As above, this gives a diffeomorphism $U(n)/U(n-1) \cong S^{2n-1}$ and the $U(n-1)$-principal bundle

$$U(n-1) \to U(n) \to S^{2n-1}.$$

We also find, as a consequence, that $U(n)$ is connected.

As a matter of fact, bundles can be exploited to obtain more delicate topological information relating the homotopy types of the total space, the base space, and the fiber.

By the same methods used above, the operation of $SU(n)$ on the unit sphere leads to the principal bundle

$$SU(n-1) \to SU(n) \to S^{2n-1},$$

and the quaternion algebra $\mathbb{H}$ gives rise to a principal bundle

$$\mathrm{Sp}(n-1) \to \mathrm{Sp}(n) \to S^{4n-1}.$$

(4.9) The group $\mathrm{SU}(n)$ operates linearly on $\mathbb{C}^n$ and the induced operation on the projective space $\mathbb{C}P^{n-1}$ of lines through the origin is transitive. The isotropy group of the point $p = \mathbb{C} \cdot e_n \in \mathbb{C}P^{n-1}$ consists of matrices of the form

$$\left[\begin{array}{c|c} A & \begin{matrix} 0 \\ \vdots \\ 0 \end{matrix} \\ \hline & \alpha \end{array}\right]$$

in $\mathrm{SU}(n)$. Thus this isotropy group is the image of the embedding

$$\mathrm{U}(n-1) \to \mathrm{SU}(n), \qquad A \mapsto \left[\begin{array}{c|c} A & \\ \hline & \alpha \end{array}\right], \qquad \alpha = \det A^{-1}.$$

Interpreting $\mathrm{U}(n-1)$ as a closed subgroup of $\mathrm{SU}(n)$ in this way, we get a principal bundle

$$\mathrm{U}(n-1) \to \mathrm{SU}(n) \to \mathbb{C}P^{n-1}.$$

The same construction yields an embedding $\mathrm{O}(n-1) \to \mathrm{SO}(n)$ and a principal bundle over the real projective space $\mathbb{R}P^{n-1}$ of lines through the origin in $\mathbb{R}^n$

$$\mathrm{O}(n-1) \to \mathrm{SO}(n) \to \mathbb{R}P^{n-1}.$$

(4.10) ***The Hopf fibration of the three-sphere*** was a surprising and important discovery in algebraic topology. It arises from (4.9) in the case $n = 2$. One can identify all three manifolds involved as spheres, namely $\mathrm{U}(1) = S^1$, $\mathrm{SU}(2) = \mathrm{Sp}(1) = S^3$ is the quaternion group, and $\mathbb{C}P^1 = S^2$ is the Riemann sphere. Thus one has a principal bundle

$$S^1 \to S^3 \to S^2.$$

As previously hinted, such a fibration gives a long exact sequence of homotopy groups, and, in this case, one obtains an isomorphism $\mathbb{Z} = \pi_3(S^3) \to \pi_3(S^2)$. In this way H. Hopf [1] discovered the first nontrivial higher homotopy group of a sphere. To this day higher homotopy groups of spheres are an active area of investigation.

(4.11) The manifold of orthonormal k-tuples of vectors in $\mathbb{R}^n$ is called the ***Stiefel manifold*** $\mathrm{V}_k(\mathbb{R}^n)$. Elements of $\mathrm{V}_k(\mathbb{R}^n)$ are often called orthonormal ***k-frames*** in $\mathbb{R}^n$. The orthogonal group $\mathrm{O}(n)$ acts transitively on the set $\mathrm{V}_k(\mathbb{R}^n)$ by

$$A \cdot (v_1, \ldots, v_k) = (Av_1, \ldots, Av_k),$$

and the isotropy group of the k-frame comprised of the last k unit vectors $p = (e_{n-k+1}, \ldots, e_n)$ is $O(n-k)$. Thus there is a bijection

$$O(n)/O(n-k) \to V_k(\mathbb{R}^n), \qquad [A] \mapsto Ap.$$

We use this bijection to define a differentiable structure on $V_k(\mathbb{R}^n)$ and obtain the principal bundle

$$O(n-k) \to O(n) \to V_k(\mathbb{R}^n).$$

(4.12) The manifold of all k-dimensional linear subspaces of $\mathbb{R}^n$ is called the ***Grassmann manifold*** $G_k(\mathbb{R}^n)$. Elements of $G_k(\mathbb{R}^n)$ are sometimes called k-planes in $\mathbb{R}^n$. The group $O(n)$ operates transitively on $G_k(\mathbb{R}^n)$ and the isotropy group of the element $\mathbb{R}^k \in G_k(\mathbb{R}^n)$ is $O(k) \times O(n-k) \subset O(n)$ via the embedding $(A, B) \mapsto \begin{bmatrix} A & 0 \\ 0 & B \end{bmatrix}$. Hence one may provide $G_k(\mathbb{R}^n)$ with a differentiable structure such that $G_k(\mathbb{R}^n) \cong O(n)/(O(k) \times O(n-k))$. This yields the principal bundle

$$O(k) \times O(n-k) \to O(n) \to G_k(\mathbb{R}^n).$$

Analogously there are Grassmann manifolds

$$G_k(\mathbb{C}^n) \cong U(n)/(U(k) \times U(n-k))$$

and

$$G_k(\mathbb{H}^n) \cong Sp(n)/(Sp(k) \times Sp(n-k)).$$

Their descriptions as quotients yield obvious diffeomorphisms $G_k(K^n) \cong G_{n-k}(K^n)$, $K = \mathbb{R}$, $\mathbb{C}$, or $\mathbb{H}$, corresponding to the pairing of a subspace with its orthogonal complement. Grassmann manifolds play an important role in the theory of vector bundles, and lead to yet another encounter with the projective spaces $G_1(\mathbb{R}^n) = \mathbb{R}P^{n-1}$, $G_1(\mathbb{C}^n) = \mathbb{C}P^{n-1}$.

We end this section by tacking on an important and oft-used theorem concerning the torus $T^n = \mathbb{R}^n/\mathbb{Z}^n$. An element $t \in T^n$ is called a ***generator*** if the group $\{t^k \mid k \in \mathbb{Z}\}$, algebraically generated by t, is dense in T^n.

(4.13) Kronecker's Theorem. *A vector $v \in \mathbb{R}^n$ represents a generator of T^n if and only if 1 and the components $v_1, \ldots, v_n$ of v are linearly independent over the rational numbers $\mathbb{Q}$. Thus almost every element is a generator and generators form a dense subset of the torus.*

PROOF. We write group operations additively. The exact sequence

$$0 \to \mathbb{Z}^n \to \mathbb{R}^n \to T^n \to 0$$

defining the torus canonically identifies $\mathbb{R}^n = \mathrm{L}\mathbb{R}^n = \mathrm{L}T^n$, thereby identifying the projection $\mathbb{R}^n \to T^n$ with the exponential map. Thus a homomorphism $f\colon T^n \to S^1$ induces a commutative diagram

$$\begin{array}{ccccccc} 0 & \longrightarrow & \mathbb{Z}^n & \longrightarrow & \mathbb{R}^n & \longrightarrow & T^n \\ & & \vdots\, {\scriptstyle \mathrm{L}f|\mathbb{Z}^n} & & \downarrow {\scriptstyle \mathrm{L}f} & & \downarrow {\scriptstyle f} \\ 0 & \longrightarrow & \mathbb{Z} & \longrightarrow & \mathbb{R} & \longrightarrow & S^1 \end{array}$$

Therefore $\mathrm{L}f(v_1, \ldots, v_n) = \alpha_1 v_1 + \cdots + \alpha_n v_n$ with $\alpha_\nu \in \mathbb{Z}$. Now the following statements are equivalent:

(i) $1, v_1, \ldots, v_n$ are linearly dependent over $\mathbb{Q}$.
(ii) $\sum_\nu q_\nu v_\nu \in \mathbb{Q}$ for some n-tuple $0 \neq (q_1, \ldots, q_n) \in \mathbb{Q}^n$.
(iii) $\sum_\nu \alpha_\nu v_\nu \in \mathbb{Z}$ for some n-tuple $0 \neq (\alpha_1, \ldots, \alpha_n) \in \mathbb{Z}^n$.
(iv) $v \bmod \mathbb{Z}^n$ is in the kernel of a nontrivial homomorphism $f\colon T^n \to S^1$.
(v) $v \bmod \mathbb{Z}^n$ is not a generator.

The equivalences (i) $\Leftrightarrow$ (ii) $\Leftrightarrow$ (iii) are trivial, (iii) $\Leftrightarrow$ (iv) results from what has been said, and (iv) $\Leftrightarrow$ (v) may be seen as follows. Let $[v] \in \ker(f\colon T^n \to S^1)$. If f is nontrivial, this kernel is not all of T^n and hence is a proper closed subgroup of T^n, so $[v]$ cannot be a generator. Conversely, a nongenerator $[v]$ is contained in a proper closed subgroup $H \subset T^n$, and the quotient group T^n/H is a nontrivial compact connected abelian Lie group. Thus T^n/H is a torus T^k, $k > 0$, and $[v]$ is in the kernel of the nontrivial homomorphism

$$T^n \to T^n/H \cong T^k = S^1 \times \cdots \times S^1 \xrightarrow{\mathrm{pr}_1} S^1. \qquad \square$$

(4.14) Corollary. *A compact Lie group contains a dense cyclic subgroup if and only if the group is isomorphic to $T^k \times \mathbb{Z}/l$ for some $k \in \mathbb{N}_0$, $l \in \mathbb{N}$.*

PROOF. If t is a generator of T^k and we choose $\tau \in T^k$ such that (in additive notation) $l \cdot \tau = t$, then $l(\tau, 1) = (t, 0)$ is a generator of $T^k \times 0$ and $\mathrm{pr}_2(\tau, 1) = 1$ generates $\mathbb{Z}/l$. Thus $(\tau, 1)$ generates a dense cyclic subgroup of $T^k \times \mathbb{Z}/l$. Conversely, if $\langle a \rangle$ is dense in G, then G is abelian, so $G \cong T^k \times B$ for some finite abelian group B. But B is generated by $\mathrm{pr}_2(a)$ and hence is cyclic. $\square$

A Lie group is called ***topologically cyclic*** if it contains an element, called a ***generator*** or ***generating element***, whose powers are dense. Thus a topologically cyclic compact Lie group is isomorphic to $T^k \times \mathbb{Z}/l$.

(4.15) Exercises

1. Exhibit two closed sets A and B in $\mathbb{R}^2$ such that $A + B = \{a + b \mid a \in A, b \in B\}$ is not closed in $\mathbb{R}^2$.

2. A global flow on $X = \mathbb{R}^2$ gives X the structure of a left $\mathbb{R}$-space $\Phi\colon \mathbb{R} \times X \to X$. Find a global flow on X such that all integral curves $\alpha_x\colon \mathbb{R} \to X$, $t \mapsto \Phi(t, x)$ are embeddings but $X/\mathbb{R}$ is not Hausdorff.

3. Let G be a compact Lie group and $M \times G \to M$ a differentiable right G-manifold. Assume further that the maps $\alpha_p : G \to M$, $g \mapsto pg$ are injections for every $p \in M$. Show that the G-manifold has the structure of a differentiable G-principal bundle.

4. Show that a bijective immersion of differentiable manifolds is a diffeomorphism. *Hint*: An immersion $f : M \to N$ is locally an embedding. If we had $\dim N > \dim M$, then $f(M)$ would have Lebesgue measure zero (locally) in N (see Bröcker and Jänich [1], §6).

5. Show that any two fibers of the Hopf fibration (4.10) are linked in S^3; see Hopf [1]:

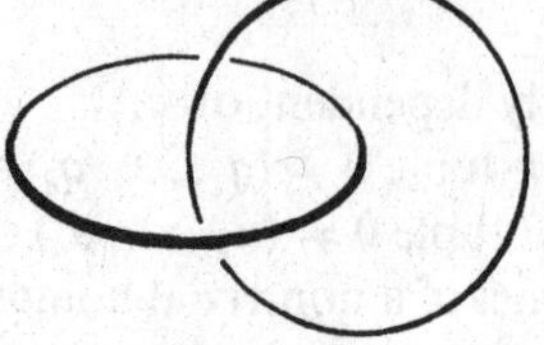

Figure 12

6. Give a diffeomorphism $SO(3) \cong \mathbb{R}P^3$, and, if you know enough about fundamental groups, show that $\pi_1(SO(n)) \cong \mathbb{Z}/2$ for $n > 2$ with a generator represented by the mapping $S^1 = SO(1) \xrightarrow{\subseteq} SO(n)$.

7. Show that $Sp(n)$ is simply connected. Show that $SU(n)$ is simply connected.

8. Show that there is a G-equivariant diffeomorphism of homogeneous spaces $G/H \xrightarrow{\cong} G/K$ if and only if H and K are conjugate in G.

9. Show that a homomorphism of tori $T^n \to T^k$ is induced by a linear map $\mathbb{R}^n \to \mathbb{R}^k$ whose associated matrix has integer coefficients.

10. Show that the only noncompact topologically cyclic Lie group is $\mathbb{Z}$.

11. Let G be a Lie group and let $X, Y \in L(G)$. Show that $[X, Y] = 0$ if and only if $\exp(sX) \cdot \exp(tY) = \exp(tY) \cdot \exp(sX)$ for all $s, t \in \mathbb{R}$. If G is connected, then the Lie algebra of the center of G is $\{X \in LG \mid [X, Y] = 0 \text{ for all } Y \in LG\}$.

5. Invariant Integration

Let X be a locally compact space and $C_c^0(X)$ be the vector space of continuous real-valued functions on X with compact support. An ***integral*** on X is a monotone linear map

$$\int : C_c^0(X) \to \mathbb{R}, \quad f \mapsto \int f.$$

"Monotone" means that if $f(x) \leq g(x)$ for all $x \in X$, then $\int f \leq \int g$. The integral $\int f$ is often denoted $\int_X f(x)\, dx$.

If $X = G$ is a Lie group, an integral on G is called ***left-invariant*** if for all $h \in G$

$$\int f \circ l_h = \int f, \quad \text{i.e.,} \quad \int_G f(h \cdot g)\, dg = \int_G f(g)\, dg.$$

If G is compact, a left-invariant integral is called ***normalized*** if $\int 1 = 1$. We will construct a left-invariant (normalized) integral on a Lie group by using the theory of differential forms on manifolds. We will recall the germane concepts, and, for detailed introductions and expositions, we refer the student to the books of Bröcker [1], Lang [1], and Spivak [1].

First recall the transformation formula for integrals on $\mathbb{R}^n$. If $U \subset \mathbb{R}^n$ is open and $\varphi: U \to V \subset \mathbb{R}^n$ is a diffeomorphism with positive Jacobian determinant, then, for every integrable function $f: V \to \mathbb{R}$,

(5.1)
$$\int_U (f \circ \varphi) \cdot \det D\varphi = \int_V f.$$

Thus the transformation of integrands

$$f \to (f \circ \varphi) \cdot \det D\varphi$$

corresponds to the transformation $\varphi: U \to V$ of domains of integration. The factor $\det D\varphi$ tells us how the oriented volume (the determinant) of an n-frame in the tangent space is transformed under the (linear) tangent map $D\varphi$: If $u_1, \ldots, u_n \in \mathrm{T}_p U = \mathbb{R}^n$, then

$$\det(u_1, \ldots, u_n) \cdot \det(D\varphi) = \det(D\varphi \cdot u_1, \ldots, D\varphi \cdot u_n).$$

So what we really can integrate, independent of coordinates, are not functions but "volume forms," where we interpret the above function f as the volume form which attaches the volume $f(p) \cdot \det(v_1, \ldots, v_n)$ to the n-frame $(v_1, \ldots, v_n) \in (\mathrm{T}_p V)^n$. The general formal definitions, which make the ideas discussed above precise, are as follows:

(5.2) Definitions. An ***alternating k-form*** on a real vector space V is a function

$$\alpha: V^k \to \mathbb{R}, \qquad (v_1, \ldots, v_k) \mapsto \alpha(v_1, \ldots, v_k),$$

which is linear in each variable (multilinear), and ***alternating***, i.e.,

$$\alpha(v_1, \ldots, v_k) = \operatorname{sign}(\sigma) \cdot \alpha(v_{\sigma(1)}, \ldots, v_{\sigma(k)})$$

for every permutation σ of $(1, \ldots, k)$. The sign of a permutation is given by $\operatorname{sign}(\sigma) = \prod_{i<j} (i - j)/(\sigma(i) - \sigma(j))$.

Let $\mathrm{Alt}^k V$ be the real vector space of alternating k-forms on V. An ***alternating differential form*** α of degree k (also called ***alternating k-form*** for short) on a manifold M is a map which attaches an alternating k-form $\alpha_p \in \mathrm{Alt}^k \mathrm{T}_p M$ to each point $p \in M$. We will usually consider ***differentiable*** forms, i.e., we require that, if $X_1, \ldots, X_k$ are differentiable vector fields on M, then the function

$$M \to \mathbb{R}, \qquad p \mapsto \alpha_p(X_1(p), \ldots, X_k(p))$$

is also differentiable.

The collection of alternating k-forms on M, viewed as a module over the ring $C^\infty(M)$ of differentiable functions on M, is denoted by $\Omega^k M$. A differentiable map $\varphi: M \to N$ induces a map

$$\varphi^* = \Omega^k(\varphi): \Omega^k(N) \to \Omega^k(M),$$
$$\varphi^*\alpha(X_1, \ldots, X_k) = \alpha(\mathrm{T}\varphi(X_1), \ldots, \mathrm{T}\varphi(X_k)),$$

making Ω^k into a contravariant functor, i.e., $\mathrm{id}^* = \mathrm{id}$ and $(\varphi \circ \psi)^* = \psi^* \circ \varphi^*$.

For any V there is an ***exterior product***

$$\wedge: \mathrm{Alt}^k V \otimes \mathrm{Alt}^l V \to \mathrm{Alt}^{k+l} V, \qquad \alpha \wedge \beta(X_1, \ldots, X_{k+l})$$
$$= \frac{1}{k!\,l!} \sum_\sigma \mathrm{sign}(\sigma)\alpha(X_{\sigma(1)}, \ldots, X_{\sigma(k)})\beta(X_{\sigma(k+1)}, \ldots, X_{\sigma(k+l)}).$$

In particular, this induces an exterior product (occasionally called "***wedge product***")

$$\wedge: \Omega^k(M) \otimes \Omega^l(M) \to \Omega^{k+l}(M).$$

It should be clear that $\mathrm{Alt}^n \mathbb{R}^n$ is one-dimensional, generated by the determinant, and that locally, after introducing coordinates $(x_1, \ldots, x_n)$, an alternating n-form on an n-dimensional manifold N^n may be written in the form $q \mapsto \alpha_q$, with

$$\alpha_q = f(q) \cdot dx_1 \wedge \cdots \wedge dx_n.$$

Here $dx_1 \wedge \cdots \wedge dx_n$ is simply the determinant corresponding to the basis $(\partial/\partial x_1, \ldots, \partial/\partial x_n)$ of the tangent space—for $dx_i(\partial/\partial x_j)$ is 1 if $i = j$ and 0 if $i \neq j$, and hence

$$dx_1 \wedge \cdots \wedge dx_n\left(\frac{\partial}{\partial x_1}, \ldots, \frac{\partial}{\partial x_n}\right) = 1.$$

If $\varphi: (U, p) \to (V, q)$ is a differentiable map of open subsets of $\mathbb{R}^n$, or the representation of a differentiable map of n-manifolds in local coordinates, then for a form $\alpha = f \cdot dx_1 \wedge \cdots \wedge dx_n$ we have

(5.3) $$(\varphi^*\alpha)_p = f(\varphi(p)) \cdot \det D\varphi_p \cdot dx_1 \wedge \cdots \wedge dx_n.$$

Thus an alternating n-form transforms like the integrand in the transformation formula—namely, like a volume in the tangent space—as long as $\det D\varphi > 0$. (In general, in fact $|\det D\varphi|$ appears in the integration formula.)

The set $\text{Alt}^n(V)\setminus\{0\}$ of nonzero alternating n-forms on the n-dimensional real vector space V splits into two classes $\mathbb{R}_+ \cdot \alpha$ and $\mathbb{R}_- \cdot \alpha$, where α is any element of $\text{Alt}^n(V)\setminus\{0\}$. These classes are called the two ***orientations*** of V.

(5.4) Definition. A ***volume form*** on an n-dimensional manifold M is an alternating differential form $\omega \in \Omega^n(M)$ such that $\omega_p \neq 0$ for all $p \in M$. An ***orientation*** of M is a class $\{f \cdot \omega \mid f: M \to \mathbb{R}_+\}$, where ω is a differentiable volume form. An ***oriented manifold*** is a manifold with an orientation which, when necessary, we indicate by giving the defining volume form. If M_1 and M_2 are oriented by ω_1 and ω_2, then a differentiable map $g: M_1 \to M_2$ with the property that

$$T_p g: T_p M_1 \cong T_{g(p)} M_2 \quad \text{for all } p \in M_1$$

is called ***orientation preserving*** if ω_1 and $g^*\omega_2$ define the same orientation.

The Euclidean space $\mathbb{R}^n$ is always oriented by $dx_1 \wedge \cdots \wedge dx_n = \det$. An open subset $U \subset M$ of an oriented manifold inherits an orientation such that the inclusion map is orientation preserving. An atlas of an oriented manifold is called ***oriented*** if all its charts (and hence all its changes of coordinates) are orientation preserving.

The ***support*** $\text{Supp}(\alpha)$ of a differential form α on M (or of a function or a section of a vector bundle) is the closure of the set $\{p \in M \mid \alpha(p) \neq 0\}$. Thus $p \notin \text{Supp}(\alpha)$ if and only if the germ of α at p vanishes.

Now let M be an n-dimensional oriented manifold and

$$\Omega_c^n(M) = \{\alpha \in \Omega^n(M) \mid \text{Supp}(\alpha) \text{ is compact}\}.$$

Let us assume, for the sake of simplicity, that the forms we consider are continuous.

(5.5) Proposition. *There is a unique integral*

$$\int_M : \Omega_c^n M \to \mathbb{R}, \qquad \alpha \mapsto \int_M \alpha$$

determined by the following properties:

(i) *If* $h: U \to U' \subset \mathbb{R}^n$ *is an orientation preserving chart of* M *and* $\text{Supp}(\alpha) \subset U$, *then*

$$\int_M \alpha = \int_U \alpha = \int_{U'} h^{-1*}\alpha = \int_{U'} a(x) \cdot dx_1 \ldots dx_n,$$

where $h^{-1*}\alpha = a \cdot dx_1 \wedge \cdots \wedge dx_n$.

(ii) *The map* $\int_M: \Omega_c^n M \to \mathbb{R}$ *is linear.*

PROOF. The definition (i) of the integral is independent of the choice of chart by (5.1) and (5.3)—that is to say the integral remains invariant under orientation preserving changes of coordinates. Now if $\alpha \in \Omega_c^n(M)$ is arbitrary, we can cover Supp(α) by finitely many chart domains U_i, $i = 1, \ldots, k$, and split α into $\alpha = \alpha_1 + \cdots + \alpha_k$ with $\text{Supp}(\alpha_i) \subset U_i$ (see (5.6) below). Then $\int \alpha = \sum_i \int \alpha_i$ by (ii) and each $\int \alpha_i$ is determined by (i), so we need only argue that the value of $\int \alpha$ is independent of the splitting of α and the choice of chart domains. But if we cover U by different chart domains $V_1, \ldots, V_l$, we have a splitting $\alpha_i = \sum_j \alpha_{ij}$ with $\text{Supp}(\alpha_{ij}) \subset U_i \cap V_j$, and the result follows from summing $\int \alpha_{ij}$ in two different orders. □

We pause to say a few words about the technical device needed to obtain the splittings of forms in the proof above.

(5.6) Definition and Notation. Let M be a manifold and $\mathfrak{U} = (U_\lambda \mid \lambda \in \Lambda)$ an open cover of M. A ***partition of unity subordinate to the cover*** $\mathfrak{U}$ is a family of functions $\varphi_j \colon M \to [0, 1]$, $j \in \mathbb{N}$ with the properties:

(i) The family is locally finite. This means that each point $p \in M$ possesses a neighborhood U such that $U \cap \text{Supp}(\varphi_j) = \emptyset$ for all but finitely many j.
(ii) For every $j \in \mathbb{N}$ there is a $\lambda \in \Lambda$ with $\text{Supp}(\varphi_j) \subset U_\lambda$.
(iii) $\sum_{j=1}^{\infty} \varphi_j(p) = 1$ for every $p \in M$. Note that, due to (i), this is always a finite sum.

Given any open cover $\mathfrak{U}$ of M, there is a differentiable partition of unity subordinate to $\mathfrak{U}$ (see Bröcker and Jänich [1]).

For the proof above we may choose a partition of unity (φ_j) subordinate to the cover of M formed by the U_i and $M \setminus \text{Supp}(\alpha)$. Then we obtain the splitting $\alpha = (\sum \varphi_j) \cdot \alpha = \sum \varphi_j \alpha = \sum \alpha_j$, with $\alpha_j = \varphi_j \cdot \alpha$. We need not insist on differentiability in this case. All we need is that the function $a(x)$ in (5.5)(i), which is the function to be integrated in the end, is really integrable. This is easy to do.

In most cases one prefers to integrate functions, not forms, and this is carried out as follows:

(5.7) Definition. Let M be a manifold with a volume form ω and $f \colon M \to \mathbb{R}$ a function—say, continuous with compact support. Then

$$\int_M f := \int_M f \cdot \omega.$$

In other words, the integral of f is defined to be the integral of the form $f \cdot \omega$ on M with orientation defined by ω.

Thus to integrate functions on a manifold, we need the additional structure ω which attaches a volume to each n-frame of the tangent space T_pM at every point $p \in M$. Observe that the volume forms ω and $-\omega$ yield the same *integral of functions*, because $f \cdot \omega$ and the orientation $[\omega]$ get simultaneously replaced by $-f\omega$ and $[-\omega]$.

(5.8) Remark. Let M be oriented by ω and let $-M$ denote the manifold M, oriented by $-\omega$. Then for $\alpha \in \Omega_c^n(M)$

$$-\int_M \alpha = \int_{-M} \alpha.$$

Thus if $\varphi: N \to M$ is a diffeomorphism,

$$\int_M \alpha = \int_N \varepsilon \cdot \varphi^* \alpha,$$

where ε is locally constant with value 1 or -1 according to whether φ locally preserves or reverses orientation.

Proof. If $\varphi: N \to M$ is an orientation preserving diffeomorphism, and $\varepsilon = 1$, the second formula follows from computing the left-hand side using charts h_λ of M and the right-hand side using the charts $h_\lambda \circ \varphi$ of N. Hence we may deduce the second formula in full generality by applying the first formula separately to connected components. To prove the first formula, we may suppose that we have a chart $h: U \to U' \subset \mathbb{R}^n$, oriented for the ω-orientation, with $\operatorname{Supp}(\alpha) \subset U$. Then

$$\int_M \alpha = \int_U \alpha = \int_{U'} h^{-1*}\alpha = \int_{U'} a\, dx_1 \wedge \cdots \wedge dx_n = \int_{U'} a(x)\, dx_1 \ldots dx_n$$

for some function $a: U' \to \mathbb{R}$. The transformation

$$\psi: U' \to \mathbb{R}^n, x \mapsto (-x_1, x_2, \ldots, x_n)$$

is orientation reversing, so

$$\begin{aligned}\int_{-M} \alpha &= \int_{\psi U'} (\psi h)^{-1*}\alpha = \int_{\psi U'} \psi^{-1*} h^{-1*}\alpha \\ &= -\int_{\psi U'} a(-x_1, x_2, \ldots, x_n)\, dx_1 \wedge \cdots \wedge dx_n \\ &= -\int_{\psi U'} a(-x_1, x_2, \ldots, x_n)\, dx_1 \ldots dx_n \\ &= -\int_{U'} a(x)\, dx_1 \ldots dx_n = -\int_M \alpha.\end{aligned}$$ □

The key to this computation is that a factor of $\det D\psi^{-1}$ enters into ψ^{-1*}, but the usual integral of functions on $\mathbb{R}^n$ transforms using the absolute value $|\det D\psi^{-1}|$.

If Supp(f) is contained in the domain U of some local coordinates $x_1, \ldots, x_n$, we express the above a bit imprecisely as

$$\text{(5.9)} \qquad \int f \cdot \omega = \int f(x) \cdot \left| \omega\left(\frac{\partial}{\partial x_1}, \ldots, \frac{\partial}{\partial x_n}\right) \right| dx_1 \ldots dx_n.$$

By the way, if $\dim M = 0$, then $T_pM = 0$, $\mathrm{Alt}^n T_pM = \mathrm{Alt}^0 T_pM = \mathbb{R}$, $\Omega^n(M)$ is the vector space of functions $M \to \mathbb{R}$, and a volume form is a function $M \to \mathbb{R}\backslash\{0\}$. The integral of a function $f\colon M \to \mathbb{R}$ with respect to ω is

$$\text{(5.10)} \qquad \int_M f \cdot \omega = \sum_{p \in M} f(p) \cdot |\omega(p)|.$$

In the case of a Lie group, there is, up to a constant factor, a canonical choice of a volume form:

(5.11) Definition. Let G be a Lie group. A form $\alpha \in \Omega^k G$ is called ***left-invariant*** if $l_g^*\alpha = \alpha$ for every left translation l_g with $g \in G$. The correspondence $\alpha \mapsto \alpha_e$ defines a canonical bijection between the space of left-invariant forms in $\Omega^k G$ and $\mathrm{Alt}^k\, LG$. From now on we will identify these spaces in this fashion.

If G is compact, then, up to sign, there is precisely one left-invariant volume form dg such that $\int_G dg = 1$. This determines a well-defined normalized integral of functions, the ***invariant*** (Haar-) ***integral***. We will only consider integrals on compact groups, where we write the invariant integral using familiar notation

$$\int_G f\, dg = \int_G f(g)\, dg$$

(5.12) Theorem. *Let G be a compact Lie group, $\varphi\colon G \to G$ be any automorphism, and $h \in G$ be any element. Then (referring to the invariant integral)*

$$\int f(g)\, dg \underset{\text{(i)}}{=} \int f(hg)\, dg \underset{\text{(ii)}}{=} \int f(gh)\, dg \underset{\text{(iii)}}{=} \int f(g^{-1})\, dg \underset{\text{(iv)}}{=} \int f \circ \varphi(g)\, dg.$$

Proof. Left invariance, i.e., equality (i), follows from the definition of dg and the observation that l_h is orientation preserving: $\int f \cdot dg = \int f \circ l_h \cdot l_h^*\, dg = \int f \circ l_h\, dg$. Now we will show below, (5.13), that there is only one left-invariant normalized integral on G.

If $\varphi\colon G \to G$ is a diffeomorphism, then the map $f \mapsto \int f \circ \varphi\, dg$ is linear, monotone, and normalized. It follows that if this map is also left-invariant, i.e., if $\int f \circ l_h \circ \varphi \cdot dg = \int f \circ \varphi \cdot dg$ or, in other notation, if $\int f(h \cdot \varphi(g))\, dg = \int f\varphi(g)\, dg$, then $\int f \cdot dg = \int f \circ \varphi \cdot dg$. We apply this principle to the remaining equalities in the theorem.

(ii) Right invariance: Choose $\varphi(g) = gh$; left invariance follows from (i) since $\int f(kgh)\, dg = \int f(gh)\, dg$.

(iii) Choosing $\varphi(g) = g^{-1}$, $\int f(hg^{-1})\,dg = \int f((gh^{-1})^{-1})\,dg = \int f(g^{-1})\,dg$ the last equality coming from (ii) applied to the composition of f with group inversion.

(iv) Choosing $\varphi = \varphi$ and letting $k = \varphi^{-1}(h)$ we have

$$\int f(h \cdot \varphi(g))\,dg = \int f \circ \varphi(kg)\,dg = \int f \circ \varphi(g)\,dg \text{ by (i).} \qquad \square$$

(5.13) Theorem. *Let G be a compact Lie group and C(G) be the real vector space of continuous functions on G. The invariant integral* $C(G) \to \mathbb{R}$, $f \mapsto \int f(g)\,dg$ *is uniquely determined by the following properties*:

(i) *It is linear, monotone, and normalized* ($\int 1 = 1$).
(ii) *Left invariance*: $\int f \circ l_h = \int f$ *for any* $h \in G$.

Proof. Suppose $f \mapsto \int f(h)\,\delta h$ denotes an integral with the given properties. Let $f \mapsto \int f(g)\,dg$ be a particular ***right-invariant*** integral, so $\int f(gh)\,dg = \int f(g)\,dg$. It is our intention to compare all possible left-invariant integrals with the given right-invariant integral, but we first remark that the concept of a group is sufficiently symmetric to guarantee the existence of a right-invariant integral, assuming the existence of a left-invariant integral. The two integrals given so far define a monotone, linear, normalized integral for continuous functions on $G \times G$: If $f: G \times G \to \mathbb{R}$, then

$$\int f(g, h) = \int\left(\int f(g, h)\,dg\right)\delta h = \int\left(\int f(g, h)\,\delta h\right)dg.$$

The second equality (a version of Fubini's theorem) may be seen as follows: If $f(g, h) = \varphi(g) \cdot \psi(h)$, the equality is clear. Next, by using approximation techniques from analysis or partitions of unity, one shows that the functions $(g, h) \mapsto \varphi(g) \cdot \psi(h)$, $\varphi, \psi \in C(G)$, span a dense subspace of $C(G \times G)$ in the topology of uniform convergence. But a linear, monotone, normalized integral commutes with uniform limits, since from $-\varepsilon \le f \le \varepsilon$ one gets $-\varepsilon \le \int f \le \varepsilon$. It follows that the equality in question is valid in general.

Now if $f: G \to \mathbb{R}$ is continuous, we can apply the above to the function $(g, h) \mapsto f(gh)$. This, together with the invariance and normalization properties of our integrals, yields

$$\begin{aligned}\int f(g)\,dg &= \int\left(\int f(g)\,dg\right)\delta h = \int\left(\int f(gh)\,dg\right)\delta h\\ &= \int\left(\int f(gh)\,\delta h\right)dg = \int f(h)\,\delta h,\end{aligned}$$

showing uniqueness. $\square$

Once one can integrate real-valued functions on G, there is a well-defined integral $\int f\, dg \in B$ for continuous functions $f\colon G \to B$ taking values in a Banach space (see, for example, Lang [1], Ch. XII). If B is a Hilbert space, there is a straightforward definition of this integral:

Let $f\colon G \to B$ be continuous and $b \in B$. Then

$$\langle f, b\rangle\colon G \to \mathbb{R}, \qquad g \mapsto \langle f(g), b\rangle$$

is a continuous real-valued function, and the map from B to $\mathbb{R}$ sending b to $\int \langle f(g), b\rangle\, dg$ is a continuous linear functional. But every such functional is given by scalar product with an element of B. We define $\int f(g)\, dg$ to be this element, so

$$\left\langle \int f(g)\, dg, b \right\rangle = \int \langle f(g), b\rangle\, dg.$$

More generally, if B is a space of functions $M \to \mathbb{C}$ (or $\mathbb{R}$), we may define the integral of a map $f\colon G \to B$ by

$$\int f\, dg\colon M \to \mathbb{C}, \qquad x \mapsto \int f(g)(x)\, dg.$$

In each specific case one must check whether this integral is contained in B. For a well-behaved example, if $f\colon G \times M \to \mathbb{C}$ is continuous, and $f\,|\,g \times M$ is holomorphic for each $g \in G$, then the map $M \to \mathbb{C}$, $x \mapsto \int f(g, x)\, dg$ is also holomorphic, since we have assumed that G is compact.

The invariant integral on a group is a tool of fundamental importance: integration is used to average functions over G, thereby making them invariant. For most applications which concern us, it suffices to know that an invariant integral exists. Thus one could assume this and forget about differential forms—in fact, a left-invariant integral, uniquely determined up to a constant factor, may be constructed for any locally compact topological group, see Hewitt and Ross [1] or the classical monograph of A. Weil [2]. But we will use the theory of differential forms as a geometric tool.

We will also integrate functions on homogeneous spaces. As long as we only deal with left-invariant integration, no problem arises.

Let H be a closed subgroup of the compact Lie group G and $\pi\colon G \to G/H$ be the canonical projection. For continuous $f\colon G/H \to \mathbb{R}$ we may define a normalized integral

$$\int_{G/H} f = \int_G f \circ \pi\, dg$$

which is invariant under the left operation of G on G/H. We would like to describe this integral by a volume form on G/H.

Now, in general, this is not possible, since a homogeneous space need not be orientable—examples include the projective spaces $\mathbb{R}P^{2n}$ ((4.9) and (5.20), Ex. 2). Nonetheless, we recall that the subgroup H acts on the left on G by conjugation and on G/H by left translation. Furthermore, the projection $\pi\colon G \to G/H$ is equivariant for these H-operations: $(hgh^{-1})H = hgH$. The coset containing the unit $H \in G/H$ is fixed by the H-operation on G/H, and the induced operation on the tangent space is denoted by $\mathrm{Ad}_{G/H}$, so

$$\text{(5.14)}\qquad \begin{aligned} l_g\colon G/H &\to G/H, & xH &\mapsto gxH,\\ \mathrm{Ad}_{G/H}\colon H &\to \mathrm{Aut}\, \mathrm{T}_H G/H, & h &\mapsto \mathrm{T}_H l_h. \end{aligned}$$

If $\dim G/H = k$, applying the contravariant functor Alt^k to the linear operation $\mathrm{Ad}_{G/H}$ yields a linear operation

$$\mathrm{Ad}^*_{G/H}\colon H \to \mathrm{Aut}(\mathrm{Alt}^k\, \mathrm{T}_H G/H), \qquad h \mapsto l_h^{-1*}.$$

(5.15) Proposition. *Let G be a compact Lie group and H a closed subgroup. If the operation $\mathrm{Ad}^*_{G/H}$ is trivial, i.e., if l_h^* is the identity of $\mathrm{Alt}^k\, \mathrm{T}_H G/H$ for all $h \in H$, then there is a volume form $d(gH)$ on the homogeneous space G/H that, up to sign, is determined by the properties*

(i) *$d(gH)$ is left-invariant, i.e., $l_h^*\, d(gH) = d(gH)$ for all $h \in G$.*
(ii) *$d(gH)$ is normalized, i.e., $\int_{G/H} d(gH) = 1$.*

In particular, the hypothesis and hence the conclusion holds if H is connected.

Proof. Since $\mathrm{Alt}^k\, \mathrm{T}_H G/H$ is one-dimensional, its automorphism group is the multiplicative group $\mathbb{R}^*$, and since the image of $\mathrm{Ad}^*_{G/H}$ is a compact subgroup, it is a subgroup of $\{1, -1\} \cong \mathbb{Z}/2$. Thus if H is connected, $\mathrm{Ad}^*_{G/H}$ is trivial, verifying the last sentence.

Now a left-invariant volume form on G/H is determined by its value at the point $H \in G/H$, because

$$\lambda_{gH} = (l_g^{-1})^* \lambda_H$$

if $\lambda \in \Omega^k G/H$ is left-invariant. The only question is whether or not the right-hand side is independent of the choice of representative $g \in gH$. But for $h \in H$,

$$(l_{gh}^{-1})^* \lambda_H = (l_h^{-1} \circ l_g^{-1})^* \lambda_H = l_g^{-1*} \circ l_h^{-1*} \lambda_H = l_g^{-1*} \lambda_H,$$

the last equality holding due to the assumption that $\mathrm{Ad}^*_{G/H}$ is trivial. Thus assigning a value at $H \in G/H$ determines a volume form and a volume form is uniquely determined by this value, and the proposition follows. □

(5.16) Proposition ("Fubini"). *Let G be a compact Lie group, H a closed subgroup, and $d(gH)$ a left-invariant normalized volume form on G/H. For any continuous real-valued function f on G,*

$$\int_G f(g)\, dg = \int_{G/H} \left(\int_H f(gh)\, dh \right) d(gH).$$

Proof. Evidently the right-hand side defines a normalized left-invariant integral on G, so the proposition follows from the uniqueness of such an integral (5.13). □

Thus for a function $f: G/H \to \mathbb{R}$ we get

$$\int_G f \circ \pi \, dg = \int_{G/H} \left(\int_H f \circ \pi(gh) \, dh \right) d(gH) = \int_{G/H} f \, d(gH),$$

giving the desired description of $\int_{G/H} f$ using a volume form.

In order to clarify the geometric result below, (5.19), we need to quote a bit more from the theory of differential forms and integration on manifolds.

First recall the ***exterior derivative*** of (differentiable) differential forms

$$d: \Omega^k M \to \Omega^{k+1} M.$$

This may be defined in local coordinates $x_1, \ldots, x_n$ by

$$d(\varphi \cdot dx_{i_1} \wedge \cdots \wedge dx_{i_k}) = d\varphi \wedge dx_{i_1} \wedge \cdots \wedge dx_{i_k},$$

but we do not need to delve into local coordinates here. The reader should also be acquainted with ***manifolds with boundary***. They possess an atlas with charts $M \supset U \to U' \subset \mathbb{R}^n_- = \{x \in \mathbb{R}^n \mid x_1 \leq 0\}$ with differentiable changes of coordinates. The ***boundary*** ∂M consists of all points which are mapped by charts to $\mathbb{R}^{n-1} = \{x \in \mathbb{R}^n \mid x_1 = 0\}$.

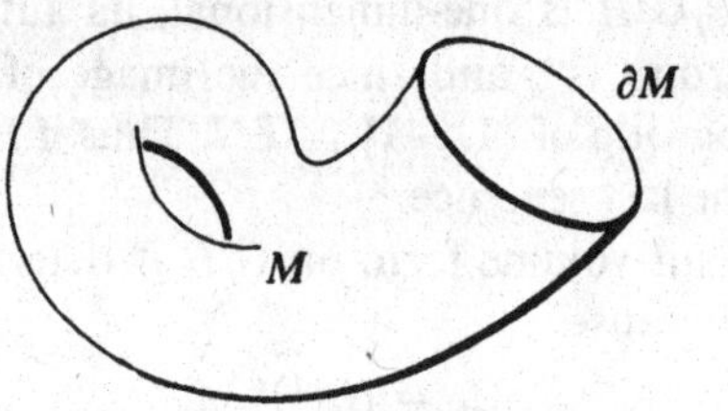

Figure 13

An orientation of M induces a canonical orientation of the boundary ∂M, since the hyperplane $T_p \partial M \subset T_p M$ for $p \in \partial M$ has a canonically defined "side" corresponding to normal vectors pointing "away from M." The following is a beautiful and fundamental result relating these concepts.

(5.17) Stokes' Theorem. *Let M be an oriented n-dimensional manifold with boundary, and let α be an alternating differential form on M of degree $n - 1$ with compact support. Then*

$$\int_{\partial M} \alpha = \int_M d\alpha.$$

See Bröcker [1], V,3, or Lang [1], XVIII, §5.

(5.18) Corollary. *Let M be a compact oriented n-dimensional manifold and let N be an arbitrary n-dimensional manifold (manifolds without boundary). Let $\alpha \in \Omega^n N$. If $f_0, f_1: M \to N$ are differentiably homotopic, then*

$$\int_M f_0^*\alpha = \int_M f_1^*\alpha.$$

PROOF. The homotopy is a map $f: M \times I \to N$, where I is the unit interval and f coincides with f_0 on $M = M \times \{0\}$ and with f_1 on $M = M \times \{1\}$. Since $d\alpha \in \Omega^{n+1}N = 0$, Stokes' theorem and (5.8) give

$$0 = \int_{M \times I} f^* \, d\alpha = \int_{M \times I} df^*\alpha = \int_{\partial(M \times I)} f^*\alpha = \int_M f_1^*\alpha + \int_{-M} f_0^*\alpha$$

$$= \int_M f_1^*\alpha - \int_M f_0^*\alpha.$$ □

(5.19) Theorem on Mapping Degrees. *Let M, N be compact, connected, oriented, n-dimensional manifolds. There is an integer $\deg(f)$ assigned to each homotopy class of (differentiable) maps $f: M \to N$, called the* ***mapping degree*** *of f, such that, for every form $\alpha \in \Omega^n(N)$,*

$$\int_M f^*\alpha = \deg(f) \cdot \int_N \alpha.$$

If $q \in N$ is a point with $f^{-1}\{q\}$ consisting of $k + l$ points $p_1, \ldots, p_{k+l}$ such that f is regular (i.e., Tf is bijective) at each p_i and preserves orientation at $p_1, \ldots, p_k$ but reverses orientation at $p_{k+1}, \ldots, p_{k+l}$, then $\deg(f) = k - l$. In particular, if $\deg(f) \neq 0$, f is surjective.

PROOF. We know that $\int_M f^*\alpha$ depends only on the homotopy class of f. Sard's theorem (Bröcker and Jänich [1], Milnor [1]) insures that a given differentiable map $f: M \to N$ has a regular value q, i.e., there is a point $q \in N$ such that Tf is an isomorphism at every $p \in f^{-1}\{q\}$. In fact, Sard's theorem says that almost every point in N is regular. So let q be regular with $f^{-1}\{q\} = \{p_1, \ldots, p_{k+l}\}$ as described in the statement of the theorem. Then f is a local diffeomorphism around each p_i, and the complement of an open neighborhood of $f^{-1}\{q\}$ is mapped by f to a compact set not containing q. Hence we may choose small open balls B about q and $B_1, \ldots, B_{k+l}$ about $p_1, \ldots, p_{k+l}$ such that

$$f \mid B_i: B_i \xrightarrow{\cong} B$$

and $f^{-1}(B) = \bigcup_i B_i$ (disjoint union). The orientation of f is constant on each B_i and hence equal to its value at p_i. Now if we had $\mathrm{Supp}(\alpha) \subset B$, the theorem would hold by (5.8) for this particular choice of q.

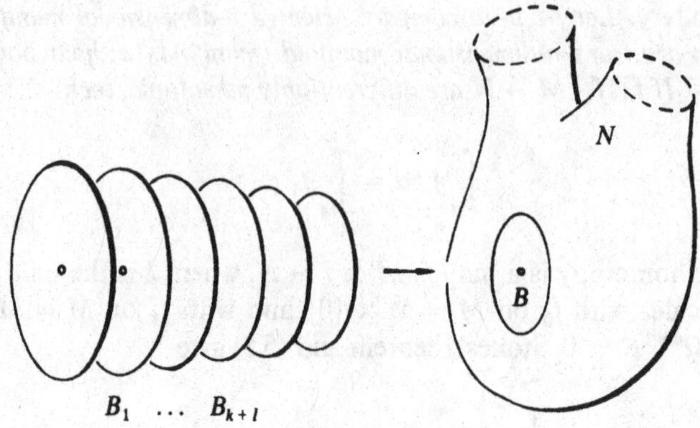

Figure 14

Consequently, if we could find a diffeomorphism $\varphi: N \to N$ homotopic to the identity on N with $\operatorname{Supp}(\alpha) \subset \varphi(B)$, the theorem would be proved, since $\operatorname{Supp}(\varphi^*\alpha) \subset B$, and the integrals are homotopy invariant. But we will show below that the sets $\varphi(B)$ cover N, where φ runs through diffeomorphisms $N \to N$ homotopic to the identity. Thus we choose a partition of unity $(\psi_j | j \in \mathbb{N})$ subordinate to this covering, and the theorem follows since it is valid for each summand of the splitting $\alpha = \sum_j \psi_j \cdot \alpha$.

To show that the $\varphi(B)$ cover N we show that, given $x \in N$, there is a φ as above with $\varphi(q) = x$. If x and q are both contained in a compact ball of a chart domain, it is fairly easy to construct such a diffeomorphism φ, for example by integrating an appropriate vector field which vanishes outside the ball. And from this case we derive the general case by joining q and x with a chain

$$q = x_0, x_1, \ldots, x_r = x$$

such that x_j and x_{j+1} are always contained in a compact ball in some chart domain. □

We intend to apply this theorem in IV, (1.7), where we will explicitly be given a regular value (a generator of a torus). Thus, for our purposes, we could incorporate the existence of a regular value as an assumption in our theorem. Then we would not need to resort to Sard's theorem.

(5.20) Exercises

1. Formulate and prove the converse of proposition (5.15).
2. Show that $\mathbb{R}P^{2n}$ is not orientable. Use the facts that the antipodal map $\tau: S^{2n} \to S^{2n}$, $x \mapsto -x$ is orientation reversing and that $\mathbb{R}P^k = S^k/\tau$.
3. Prove that if a manifold has an atlas all of whose changes of coordinates are orientation preserving then it is orientable. *Hint*: Partitions of unity.

4. Construction of partitions of unity: Show that there is a C^∞-function $\varphi: \mathbb{R}^n \to \mathbb{R}$ such that $\varphi(x) > 0$ for $|x| < 1$ and $\varphi(x) = 0$ for $|x| \geq 1$. One way to do this is to start with the function $\lambda: \mathbb{R} \to \mathbb{R}$ given by

$$\lambda(t) = \begin{cases} \exp(-t^{-1}) & \text{for } t > 0, \\ 0 & \text{for } t \leq 0. \end{cases}$$

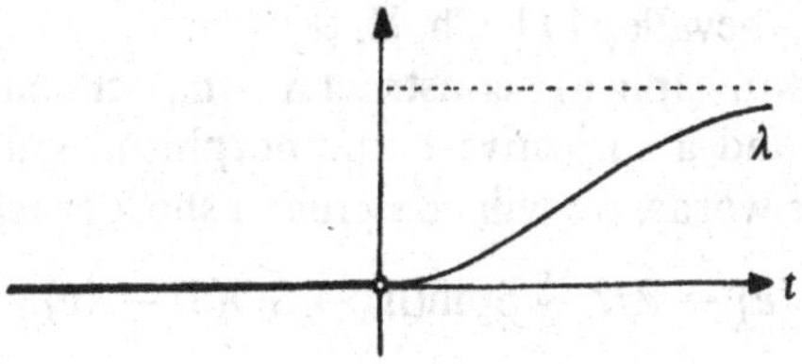

Figure 15

Show that if M is compact and $\{U_\alpha\}$ is an open cover of M, then there is a partition of unity subordinate to the cover.

5. The manifold $\mathbb{R} \times \mathbb{R}$ with the group structure

$$(a_1, b_1) \cdot (a_2, b_2) = (a_1 + \exp(b_1) \cdot a_2, b_1 + b_2)$$

is a Lie group (in fact it is a semidirect product of $\mathbb{R}$ with itself). Show that a left-invariant integral on this group is not right-invariant, and that the left-invariant integral is not invariant under conjugation.

6. Euler angles: Let α, β be the one-parameter groups of SO(3) given by

$$\alpha(t) = \begin{bmatrix} \cos t & -\sin t & 0 \\ \sin t & \cos t & 0 \\ 0 & 0 & 1 \end{bmatrix}, \qquad \beta(t) = \begin{bmatrix} 1 & 0 & 0 \\ 0 & \cos t & -\sin t \\ 0 & \sin t & \cos t \end{bmatrix}.$$

Show that the map $\gamma: T^3 \to \mathrm{SO}(3)$, $(\varphi, \vartheta, \psi) \mapsto \alpha(\varphi) \cdot \beta(\vartheta) \cdot \alpha(\psi)$, $0 \leq \varphi, \vartheta, \psi \leq 2\pi$ is surjective. Which points of T^3 are sent to the same place by γ, i.e., how can SO(3) be described as a quotient of T^3 using γ? Let dg be the normalized invariant volume form on SO(3). Then

$$\gamma^* dg = \pm \frac{1}{8\pi^2} \sin \vartheta \cdot d\varphi \wedge d\vartheta \wedge d\psi.$$

The invariant integral on SO(3) is given by

$$f \mapsto \frac{1}{8\pi^2} \int_0^{2\pi} \int_0^{\pi} \int_0^{2\pi} \sin \vartheta \cdot f \circ \gamma(\varphi, \vartheta, \psi)\, d\varphi\, d\vartheta\, d\psi.$$

The parameters φ, ϑ, and ψ are called ***Euler angles***.

7. Let $f: G \to H$ be a surjective homomorphism of compact Lie groups of equal dimension. Show that every map homotopic to f is surjective.

8. Show that given two points $p, q \in \mathbb{R}^n$ there is a diffeomorphism $\varphi: \mathbb{R}^n \to \mathbb{R}^n$ with $\varphi(p) = q$ such that φ is homotopic to the identity via φ_t, where outside of a compact ball $\varphi_t(x) = x$ for all t and x.

6. Clifford Algebras and Spinor Groups

Assuming a little knowledge about fundamental groups and covering spaces, it is not hard to see that, for $n > 2$, SO(n) has a connected double cover. Furthermore, a covering space of a connected Lie group always has a canonical Lie group structure such that the projection is a homomorphism (see Tits [2], II, or Chevalley [1], Ch. II, §8).

In any case, we will explicitly construct a compact connected (for $n \geq 2$) Lie group Spin(n) and a surjective homomorphism Spin(n) → SO(n) with kernel $\mathbb{Z}/2$. In other words, we will construct a short exact sequence

$$\{e\} \to \mathbb{Z}/2 \to \mathrm{Spin(n)} \to \mathrm{SO}(n) \to \{e\}.$$

The construction follows Atiyah, Bott, and Shapiro [1] and uses the theory of real Clifford algebras corresponding to the quadratic form $Q\colon \mathbb{R}^n \to \mathbb{R}$, $x \mapsto -|x|^2$.

(6.1) Definition. Let V be a real finite-dimensional vector space and $Q\colon V \to \mathbb{R}$ a quadratic form. The ***Clifford Algebra*** C(Q) is an $\mathbb{R}$-algebra with unit 1 together with a linear map $i = i_Q\colon V \to \mathrm{C}(Q)$ called the ***structure map***. The structure map is required to satisfy $(i(x))^2 = Q(x) \cdot 1$ for all $x \in V$, and C(Q) is required to satisfy the following ***universal property***:

If A is any $\mathbb{R}$-algebra with 1 and a linear map

$$j\colon V \to A, \qquad (j(x))^2 = Q(x) \cdot 1,$$

then there is a unique homomorphism κ_j of real algebras with 1, called the ***universal*** homomorphism, which makes the following diagram commutative:

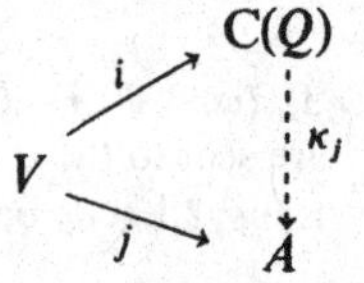

Note that in this book we use the word "algebra" to mean "associative algebra" except when we specify "Lie algebra."

(6.2) A familiar argument shows that there is, up to isomorphism, only one pair (C(Q), i) for a given (V, Q): if (A, j) has the same universal property, we get universal homomorphisms $\lambda_i\colon A \to \mathrm{C}(Q)$ and $\kappa_j\colon \mathrm{C}(Q) \to A$ and conclude that $\lambda_i \circ \kappa_j$ and $\kappa_j \circ \lambda_i$ are the identity by the uniqueness of universal homomorphisms. It is also easy to "construct" the Clifford algebra C(Q) with its canonical map $i\colon V \to \mathrm{C}(Q)$:

We start with the tensor algebra T of V:

$$T = \bigoplus_{\nu=0}^{\infty} V^{(\nu)}, \qquad V^{(0)} = \mathbb{R}, \qquad V^{(1)} = V, \qquad V^{(\nu)} = V \otimes \cdots \otimes V$$

($V^{(\nu)}$ has ν factors) and the inclusion map $V = V^{(1)} \xrightarrow{\subset} T$. Multiplication in T is induced by $a \cdot b = a \otimes b$ for $a \in V^{(\nu)}, b \in V^{(\mu)}$. Then $a \otimes b \in V^{(\nu+\mu)}$.

Let $\mathfrak{a} \subset T$ be the ideal of T generated by the elements

$$\{x \otimes x - Q(x) \cdot 1 \,|\, x \in V\}.$$

Then $\mathrm{C}(Q) = T/\mathfrak{a}$, and the structure map i is the composition

$$i\colon V \xrightarrow{\subset} T \to T/\mathfrak{a}.$$

(6.3) Remark and Notation. If $V = \mathbb{R}^n$ with the quadratic form $Q\colon \mathbb{R}^n \to \mathbb{R}$, $x \mapsto -|x|^2$ and the canonical basis $e_1, \dots, e_n$, then the corresponding Clifford algebra $\mathrm{C}(Q)$ is denoted by C_n. The relations in the ideal $\mathfrak{a}$ then say that

$$e_\nu^2 = -1 \quad \text{and} \quad e_\nu \cdot e_\mu = -e_\mu \cdot e_\nu \quad \text{for } \nu \neq \mu.$$

We have the canonical isomorphisms

$$\mathrm{C}_0 = \mathbb{R}, \qquad \mathrm{C}_1 = \mathbb{C}, \qquad \mathrm{C}_2 = \mathbb{H}.$$

PROOF. The relations are easily derived from $(\lambda_1 e_1 + \cdots + \lambda_k e_k)^2 = -(\lambda_1^2 + \cdots + \lambda_k^2)$. The isomorphism $\mathrm{C}_1 \to \mathbb{C}$ maps e_1 to $i = \sqrt{-1}$, and the isomorphism $\mathrm{C}_2 \to \mathbb{H}$ is given by $e_1 \mapsto i$, $e_2 \mapsto j$ (see (1.9)). We will see below, (6.16), that $\dim C_n = 2^n$, so the relations in the remark generate all the relations. □

(6.4) Another Remark and More Notation. The algebra $\mathrm{C}(Q)$ has a canonical ***anti-automorphism*** denoted

$$t\colon \mathrm{C}(Q) \to \mathrm{C}(Q), \qquad x \mapsto x^t$$

satisfying $(x \cdot y)^t = y^t \cdot x^t$ and $t^2 = \mathrm{id}$. It is uniquely determined by $x^t = x$ for $x \in i(V)$.

PROOF. Given $\mathrm{C}(Q)$ we define the opposite algebra $\mathrm{C}(Q)^{\mathrm{op}}$ by letting $\mathrm{C}(Q)^{\mathrm{op}} := \mathrm{C}(Q)$ as a real vector space and setting $x \cdot y$ in $\mathrm{C}(Q)^{\mathrm{op}}$ equal to $y \cdot x$ in $\mathrm{C}(Q)$. In $\mathrm{C}(Q)^{\mathrm{op}}$ we still have $x^2 = Q(x) \cdot 1$ for $x \in i(V)$, and from (6.1) we conclude that $\mathrm{C}(Q)^{\mathrm{op}}$ has the universal property for all opposite algebras. But the collection of all opposite algebras is the same as the collection of all algebras, so $\mathrm{C}(Q)^{\mathrm{op}}$ is universal and there is a unique isomorphism $t\colon \mathrm{C}(Q) \to \mathrm{C}(Q)^{\mathrm{op}}$ making the following diagram commutative:

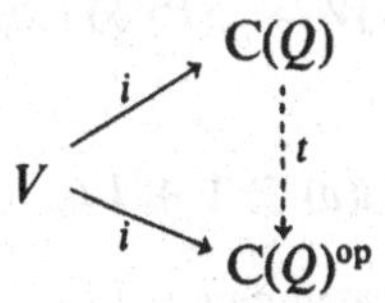

We also conclude from uniqueness of the universal homomorphism that the composition $\mathrm{C}(Q) \xrightarrow{t} \mathrm{C}(Q)^{\mathrm{op}} \to \mathrm{C}(Q)^{\mathrm{op\,op}} = \mathrm{C}(Q)$ is the identity. □

Each element from C(Q) may be expressed (nonuniquely) as a linear combination of elements of the form $x = x_1 \dots x_k$ with $x_\nu \in i(V)$. The diagram shows that

$$(x_1 \dots x_k)^t = x_k \cdot x_{k-1} \dots x_1,$$

which completely determines t.

Analogously, there is a ***canonical automorphism***

(6.5) $\alpha: C(Q) \to C(Q), \quad \alpha^2 = \mathrm{id}, \quad \alpha(x) = -x \quad \text{for } x \in i(V).$

Thus $\alpha(x_1 \dots x_k) = (-1)^k(x_1 \dots x_k)$ for $x_j \in i(V)$.

For $\nu = 0, 1$ let $C(Q)^\nu$ be the eigenspace for the eigenvalue $(-1)^\nu$ of α. Thus $C(Q)^\nu = \{x \in C(Q) \mid \alpha(x) = (-1)^\nu x\}$ and we have

$$C(Q) = C(Q)^0 \oplus C(Q)^1$$

as vector spaces. This is what is called a $\mathbb{Z}/2$-***grading***, which means that if $x \in C(Q)^\nu$ and $y \in C(Q)^\mu$, then $x \cdot y \in C(Q)^{\nu+\mu}$ where we reduce exponents modulo 2.

Now if we have two algebras which are $\mathbb{Z}/2$-graded in this sense

$$A = A^0 \oplus A^1, \qquad B = B^0 \oplus B^1,$$

there is a ***graded tensor product*** $A \hat{\otimes} B$ defined by

$$(A \hat{\otimes} B)^0 = (A^0 \otimes B^0) \oplus (A^1 \otimes B^1),$$
$$(A \hat{\otimes} B)^1 = (A^0 \otimes B^1) \oplus (A^1 \otimes B^0),$$

with multiplication

$$(a' \otimes b) \cdot (a \otimes b') = (-1)^{\nu\mu}(a' \cdot a) \otimes (b \cdot b')$$

for $a \in A^\mu$ and $b \in B^\nu$. It is easy to verify that $A \hat{\otimes} B$ is again a $\mathbb{Z}/2$-graded algebra.

(6.6) Proposition. *Let V and W be finite-dimensional vector spaces with quadratic forms P and Q. Then there is a quadratic form $P \oplus Q$ on $V \oplus W$ defined by $(P \oplus Q)(v, w) = P(v) + Q(w)$. Let $i = i_P: V \to C(P)$ and $j = j_Q: W \to C(Q)$ be the structure maps of the corresponding Clifford algebras, and define a linear map*

$$f: V \oplus W \to C(P) \hat{\otimes} C(Q)$$

by

$$(v, w) \mapsto i(v) \otimes 1 + 1 \otimes j(w).$$

Then f induces an isomorphism (also denoted by f)

$$f: C(P \oplus Q) \to C(P) \hat{\otimes} C(Q).$$

PROOF. For $(v, w) \in V \oplus W$ we have

$$\begin{aligned}(f(v, w))^2 &= (i(v) \otimes 1)^2 + (1 \otimes j(w))^2 + (-1)^0(i(v) \otimes j(w)) \\ &\quad + (-1)^1(i(v) \otimes j(w)) \\ &= (P \oplus Q)(v, w) \cdot (1 \otimes 1).\end{aligned}$$

The universal property of the Clifford algebra $C(P \oplus Q)$ then gives us an algebra homomorphism extending the given f

$$f: C(P \oplus Q) \to C(P) \hat{\otimes} C(Q).$$

The inverse homomorphism may be obtained as follows: Let $\varphi: C(P) \to C(P \oplus Q)$ and $\psi: C(Q) \to C(P \oplus Q)$ be the maps induced by the inclusions $V \to V \oplus W$ and $W \to V \oplus W$. Then

$$f^{-1}: C(P) \hat{\otimes} C(Q) \to C(P \oplus Q), \qquad v \otimes w \mapsto \varphi(v) \cdot \psi(w). \qquad \square$$

(6.7) Corollary. *The structure map $i_Q: V \to C(Q)$ is injective. Thus we will view V as a subspace of $C(Q)$ via this map. If $e_1, \ldots, e_n$ form a basis of V, then the products*

$$e_{\nu_1} \ldots e_{\nu_k}, \qquad 1 \le \nu_1 < \cdots < \nu_k \le n$$

and 1 form a basis of the real vector space $C(Q)$. In particular, (6.3) contains a complete set of relations for these products, and $C(Q)$ has dimension 2^n.

PROOF. If $\dim V = 1$, the tensor algebra T is just the polynomial ring $\mathbb{R}[X]$ where $i(e_1) = X$. Thus $C(Q) = \mathbb{R}[X]/\langle X^2 - Q(e_1)\rangle$ and in this case the corollary is clear. In general the relation $Q(e_\nu + e_\mu) - Q(e_\nu) - Q(e_\mu) = e_\nu e_\mu + e_\mu e_\nu$ shows that the stated elements generate $C(Q)$, so all that is left is the calculation of $\dim C(Q)$. For this purpose, we may pick any basis of V we so desire. But we may diagonalize any quadratic form over $\mathbb{R}$, which means there is an orthogonal splitting $(V, Q) \cong \bigoplus_{\nu=1}^n (V_\nu, Q_\nu)$ with $\dim V_\nu = 1$. Choosing a basis element $e_\nu \in V_\nu \subset V$, the corollary follows from (6.6) by induction on $n = \dim V$. $\square$

(6.8) Still More Remarks and Notation. There is a ***conjugation*** defined on $C(Q)$ by

$$t\alpha = \alpha t: C(Q) \to C(Q), \quad \text{denoted } x \mapsto \bar{x}$$

with α and t as in (6.4) and (6.5). This is an algebra anti-automorphism. Let $C(Q)^*$ be the group of units of $C(Q)$ and

$$\Gamma(Q) = \{x \in C(Q)^* \mid \alpha(x) \cdot v \cdot x^{-1} \in V \text{ for all } v \in V\}.$$

Then $\Gamma(Q)$ is a subgroup of $C(Q)^*$ and is called the ***Clifford group*** of Q. The map

$$\mathrm{N}: C(Q) \to C(Q), \qquad x \mapsto x \cdot \bar{x}$$

is called the ***norm*** of $C(Q)$. Thus for $x \in V$ we have $\mathrm{N}(x) = -Q(x) \cdot 1$.

By the way, units are those $x \in C(Q)$ for which there is a $y \in C(Q)$ with $xy = yx = 1$. Also note that the equation $\alpha t = t\alpha$ holds on V and hence all of $C(Q)$, and that $\Gamma(Q)$ is a group because α is an automorphism (and V is finite dimensional).

(6.9) Lemma. *The maps* α *and* t *from* (6.4) *and* (6.5) *induce an automorphism and anti-automorphism of* $\Gamma(Q)$.

PROOF. The maps are $\pm\mathrm{id}$ on V. Hence if $\alpha(x)vx^{-1} \in V$ for all $v \in V$ then

$$\begin{aligned}\alpha(\alpha(x)) \cdot v \cdot \alpha(x)^{-1} &= -\alpha(\alpha(x)) \cdot \alpha(v) \cdot \alpha(x)^{-1} \\ &= -\alpha(\alpha(x) \cdot v \cdot x^{-1}) \\ &= \alpha(x) \cdot v \cdot x^{-1} \in V,\end{aligned}$$

so $\alpha(x) \in \Gamma(Q)$. Analogously $\alpha t(x) \cdot v \cdot t(x)^{-1} = t\alpha x \cdot tv \cdot t(x)^{-1} = t(x^{-1}v\alpha(x)) \in V$ for all v because $\Gamma(Q)$ contains inverses and α leaves $\Gamma(Q)$ invariant. □

We now turn to the algebras C_n corresponding to the vector spaces $V = \mathbb{R}^n$ with quadratic form $Q: \mathbb{R}^n \to \mathbb{R}$, $x \mapsto -|x|^2$. By (6.2) and (6.6) we have

$$C_n = \mathbb{C} \hat{\otimes} \cdots \hat{\otimes} \mathbb{C} \quad (n\text{-factors})$$

with basis and relations as in (6.2). The Clifford group Γ_n of the algebra C_n comes with a ready-made homomorphism

(6.10) $\rho: \Gamma_n \to \mathrm{Aut}(\mathbb{R}^n)$, $\quad \rho(x)v = \alpha(x)vx^{-1}$ for $x \in \Gamma_n$ and $v \in \mathbb{R}^n$.

(6.11) Lemma. *The kernel of* $\rho: \Gamma_n \to \mathrm{Aut}(\mathbb{R}^n)$ *is* $\mathbb{R}^*$, *the multiplicative group of nonzero real multiples of* $1 \in C_n$.

PROOF. Let $x \in \ker(\rho)$. Then by definition

(i) $$\alpha(x) \cdot v = v \cdot x \quad \text{for all } v \in \mathbb{R}^n.$$

Expressing $x = x^0 + x^1$, where $C_n = C_n^0 \oplus C_n^1$ and $x^\nu \in C_n^\nu$, equation (i) says

(ii) $$x^0 v = vx^0 \quad \text{and} \quad -x^1 v = vx^1.$$

Next we use (6.7) to write x^0 as a linear combination of monomials in the canonical basis of $\mathbb{R}^n$, so

$$x^0 = a^0 + e_1 b^1, \qquad a^0 \in C_n^0, \qquad b^1 \in C_n^1,$$

where neither a^0 nor b^1 contains a summand with a factor e_1. Applying the first relation in (ii) to $v = e_1$ yields

(iii) $$a^0 + e_1 b^1 = e_1^{-1}(a^0 + e_1 b^1)e_1.$$

Since each monomial in a^0 is of even degree and contains no factor e_1, the relations (6.3) show that $a^0e_1 = e_1a^0$. Similarly $e_1b^1 = -b^1e_1$, so

$$a^0 + e_1b^1 = a^0 - e_1b^1.$$

We conclude that $e_1b^1 = 0$ and x^0 contains no monomial with a factor e^1. The same argument applied successively to the other basis elements demonstrates that we can write x^0 as a linear combination of monomials with no e_v as a factor. In other words, $x^0 \in \mathbb{R} \cdot 1$. Proceeding in a like fashion with the second relation from (ii), $x^1 = a^1 + e_1b^0$ and $v = e_1$ shows

$$a^1 + e_1b^0 = -e_1^{-1}a^1e_1 - b^0e_1 = a^1 - e_1b^0.$$

Thus $b^0 = 0$ and $x^1 \in \mathbb{R}$. But $\mathbb{R} \subset C_n^0$, so $x^1 = 0$. Thus $x = x^0 \in \mathbb{R} \cap \Gamma_n = \mathbb{R}^*$. □

The norm $N(x) = x \cdot \bar{x} = x \cdot t\alpha(x)$ agrees with $-x^2 = -Q(x) = |x|^2$ on $\mathbb{R}^n$, so there it is the customary square of the Euclidean absolute value. And by (6.9) we have $N(\Gamma_n) \subset \Gamma_n$. As a matter of fact, the following stronger statement holds:

(6.12) Lemma. *If* $x \in \Gamma_n$, *then* $N(x) \in \mathbb{R}^*$.

Proof. We compute that $N(x)$ is in the kernel of ρ, i.e., that $N(x)$ acts trivially on $\mathbb{R}^n$. To say $x \in \Gamma_n$ means

$$\alpha(x)vx^{-1} \in \mathbb{R}^n \quad \text{for all } v \in \mathbb{R}^n.$$

Applying t gives $t(x)^{-1}vt\alpha(x) = \alpha(x)vx^{-1}$ since t is the identity on $\mathbb{R}^n$. Thus $v = t(x)\alpha(x)v(t\alpha(x) \cdot x)^{-1} = \alpha(\bar{x}x)v(\bar{x}x)^{-1}$, so $\bar{x}x \in \ker(\rho)$ and the same is true for $x\bar{x} = \bar{\bar{x}}\bar{x}$ by (6.9). □

(6.13) Lemma. $N|\Gamma_n\colon \Gamma_n \to \mathbb{R}^*$ *is a homomorphism and* $N(\alpha(x)) = N(x)$.

Proof. $N(xy) = xy\bar{y}\bar{x} = xN(y)\bar{x} = x\bar{x}N(y) = N(x)N(y)$, the third equality holding because $N(y) \in \mathbb{R}^*$.

$$N(\alpha x) = \alpha(x)\alpha(\bar{x}) = \alpha(x\bar{x}) = \alpha N(x) = N(x). \qquad \square$$

(6.14) Lemma. $\mathbb{R}^n\backslash\{0\} \subset \Gamma_n$, *and if* $x \in \mathbb{R}^n\backslash\{0\}$, *then* $\rho(x)$ *is the reflection in the hyperplane orthogonal to* x. *Also,* $\rho\Gamma_n \subset O(n)$.

Proof. Given $x \in \mathbb{R}^n\backslash\{0\}$, we may choose a basis of $\mathbb{R}^n$ such that $x = |x|e_1$. By (6.11), $\rho(|x|e_1) = \rho(e_1)$, so we may assume that $x = e_1$. Then we have

$$\rho(e_1)e_1 = \alpha(e_1)e_1e_1^{-1} = -e_1,$$

$$\rho(e_1)e_v = \alpha(e_1)e_ve_1^{-1} = -e_1e_ve_1^{-1} = e_ve_1e_1^{-1} = e_v, \qquad v \neq 1.$$

This demonstrates the first sentence. Now let x be any element of Γ_n and let $v \in \mathbb{R}^n\backslash\{0\}$. Then $N(\rho(x)v) = N(\alpha(x)vx^{-1}) = N(\alpha(x)) \cdot N(v) \cdot N(x^{-1}) = N(x) \cdot N(v) \cdot N(x)^{-1} = N(v)$, and since N is simply the square of the Euclidean absolute value, this says precisely that $\rho(x) \in O(n)$. □

We may now achieve our original goal:

(6.15) Theorem. *Let* Pin(n) *be the kernel of* N: $\Gamma_n \to \mathbb{R}^*$ *for* $n \geq 1$. *Then the map* $\rho|\text{Pin}(n)$ *has image* O(n) *and kernel generated by* $\{-1\} \in C_n$. *Thus we have an exact sequence of groups*

$$\{e\} \to \mathbb{Z}/2 \to \text{Pin}(n) \xrightarrow{\rho} \text{O}(n) \to \{e\}.$$

PROOF. Every orthogonal transformation $A \in \text{O}(n)$ is a product of k reflections with $k \leq n$. This is easily verified by induction on n, for given $A \in \text{O}(n)$, $A \neq E$, we can find a reflection σ such that $\sigma(Ae_n) = e_n$, and then $\sigma \circ A \in \text{O}(n-1)$. But all reflections are in the image of $\rho|\text{Pin}(n)$ by (6.14), and the kernel of $\rho|\text{Pin}(n)$ is $\ker(\rho) \cap \ker(\text{N}) = \{t \in \mathbb{R}^* | \text{N}(t) = 1\} = \{1, -1\}$. □

The group Pin(n) has a well-defined structure as a Lie group such that ρ is a homomorphism of Lie groups and a double cover. The Lie group structure may be obtained as follows:

The group of units C_n^* is open in C_n because an element x is a unit if and only if left multiplication by x is a linear isomorphism, and linear isomorphisms are open in the space of linear endomorphisms of C_n. Thus C_n^* is a Lie group, and since Γ_n and Pin(n) are closed in C_n^*, they are also Lie groups. Actually, one may apply a more general argument to show that Pin(n) is a Lie group (see (6.22), Ex. 5).

(6.16) Definition. We define $\text{Spin}(n) \subset \text{Pin}(n)$ to be the inverse image of SO(n) under ρ: Pin(n) $\to$ O(n).

Thus we have an exact sequence of Lie groups

$$\{e\} \to \mathbb{Z}/2 \to \text{Spin}(n) \to \text{SO}(n) \to \{e\}.$$

For $n = 1$ we have

$$C_1 = \mathbb{C}, \qquad \Gamma_1 = \{z \in \mathbb{C}^* | \bar{z}(i\mathbb{R})z^{-1} \subset i\mathbb{R}\} = (\mathbb{R} \cup i\mathbb{R})\backslash\{0\}.$$

Thus Pin(1) $= \mathbb{Z}/4$ and is generated by $i \in \mathbb{C}$, and Spin(1) $= \{-1, 1\} = \mathbb{Z}/2$.

(6.17) Proposition. *For* $n \geq 2$ *the homomorphism* ρ: Spin(n) $\to$ SO(n) *is a nontrivial double covering.*

PROOF. We need to show that there is an arc in Pin(n) connecting the elements 1 and -1 which constitute the kernel of ρ. Such an arc is given by

$$w: t \mapsto \cos(t) + \sin(t)e_1 \cdot e_2, \qquad 0 \leq t \leq \pi.$$

We have $w(t)^{-1} = \bar{w}(t) = \cos(t) - \sin(t) \cdot e_1 \cdot e_2$, so $w(t) \in \text{Pin}(n)$. Since $\rho w(t)$ must stay in a connected component of SO(n), and $\rho w(0) = E \in \text{SO}(n)$, $w(t) \in \text{Spin}(n)$ for all t. □

(6.18) Remark. There is a standard isomorphism of groups

$$\kappa\colon \mathrm{SU}(2) = \mathrm{Sp}(1) \xrightarrow{\cong} \mathrm{Spin}(3)$$

and a diffeomorphism of manifolds $\mathrm{SO}(3) \cong \mathbb{R}P^3$ which may be described as follows. The group of units $\mathbb{H}^*$ operates orthogonally on the space $\mathbb{H}$ of quaternions by

$$\mathbb{H}^* \times \mathbb{H} \to \mathbb{H}, \qquad (q, x) \mapsto qxq^{-1},$$

since $\mathrm{N}(qxq^{-1}) = \mathrm{N}(q)\mathrm{N}(x)\mathrm{N}(q)^{-1} = \mathrm{N}(x)$ for the norm N on $\mathbb{H}$. This operation leaves the subspace $\mathbb{R} \subset \mathbb{H}$ invariant. Hence it also fixes the orthogonal complement $\mathbb{R}^3$, which is the group of pure quaternions. Thus we get an induced operation of $\mathbb{H}^*$ on $\mathbb{R}^3$, and restricting this operation to the subgroup $\mathrm{Sp}(1) \subset \mathbb{H}^*$ of elements of norm one gives a projection

$$\pi\colon \mathrm{Sp}(1) \to \mathrm{SO}(3), \qquad \pi(q)(x) = qxq^{-1} \quad \text{for } q \in \mathrm{Sp}(1), \qquad x \in \mathbb{R}^3.$$

This projection is essentially the same as the projection $\rho\colon \mathrm{Spin}(3) \to \mathrm{SO}(3)$.
Now one has an inclusion of algebras (!)

$$\kappa\colon \mathbb{H} \to C_3, \qquad (i, j, k) \mapsto (e_2 e_3, e_3 e_1, e_1 e_2),$$

which restricts to an injective homomorphism $\kappa\colon \mathbb{H}^* \to \Gamma_3$ preserving the respective norms (!) and compatible with the respective operations on $\mathbb{R}^3$. In other words, the following diagram commutes (!)

$$\begin{array}{ccccc} \mathbb{H}^* \times \mathbb{R}^3 & \longrightarrow & \mathbb{R}^3 & \qquad & (i, j, k) \\ \Big\downarrow{\scriptstyle \kappa \times \lambda} & & \Big\downarrow{\scriptstyle \lambda} & & \Big\downarrow{\scriptstyle \lambda} \\ \Gamma_3 \times \mathbb{R}^3 & \longrightarrow & \mathbb{R}^3 & & (e_1, e_2, e_3) \end{array}$$

where the horizontal arrows come from the appropriate operations on $\mathbb{R}^3$. Restricting κ to elements of norm one again, we get an injection $\kappa\colon \mathrm{Sp}(1) \to \mathrm{Spin}(3)$. Since the groups have the same dimension, this injection is an isomorphism. Note also that $\pi = \rho \circ \kappa$, so $\ker(\pi) = \{1, -1\}$ and $\mathrm{SO}(3) \cong \mathrm{Sp}(1)/\{1, -1\} = S^3/(x \sim -x) = \mathbb{R}P^3$. The reader is invited to check the details marked by a (!) in (6.22), Ex. 9.

(6.19) Corollary. *The fundamental group* $\pi_1 \mathrm{SO}(n)$ *is isomorphic to* $\mathbb{Z}/2$ *for* $n \geq 3$, *so* $\mathrm{Spin}(n)$ *is simply connected and* $\rho\colon \mathrm{Spin}(n) \to \mathrm{SO}(n)$ *is the universal covering (for* $n \geq 3$).

PROOF. We use induction on n and consider the principal bundle (4.7)

$$\mathrm{SO}(n) \to \mathrm{SO}(n+1) \to S^n.$$

Since $\pi_1 S^n = \pi_2 S^n = \{e\}$ for $n \geq 3$, the inclusion $\mathrm{SO}(n) \to \mathrm{SO}(n+1)$ induces an isomorphism of fundamental groups (see Hu [1], V, 6, p. 152 or G. W. Whitehead [1], IV, 8), and the corollary follows. □

We will not use this result in this book, but we will come back to the problem of computing the fundamental group of a compact Lie group in V, §7.

The Lie algebra su(2) = sp(1) ≅ so(3) of Sp(1), considered as a vector space, is the tangent space of $S^3 \subset \mathbb{H} = \mathbb{R}^4$ at the point 1. Thus it is the space of pure quaternions—the orthogonal complement $\mathbb{R}^3$ of $\mathbb{R}$ in $\mathbb{H}$. As a real vector space it is generated by the three quaternions

$$i = \begin{bmatrix} i & 0 \\ 0 & -i \end{bmatrix}, \quad j = \begin{bmatrix} 0 & 1 \\ -1 & 0 \end{bmatrix}, \quad \text{and} \quad k = \begin{bmatrix} 0 & i \\ i & 0 \end{bmatrix}.$$

The map π: Sp(1) → SO(3) therefore describes the operation of Sp(1) on its Lie algebra by conjugation. Rephrasing this:

(6.20) Remark. The canonical projection π: Sp(1) = SU(2) ≅ Spin(3) → SO(3) is the adjoint representation of Sp(1).

(6.21) Physicists' Notation. Recall that physicists apply a factor of $(-i)$ to Lie algebras (2.21). Thus they also multiply the above complex matrices $i, j, k \in \operatorname{End}(\mathbb{C}^2)$ by $(-i)$ (this is not the same as multiplication in $\mathbb{H}$!). This yields the Hermitian matrices

$$\begin{bmatrix} 1 & 0 \\ 0 & -1 \end{bmatrix}, \quad \begin{bmatrix} 0 & -i \\ i & 0 \end{bmatrix}, \quad \text{and} \quad \begin{bmatrix} 0 & 1 \\ 1 & 0 \end{bmatrix}$$

as real generators of the (physicists') Lie algebra su(2), which is the same as $(-i)$ su(2) $\subset \operatorname{End}(\mathbb{C}^2)$ in our notation. These matrices are called ***Pauli spin matrices.***

We will return to Clifford algebras in VI, §6. There we will describe the structure of the algebras C_m and classify the left modules over these algebras in order to construct the "half-spin" representations.

The spin representation of the special orthogonal group by means of Clifford algebras was discovered by R. Lipschitz (*Untersuchungen über die Summen von Quadraten*, Bonn, 1886), see also van der Waerden [2], p. 14. The reader should not forgo the pleasure of reading Lipschitz through his medium . . . . . . in Correspondence [1]. There one may also find the Cayley parametrization of the orthogonal group.

(6.22) Exercises

1. The tensor algebra $T = \bigoplus_j V^{(j)}$ has a **filtration** $F^kT = \bigoplus_{j=0}^k V^{(j)}$, and this yields a filtration of $C(Q)$ by $F^kC(Q) = \kappa F^kT$, where κ is the canonical epimorphism $\kappa\colon T \to C(Q)$, $\kappa | V = \mathrm{id}_V$. Show that $F^kC(Q) \cdot F^lC(Q) \subset F^{k+l}C(Q)$. Setting $A^k = F^kC(Q)/F^{k-1}C(Q)$, we get an induced multiplication $A^k \otimes A^l \to A^{k+l}$, making $\bigoplus_{k=0}^\infty A^k$ an $\mathbb{R}$-algebra. Show that this algebra, considered as a real vector space, is isomorphic to $C(Q)$, and that as an algebra it is isomorphic to the exterior algebra $\bigoplus_{k=0}^\infty \operatorname{Alt}^k V$.

2. Show that in $C(Q)$ we have $x \cdot y = 1$ if and only if $y \cdot x = 1$.

3. Show that the map

$$f: \left(\bigtimes_{j=1}^{n} S^{n-1}\right) \cup \left(\bigtimes_{j=1}^{n-1} S^{n-1}\right) \to \mathrm{Pin}(n) \quad \text{(disjoint union)}$$

$$(x_1, \ldots, x_k) \mapsto x_1 \cdot \ldots \cdot x_k, \quad \text{where } k = n \text{ or } n-1$$

is surjective. If n is even, then $\mathrm{Spin}(n) = f(\bigtimes_{j=1}^{n} S^{n-1})$ and if n is odd, then $\mathrm{Spin}(n) = f(\bigtimes_{j=1}^{n-1} S^{n-1})$.

4. Show that $\mathrm{Spin}(n) = \mathrm{Pin}(n) \cap C_n^0$.

5. Let $f: \tilde{G} \to G$ be a covering with G a connected Lie group and let $f(\tilde{e}) = e$, the unit. Show that $\tilde{G}$ has a unique structure of a Lie group with unit $\tilde{e}$ such that f is a homomorphism of Lie groups.

6. Let w be the arc in $\mathrm{Spin}(n)$ given in (6.17). Show

$$\rho w(t) = \left[\begin{array}{cc|ccc} \cos 2t & -\sin 2t & & 0 & \\ \sin 2t & \cos 2t & & & \\ \hline & & 1 & & \\ & 0 & & \ddots & \\ & & & & 1 \end{array}\right].$$

7. The linear group $GL(2, \mathbb{C})$ acts on the Riemann sphere $\mathbb{C}P^1 = \mathbb{C} \cup \{\infty\} \cong S^2$ by projective transformations, i.e., via the projection $GL(2, \mathbb{C}) \to PGL(2, \mathbb{C})$. In the notation of complex analysis, the matrix $A = \begin{bmatrix} a & b \\ c & d \end{bmatrix}$ in $GL(2, \mathbb{C})$ acts by sending z to $(az + b)/(cz + d)$. Show that $A \in U(2)$ if and only if the corresponding transformation of $\mathbb{C}P^1 \cong S^2$ is orthogonal (isometric). This yields an isomorphism $SU(2)/(\mathbb{Z}/2) \to SO(3)$ mapping A to the transformation of S^2 induced by A, and hence we get a new description of the projection $SU(2) \to SO(3)$.

8. Compute C_2^* and Γ_2. Check explicitly which two circles in Γ_2 form the group $\mathrm{Pin}(2)$.

9. Compute Γ_3 and $\mathrm{Pin}(3)$ in terms of the canonical basis of C_3 and check all the details in the description of the isomorphism $\kappa: \mathrm{Sp}(1) \to \mathrm{Spin}(3)$ in (6.18).

CHAPTER II

Elementary Representation Theory

In this chapter we meet the objects which are the focus of this book: finite-dimensional representations of compact Lie groups. In §1 we introduce and discuss the notion of a complex representation and show that every representation is the direct sum of irreducible representations. Uniqueness of the direct sum decomposition is shown in §2 in the context of semisimple modules. The standard constructions from linear algebra, such as tensor products, symmetric powers, and exterior powers, are reviewed briefly in §3 and used to build new representations from old.

The emphasis shifts to characters of representations in §4. The character is a function on the group and determines the representation up to isomorphism. Using characters greatly simplifies computations without sacrificing vital information. The orthogonality relations for characters and for entries of matrix representations are proved in this section.

In §5 we give an elementary construction of the irreducible representations of the groups SU(2), SO(3), U(2), and O(3). This is continued in §10. In §6 we turn to representations on real and quaternionic vector spaces and the relations between these and complex representations. We do some book-keeping which keeps track of these relations.

The ring of functions generated by the characters on a group is the topic of §7. This character ring has a more abstract description as the (Grothendieck) representation ring. We also introduce the exterior power and Adams operations on the representation ring.

Representations of the abelian groups are described in §8. These yield simple examples and are important for the development of the general theory because representations are determined by their restrictions to (certain) abelian subgroups. The key notions of weights and (in Chapter V) roots are based upon representations of tori.

In §9 we briefly introduce the infinitesimal form of a representation: the representation of a Lie algebra. Although we avoid using Lie algebra techniques in this book, it seems convenient to have the infinitesimal weights at our disposal. They are later used in computations for the classical groups.

The infinitesimal aspect is furthered in §10, where the irreducible representations of the Lie algebra $\mathrm{sl}(2, \mathbb{C})$ are described. The elementary discussion of SO(3)-representations is also supplemented by spherical functions, Legendre functions, and the differential equations coming from the Lie derivative.

1. Representations

Representations of compact Lie groups are the chief objects of interest in this book. Groups are intended to describe symmetries of geometric and other mathematical objects. Representations are symmetries of some of the most basic objects in geometry and algebra, namely vector spaces.

We begin by considering finite-dimensional vector spaces over the complex numbers $\mathbb{C}$. Later we will indicate the modifications necessary for working with real vector spaces, infinite-dimensional vector spaces, and so on.

Representations have three different aspects—geometric, numerical, and algebraic—and manifest themselves in corresponding forms. We begin with the geometric form.

(1.1) Definition. A ***representation*** of the Lie group G on the (finite-dimensional complex) vector space V is a continuous action

$$\rho: G \times V \to V$$

of G on V such that for each $g \in G$ the translation $l_g: v \mapsto \rho(g, v)$ is a linear map. We call the pair (V, ρ) a ***complex representation*** and V the ***representation space***. The dimension of V (as a complex vector space) is called the ***dimension*** $\dim V$ of the representation.

Recall that saying ρ is an action means that

$$\rho(e, v) = v \quad \text{and} \quad \rho(g, \rho(h, v)) = \rho(gh, v).$$

We usually denote $\rho(g, v)$ by gv, so the defining equations of an action take on the more suggestive form

$$ev = v \quad \text{and} \quad (gh)v = g(hv).$$

Written in terms of translations, these equations become

$$l_e = \mathrm{id}_V \quad \text{and} \quad l_g \circ l_h = l_{gh}.$$

Thus l_g is a linear automorphism of V with inverse $l_{g^{-1}}$, and the map $g \mapsto l_g$ is a homomorphism

$$l: G \to \mathrm{Aut}(V) = \mathrm{Aut}_{\mathbb{C}}(V).$$

Conversely, given any such homomorphism l, we may define an action

$$\rho: G \times V \to V, \quad (g, v) \mapsto l_g v.$$

The reader may easily check that ρ is continuous if and only if l is continuous.

A choice of a basis for V determines an isomorphism $\mathrm{Aut}(V) \cong \mathrm{GL}(n, \mathbb{C})$. This leads to the numerical form of a representation.

(1.2) Definition. A ***matrix representation*** of G is a continuous homomorphism $l: G \to \mathrm{GL}(n, \mathbb{C})$.

A representation is called ***faithful*** if the associated homomorphism $G \to \mathrm{Aut}(V)$ is injective. Later we will show that every compact Lie group has a faithful representation and is therefore isomorphic to a closed subgroup of a matrix group. The desire to represent abstract groups in the concrete numerical terms of matrices and matrix multiplication is one of the origins of representation theory.

Finally, we come to the algebraic form of a representation. To begin with, let G be a finite group. Then we may form what is called the ***group ring*** $\mathbb{C}[G]$ of G over $\mathbb{C}$. Additively this is just the complex vector space with a basis consisting of the elements of G. Thus each element of $\mathbb{C}[G]$ may be uniquely expressed as a formal linear combination $\sum_{g \in G} \lambda_g \cdot g$, $\lambda_g \in \mathbb{C}$.

Two such elements are multiplied as follows:

$$\left(\sum_g \lambda_g g\right) * \left(\sum_h \mu_h h\right) = \sum_{g,h} \lambda_g \mu_h gh$$
$$= \sum_k \left(\sum_{gh=k} \lambda_g \mu_h\right) k = \sum_g \left(\sum_h \lambda_{gh^{-1}} \mu_h\right) g.$$

Given a representation $G \times V \to V$ of G, we associate to $x = \sum \lambda_g g$ the linear endomorphism $v \mapsto \sum \lambda_g gv = x * v$ of V. It is easy to verify that the map $(x, v) \mapsto x * v$ makes V into a left module over the ring $\mathbb{C}[G]$, and if V has the structure of a left $\mathbb{C}[G]$-module, we have by definition a linear action of G on V which is continuous since G is finite. Thus the theory of representations of the finite group G may be viewed as the theory of modules over the group ring $\mathbb{C}[G]$.

This module-theoretic point of view presents technical difficulties in the case of Lie groups because continuity must play a role in the definition of "group ring" and "module." Nevertheless, it is good to have at least an heuristic understanding of the concepts involved. Otherwise some of the proofs in later sections will seem artificial and unmotivated.

Note that an element of $\mathbb{C}[G]$ defines a function $G \to \mathbb{C}$ assigning the coefficient λ_g to g. Therefore it is natural to try to replace elements of the group ring with continuous functions $f: G \to \mathbb{C}$. In terms of functions, the multiplication rule reads

$$(f_1 * f_2)(g) = \sum_{h \in G} f_1(gh^{-1}) \cdot f_2(h).$$

As usual for Lie groups, such finite sums have to be replaced by integrals. This leads to the ***convolution product***

(1.3)
$$(f_1 * f_2)(g) = \int_G f_1(gh^{-1}) f_2(h)\, dh$$

of two continuous functions f_1 and f_2 on the compact Lie group G. It turns out that the space $C^0(G, \mathbb{C})$ of continuous complex-valued functions on G, with the supremum norm and the convolution product, is a Banach algebra, and that representations are suitable modules over this algebra.

We mention one final difficulty with this approach. The element $g \in G$ is contained in $\mathbb{C}[G]$ as the function whose value is 1 at g and zero otherwise. But this function is, in general, not continuous and hence not contained in $C^0(G, \mathbb{C})$. So if we want to include the group elements, we must enlarge $C^0(G, \mathbb{C})$ by "Dirac delta functions," i.e., we are forced to consider appropriate measures. Occasionally we shall have this approach in mind as a guiding viewpoint, but we will not formally develop measure theory on groups.

We will, however, abuse notation and allow ourselves to call representations V as in definition (1.1) ***complex G-modules***.

(1.4) Definition. A ***morphism*** $f: V \to W$ between representations is a linear map which is equivariant, i.e., which satisfies $f(gv) = gf(v)$ for $g \in G$ and $v \in V$. Morphisms are also called ***intertwining operators***.

We also call these ***morphisms of G-modules*** and let $\mathrm{Hom}_G(V, W)$ denote the set of all such morphisms. This defines a category, and, as usual, we have the notion of an ***isomorphism***, which is a morphism with an inverse. Isomorphic representations are also called ***equivalent*** representations.

Let α and β be two matrix representations $G \to \mathrm{GL}(n, \mathbb{C})$, and $V_\alpha = (\mathbb{C}^n, \rho_\alpha)$, $V_\beta = (\mathbb{C}^n, \rho_\beta)$ be the corresponding representations on $\mathbb{C}^n$. Then using the correspondence between linear maps $V_\alpha \to V_\beta$ and complex $(n \times n)$-matrices, we see that V_α and V_β are isomorphic if and only if there is an invertible matrix A such that

(1.5)
$$A\alpha(g)A^{-1} = \beta(g) \quad \text{for all } g \in G.$$

If two homomorphisms α and β are related as in (1.5), they are said to be ***similar*** or ***conjugate***. This should not be confused with complex conjugation.

(1.6) Definition. If V is a complex G-module, an (Hermitian) inner product $V \times V \to \mathbb{C}$, $(u, v) \mapsto \langle u, v\rangle$ is called ***G-invariant*** if $\langle gu, gv\rangle = \langle u, v\rangle$ for all $g \in G$ and $u, v \in V$. A representation together with a G-invariant inner product is called a ***unitary representation.***

If we choose an orthonormal basis for the space V of a unitary representation, then the associated matrix representation is a homomorphism $G \to \mathrm{U}(n)$. Conversely, any such continuous homomorphism defines a unitary representation on $\mathbb{C}^n$ with its standard inner product. The existence of invariant inner products for compact groups is an important application of invariant integration. In fact, the results we are about to derive are true for arbitrary compact topological groups. But the reader not familiar with general topological groups may take the term "compact group" to mean "compact Lie group."

(1.7) Theorem. *Let V be a representation of the compact group G. Then V possesses a G-invariant inner product.*

PROOF. Let $b\colon V \times V \to \mathbb{C}$ be any inner product and define

$$c(u, v) = \int_G b(gu, gv)\, dg,$$

where the integral is normalized and left-invariant. Then c is linear in u, conjugate linear in v, G-invariant since the integral is left-invariant and positive definite since the integral of a positive continuous function is positive. Thus c is a G-invariant inner product. □

(1.8) Definition. Let V be a G-module. A subspace $U \subset V$ which is G-invariant (i.e., $gu \in U$ for $g \in G$ and $u \in U$) is called a ***submodule*** of V or a ***subrepresentation***. A nonzero representation V is called ***irreducible*** if it has no submodules other than $\{0\}$ and V. A representation which is not irreducible is called ***reducible.***

(1.9) Proposition. *Let G be a compact group. If V is a submodule of the G-module U, then there is a complementary submodule W such that $U = V \oplus W$. Each G-module is a direct sum of irreducible submodules.*

PROOF. Choose a G-invariant inner product on U and let W be the orthogonal complement of V in U, then W is again a G-submodule.

The second statement now follows by induction on the dimension of U, since if $U \neq \{0\}$ is reducible then $U = V \oplus W$ with $0 < \dim V < \dim U$. □

The next theorem is an extremely useful tool in the theory of representations:

(1.10) Theorem (Schur's Lemma). *Let G be any group and let V and W be irreducible G-modules. Then*

(i) *A morphism $V \to W$ is either zero or an isomorphism.*
(ii) *Every morphism $f\colon V \to V$ has the form $f(v) = \lambda v$ for some $\lambda \in \mathbb{C}$.*
(iii) $\dim_{\mathbb{C}} \operatorname{Hom}_G(V, W) = 1$ *if* $V \cong W$, *and*

$$\dim_{\mathbb{C}} \operatorname{Hom}_G(V, W) = 0 \quad \textit{if } V \ncong W.$$

PROOF. Since V is irreducible, the kernel of f is either $\{0\}$ or V. In the latter case f is zero, and in the former f is injective. If f is injective, its image is a nonzero submodule of the irreducible G-module W, and hence is all of W. We conclude that f is an isomorphism, showing (i).

To prove (ii), assume that f is nontrivial and let λ be any eigenvalue of f and W the corresponding eigenspace. Thus $W = \{w \in V \mid f(w) = \lambda w\}$ and one easily checks that W is a G-submodule. Hence $W = V$, which gives (ii). The third part follows from (i) and (ii). □

(1.11) Definitions. The representations of $SU(n)$, $U(n)$, and $GL(n, \mathbb{C})$ on $\mathbb{C}^n$ in which elements of the stated Lie groups simply operate by matrix multiplication are called the ***standard representations***.

A representation is called ***trivial*** if each group element acts as the identity.

(1.12) If V and W are G-modules, we may form their ***direct sum*** $V \oplus W$. This becomes a G-module with the action $g(v, w) = (gv, gw)$. In terms of matrices, this corresponds to the following construction: If $G \to GL(m, \mathbb{C})$, $g \mapsto A(g)$ and $G \to GL(n, \mathbb{C})$, $g \mapsto B(g)$, then we obtain the direct sum representation $G \to GL(m + n, \mathbb{C})$ by forming the block matrices

$$g \mapsto \begin{pmatrix} A(g) & 0 \\ 0 & B(g) \end{pmatrix}.$$

(1.13) Proposition. *An irreducible representation of an abelian Lie group G is one-dimensional.*

PROOF. Since G is abelian, the translation $l_g\colon V \to V$ is a morphism of representations for each $g \in G$. By (1.10)(ii), every l_g is multiplication by $\lambda(g) \in \mathbb{C}$. But this implies that any subspace of V is G-invariant. The result follows, since if $\dim V > 1$, V would have a one-dimensional subspace, and since all subspaces are submodules, this would contradict the irreducibility of V. □

Let G be a compact group. We denote by $\operatorname{Irr}(G, \mathbb{C})$ a complete set of pairwise nonisomorphic complex G-modules, i.e., each irreducible G-module is isomorphic to exactly one element of $\operatorname{Irr}(G, \mathbb{C})$. For arbitrary representations of G we use the following terminology:

If U is isomorphic to a submodule of V we say U is ***contained*** in V. If W is irreducible we call the dimension $\dim_{\mathbb{C}} \operatorname{Hom}_G(W, V)$ the ***multiplicity*** of W

in V. The significance of this number and its name is the following: Suppose we have a decomposition $V = \bigoplus_j V_j$ of V into irreducible submodules V_j (such a decomposition exists by (1.9)). Then $\mathrm{Hom}_G(W, V) = \bigoplus_j \mathrm{Hom}_G(W, V_j)$, and so by Schur's lemma, (1.10), $\dim_{\mathbb{C}} \mathrm{Hom}_G(W, V)$ is simply the number of V_j that are isomorphic to W. In particular, this multiplicity is nonzero if and only if W is contained in V, and this happens for only those finitely many $W \in \mathrm{Irr}(G, \mathbb{C})$ isomorphic to some V_j.

The multiplicities become even more transparent in the ***canonical decomposition*** of V described in the next proposition. Given an irreducible W, we consider the map

$$d_W\colon \mathrm{Hom}_G(W, V) \otimes_{\mathbb{C}} W \to V, \qquad \varphi \otimes w \mapsto \varphi(w).$$

The group G acts on the domain of d_W by $(g, \varphi \otimes w) \mapsto \varphi \otimes gw$, and it is easy to see that d_W is a G-map with this action. We can put all the maps d_W together to form a map

$$d = (d_W)\colon \bigoplus_W \mathrm{Hom}_G(W, V) \otimes_{\mathbb{C}} W \to V,$$

where W ranges over $\mathrm{Irr}(G, \mathbb{C})$. This is again a morphism of G-modules.

(1.14) Proposition. *The map above* $d\colon \bigoplus_{W \in \mathrm{Irr}(G, \mathbb{C})} \mathrm{Hom}_G(W, V) \otimes_{\mathbb{C}} W \to V$ *is an isomorphism.*

Proof. The map d is compatible with G-isomorphisms $V \to V'$ and direct sum decompositions $V = V_1 \oplus V_2$. Thus it suffices to consider $V \in \mathrm{Irr}(G, \mathbb{C})$ (see (1.16), Ex. 11). In this case Schur's lemma, (1.10), tells us that the only nonzero summand in the domain of d is $\mathrm{Hom}_G(V, V) \otimes_{\mathbb{C}} V \cong \mathbb{C} \otimes_{\mathbb{C}} V$, and d describes the canonical isomorphism $\mathbb{C} \otimes_{\mathbb{C}} V \to V$, $\lambda \otimes v \mapsto \lambda v$. Thus the proposition is proved. □

Let $V(W)$ be the image of d_W. We call $V(W)$ the ***W-isotypical summand*** of V, and we also call $\dim_{\mathbb{C}} \mathrm{Hom}_G(W, V)$ the ***multiplicity*** of $V(W)$. The reader should show as an exercise that $V(W_1) = V(W_2)$ if $W_1 \cong W_2$, so each isomorphism class of irreducible complex G-modules yields a uniquely determined submodule of V. The W-isotypical summand enjoys the following property:

(1.15) Proposition. *$V(W)$ is generated by the irreducible submodules of V that are isomorphic to W.*

Proof. If $i\colon W \to V$ is the inclusion of a G-submodule, then $d_W(i \otimes w) = i(w) = w$. Hence $W \subset V(W)$. Thus if $W_1 \subset V$ and $W_1 \cong W \in \mathrm{Irr}(G, \mathbb{C})$, then $W_1 \subset V(W_1) = V(W)$, so $V(W)$ contains all the submodules of V isomorphic to W. On the other hand, if $\varphi \in \mathrm{Hom}_G(W, V)$, the image of φ

lies in the direct sum of the irreducible submodules of V isomorphic to W. Thus the image of d_W lies in the same direct sum, which completes the proof. □

If V is a G-module and $\alpha: H \to G$ is a homomorphism of Lie groups, we obtain an H-module $\alpha^* V$ with the same underlying vector space, but with the H-action $H \times V \to V, (h, v) \mapsto \alpha(h)v$. For an inclusion $\alpha: H \to G$ we call $\alpha^* V$ the ***restriction*** of V to H and sometimes denote it by $\mathrm{res}_H^G V$.

We have already encountered representations in Chapter I, most notably the ***adjoint representation***

$$G \times LG \to LG,$$

which is a representation on a real vector space (cf. §6). This representation is of great importance to the structure theory of G and will be analyzed in detail in later chapters.

(1.16) Exercises

1. Show that the standard representations (1.11) are irreducible.

2. Show that the matrix

$$A(t) = \begin{pmatrix} 1 & t \\ 0 & 1 \end{pmatrix}, \qquad t \neq 0$$

is not conjugate to any unitary matrix. Show that the representation of the additive group $\mathbb{R}$ given by $t \mapsto A(t)$ has no invariant inner product. Show that this representation is not a direct sum of irreducible submodules. What are the irreducible submodules? Show that $\mathbb{R}$ has an uncountable number of nonisomorphic irreducible unitary representations.

3. Show that $SL(2, \mathbb{R})$ has no nontrivial unitary representation as follows:
   (i) For any natural number m verify the identity

$$\begin{pmatrix} m & 0 \\ 0 & m^{-1} \end{pmatrix} A(t) \begin{pmatrix} m & 0 \\ 0 & m^{-1} \end{pmatrix}^{-1} = A(m^2 t) = A(t)^{m^2}$$

   with $A(t)$ as in Exercise 2.
   (ii) Let $\varphi: SL(2, \mathbb{R}) \to U(n)$ be a representation. Use (i) to show that the eigenvalues of $\varphi A(t)$ are a permutation of their m^2th powers for any m and are therefore roots of unity. Conclude that all eigenvalues must be 1.
   (iii) Show that the normal subgroup generated by the $A(t)$ is equal to the whole group.

   Note that this proof does not use the continuity of φ! For more details and references see Hewitt and Ross [1], p. 349.

4. By viewing $\mathbb{R}$ as a vector space over $\mathbb{Q}$ show that $S^1 = \mathbb{R}/\mathbb{Z}$ has many discontinuous representations $S^1 \to S^1 = U(1)$.

   It is a remarkable theorem of van der Waerden [1] that bounded representations of compact semisimple groups (like $SO(n)$, $SU(n)$, $Sp(n)$ and products of such groups) are automatically continuous.

5. Consider the representation $S^1 \to O(2) \subset U(2)$

$$\exp(it) \mapsto \begin{pmatrix} \cos t & -\sin t \\ \sin t & \cos t \end{pmatrix} = R(t).$$

Find the irreducible subspaces. Find a unitary matrix A such that $AR(t)A^{-1}$ is diagonal and consists of two one-dimensional representations.

6. Let A be an abelian subgroup of $U(n)$. Show that A is conjugate to a subgroup of the group of diagonal matrices in $U(n)$. (Use (1.13).)

7. Let V be irreducible. Show that any two G-invariant inner products on V differ by a constant factor. How can one determine all G-invariant inner products for an arbitrary V?

8. Let V and W be isomorphic unitary representations of the compact group G. Show that V and W are actually isometric. In other words, if two homomorphisms $\alpha, \beta: G \to U(n)$ are conjugate in $GL(n, \mathbb{C})$, they are already conjugate in $U(n)$.

9. Let V be a representation of the compact group G with isotypical decomposition $V = V_1 \oplus \cdots \oplus V_r$. Show that the group of G-automorphisms of V is isomorphic to $GL(n_1, \mathbb{C}) \times \cdots \times GL(n_r, \mathbb{C})$, where n_i is the multiplicity of V_i.

10. Does Schur's lemma help determine the center of $GL(n, \mathbb{C})$?

11. Verify the assertions in the first sentence of the proof of (1.14).

2. Semisimple Modules

We have seen that any complex representation of a compact Lie group is a direct sum of irreducible representations. In this section we shall see that the decomposition is unique in a certain sense. We discuss this in a slightly more general setting so that we may apply it to other situations we will encounter later.

Let Ω be any set. An additive abelian group M has the structure of a (left) ***Ω-module*** if for each $\alpha \in \Omega$ there is given an endomorphism $x \mapsto \alpha x$ of M. An Ω-module M is also called a ***group with Ω-operators***. An ***Ω-homomorphism*** between two Ω-modules M and N is a group homomorphism $f: M \to N$ such that $f(\alpha x) = \alpha f(x)$ for all $\alpha \in \Omega$. A subgroup N of an Ω-module M is called a ***submodule*** if $\alpha x \in N$ for each $x \in N$ and $\alpha \in \Omega$. Kernels, images, and cokernels of Ω-homomorphisms are Ω-modules. An Ω-module $M \neq \{0\}$ is called ***simple*** or ***irreducible*** if its only submodules are $\{0\}$ and M. The arguments used in (1.10) yield the same result:

(2.1) Schur's Lemma. *Let $f: M \to N$ be an Ω-homomorphism. If M is simple then f is either injective or zero; if N is simple then f is either surjective or zero; if both are simple then f is either an isomorphism or zero.* □

The ***direct sum*** of a family of Ω-modules is the group-theoretic direct sum with componentwise Ω-operation. An Ω-module is called ***semisimple*** if it is a direct sum of simple modules.

(2.2) Theorem. *Let M be an Ω-module that is a sum (not necessarily direct) of a family $(N_j \mid j \in J)$ of simple submodules. Let E be any submodule of M. Then there is a submodule F of M such that $M = E \oplus F$ and F is the direct sum of some subfamily $(N_j \mid j \in I)$, $I \subset J$. Furthermore, there is an index set $I' \subset J$ with $I \cap I' = \varnothing$ such that M is the direct sum of the simple submodules $(N_j \mid j \in I \cup I')$ and E is isomorphic to the direct sum of the $(N_j \mid j \in I')$. In particular, every submodule of M is semisimple.*

PROOF. Consider those subsets $K \subset J$ such that $\sum_{j \in K} N_j = N(K)$ is in fact a direct sum and such that $N(K) \cap E = \{0\}$. Applying Zorn's lemma, let $I \subset J$ be maximal with respect to these properties. Suppose $E + N(I) \neq M$. Then there is a $j \in J$ with N_j not contained in $E + N(I)$. Since N_j is simple, $N_j \cap E = N_j \cap N(I) = \{0\}$. Thus $N(I) + N_j = N(I \cup \{j\})$ is direct and $E \cap N(I \cup \{j\}) = \{0\}$, contradicting the maximality of I. This proves the first statement, and the second follows by applying the first to the submodule $N(I)$ in place of E. This yields I' with $I \cap I' = \varnothing$ and $M = N(I) \oplus N(I')$, and since $M = E \oplus N(I)$, $E \cong M/N(I) \cong N(I')$. □

A semisimple Ω-module M is called ***isotypical*** if it is a direct sum of simple submodules each of which is isomorphic to a given simple module S. We also say that M is ***S-isotypical*** in this case. If M is S-isotypical, the previous theorem (2.2) implies that each simple submodule of M is isomorphic to S and hence each submodule of M is itself S-isotypical.

(2.3) Proposition. *Let M be a semisimple Ω-module and S a simple Ω-module. Let $M(S)$ be the submodule of M generated by all simple submodules of M isomorphic to S. Then $M(S)$ is S-isotypical and the S-isotypical submodules of M are submodules of $M(S)$. If N is a submodule of M, then $N(S) = N \cap M(S)$. Finally, M is the direct sum of the submodules $M(S) \neq \{0\}$, where S ranges over a complete set of pairwise nonisomorphic simple Ω-modules (contained in M).*

PROOF. $M(S)$ is semisimple and S-isotypical by (2.2) and contains every S-isotypical submodule of M by definition. If N is a submodule of M, then $N(S)$ is S-isotypical, so $N(S) \subset M(S) \cap N$, and conversely $M(S) \cap N$ is S-isotypical, so $M(S) \cap N \subset N(S)$, showing equality. To prove the final statement, note that by collecting those simple submodules isomorphic to the same S, we may write $M = \bigoplus_S M_S$ with M_S S-isotypical. Note also that if M has no submodule isomorphic to S, we have $M_S = M(S) = \{0\}$, so we may take S to run over the index set stated in the proposition. Fixing S, let

N be a submodule of M isomorphic to S. Then by Schur's lemma the map $N \subset M \to \bigoplus_{T \neq S} M_T$ is the zero map. Hence $N \subset M_S$, so $M_S = M(S)$ and the proposition is proved. □

We call the uniquely determined submodule $M(S)$ ***the S-isotypical part*** of M.

The foregoing may be applied to the representations of a compact Lie group G as follows. Let $\Omega = G \cup \mathbb{C}$ as a set. If V is a representation of G, then we are given an action of $g \in G \subset \Omega$ on V, and $\lambda \in \mathbb{C} \subset \Omega$ acts via scalar multiplication. Thus V may be considered an Ω-module, and an Ω-submodule is clearly a subrepresentation. By (1.9), such representations are semisimple as Ω-modules. Since simple Ω-submodules of V are irreducible representations, we end up with a decomposition of V into its S-isotypical parts, where S runs through a complete set of pairwise nonisomorphic irreducible representations of G.

This section is based on Dieudonné [2], annexe. See also Cartan and Eilenberg [1], I.4.

(2.4) Exercises

1. Let M be an S-isotypical representation of the compact group G. Suppose $M = M_1 \oplus \cdots \oplus M_k$, where each M_i is irreducible. Show that k is uniquely determined by M, but that the M_i themselves are not uniquely determined if $k \geq 2$.
2. Show that the additive group $\mathbb{R}$ admits representations which are not semisimple.
3. Let V be a finite-dimensional complex vector space and let u be an endomorphism of V. Let $\Omega = \mathbb{C} \cup \{u\}$, with $\mathbb{C}$ operating on V by scalar multiplication and u by $x \mapsto u(x)$. The endomorphism u is called semisimple if V is a semisimple Ω-module. Show that V is semisimple if and only if V has a basis consisting of eigenvectors of u. Determine the isotypical components.
4. Show that the ring of Ω-endomorphisms of a simple Ω-module is a skew field.
5. Explain how Ω-modules may be considered as modules over the ring freely generated by Ω.

3. Linear Algebra and Representations

Algebraic constructions may be used to obtain new representations from old ones. We have already seen how to form ***direct sums*** $V \oplus W$, and now we consider tensor products, exterior powers, and homomorphisms.

Let V and W be representations of G. The ***tensor product*** representation

$$V \otimes W$$

has the action $g(v \otimes w) = gv \otimes gw$. If $v_1, \ldots, v_n$ is a basis of V and $w_1, \ldots, w_m$ is a basis of W, then the $v_i \otimes w_k$ form a basis of $V \otimes W$ and the map $V \times W \to V \otimes W$, $(v, w) \to v \otimes w$, is bilinear. If g acts on V and W via the matrices (r_{ij}) and (s_{kl}), then g acts on $V \otimes W$ via the matrix $(r_{ij}s_{kl})$ whose entry in the (i, k)th row and (j, l)th column is $r_{ij}s_{kl}$. More explicitly, if $gv_j = \sum_i r_{ij}v_i$ and $gw_l = \sum_k s_{kl}w_k$, then

$$g(v_j \otimes w_l) = \sum_{i,k} r_{ij}s_{kl}v_i \otimes w_k.$$

The matrix $(r_{ij}s_{kl})$ is sometimes called the ***Kronecker product*** of (r_{ij}) and (s_{kl}).

We also have an action of G on

$$\mathrm{Hom}(V, W)$$

given by $(g \cdot f)(v) = gf(g^{-1}v)$, where $f \in \mathrm{Hom}(V, W)$, $g \in G$, and $v \in V$. This makes the following diagram commute:

$$\begin{array}{ccc} V & \xrightarrow{\ f\ } & W \\ {\scriptstyle g}\downarrow & & \downarrow{\scriptstyle g} \\ V & \dashrightarrow_{g \cdot f} & W \end{array}$$

If $W = \mathbb{C}$ is the trivial representation, then

$$\mathrm{Hom}(V, \mathbb{C}) = V^*$$

is called the ***dual representation*** of V. If $gv_j = \sum_i r_{ij}v_i$ with respect to the basis $v_1, \ldots, v_n$ of V, and if $v_1^*, \ldots, v_n^*$ is the dual basis, then we may write $g \cdot v_j^* = \sum_i s_{ij}(g)v_i^*$. But

$$s_{ij}(g) = (gv_j^*)(v_i) = v_j^*(g^{-1}v_i) = v_j^*\left(\sum_k r_{ki}(g^{-1})v_k\right) = r_{ji}(g^{-1}),$$

in other words g acts via the transpose of the inverse.

If V is a complex vector space, we may define the conjugate space $\overline{V}$ which has the same additive structure as V but scalar multiplication defined by

$$\mathbb{C} \times V \to V, \qquad (z, v) \mapsto \bar{z}v.$$

If V is a G-module, then $\overline{V}$ is also a G-module, called the ***conjugate representation*** of V. If we choose an invariant inner product on V, then

$$\overline{V} \to V^*, \qquad v \mapsto \langle -, v \rangle$$

is an isomorphism.

Occasionally we must resort to fancier constructions from linear algebra; for example, the ith ***exterior power*** $\Lambda^i(V)$ of V and the ith ***symmetric power*** $S^i(V)$ of V. If $v_1, \ldots, v_n$ is a basis of V, then a basis of $\Lambda^i(V)$ is given by the symbols $v_{k_1} \wedge \cdots \wedge v_{k_i}$, $k_1 < \cdots < k_i$, and the map $V^i \to \Lambda^i(V)$, $(w_1, \ldots, w_i) \mapsto w_1 \wedge \cdots \wedge w_i$ is linear in each variable and alternating. In particular, $\Lambda^n(V)$

is one-dimensional and $g \in G$ operates via multiplication by the determinant of l_g.

There are many canonical isomorphisms between these kind of constructions, and they yield canonical isomorphisms between the corresponding representations. The examples listed below provide an excellent opportunity for the reader to check his understanding of linear algebra.

(3.1)
$$(U \otimes V) \otimes W \cong U \otimes (V \otimes W),$$
$$U \otimes V \cong V \otimes U,$$
$$U \otimes (V \oplus W) \cong (U \otimes V) \oplus (U \otimes W),$$
$$\Lambda^k(V \oplus W) \cong \bigoplus_{i=0}^{k} \Lambda^i(V) \otimes \Lambda^{k-i}(W),$$
$$\mathrm{S}^k(V \oplus W) \cong \bigoplus_{i=0}^{k} \mathrm{S}^i(V) \otimes \mathrm{S}^{k-i}(W),$$
$$V \otimes V \cong \mathrm{S}^2(V) \oplus \Lambda^2(V).$$

Of particular importance for representation theory is the isomorphism

$$\theta\colon V^* \otimes W \to \mathrm{Hom}(V, W),$$

which maps $v^* \otimes w$ to the homomorphism $u \mapsto v^*(u)w$. We remark that although this map is canonical, it is an isomorphism only if V or W is finite dimensional.

If $V = W$, then the map

(3.2) $$\mathrm{Hom}(V, V) \cong V^* \otimes V \to \mathbb{C}, \qquad v^* \otimes u \mapsto v^*(u)$$

associates to $f \in \mathrm{Hom}(V, V)$ its ***trace*** $\mathrm{Tr}(f) \in \mathbb{C}$. If $v_1, \ldots, v_n$ is a basis of V and $fv_j = \sum_i r_{ij} v_i$, then $\theta(\sum r_{ik} v_k^* \otimes v_i) = f$.

Consequently

$$\mathrm{Tr}(f) = \sum_i r_{ii}.$$

We collect some properties of the trace.

(3.3) Proposition.

(i) $\mathrm{Tr}\colon \mathrm{Hom}(V, V) \to \mathbb{C}$ *is linear.*
(ii) $\mathrm{Tr}(\varphi f \varphi^{-1}) = \mathrm{Tr}(f)$ *for each* $\mathbb{C}$*-automorphism* φ *of* V.
(iii) *For* $f\colon V \to W$ *and* $h\colon W \to V$, $\mathrm{Tr}(fh) = \mathrm{Tr}(hf)$.
(iv) $\mathrm{Tr}(f \oplus h) = \mathrm{Tr}(f) + \mathrm{Tr}(h)$.
(v) $\mathrm{Tr}(f \otimes h) = \mathrm{Tr}(f) \cdot \mathrm{Tr}(h)$.
(vi) $f\colon V \to V$ *induces a map* $f^*\colon V^* \to V^*$ *and* $\mathrm{Tr}(f^*) = \mathrm{Tr}(f)$.
(vii) *If* $f\colon V \to V$ *is idempotent, then* $\mathrm{Tr}(f)$ *is the dimension of the image of* f.
(viii) $f\colon V \to V$ *induces a map* $\bar{f}\colon \bar{V} \to \bar{V}$ *and* $\mathrm{Tr}(\bar{f}) = \overline{\mathrm{Tr}(f)}$.

PROOF. The proof of this proposition is relegated to (3.4), Ex. 3. □

(3.4) Exercises

1. Let V be a G-module. The symmetric group $S(k)$ acts on the k-fold tensor product $V^{\otimes k} = V \otimes \cdots \otimes V$ by permuting the factors. This action commutes with the action of G on $V^{\otimes k}$. The subspace of ***alternating tensors***
$$a(V^{\otimes k}) = \{x \mid \pi x = (\operatorname{sign} \pi)x \text{ for all } \pi \in S(k)\}$$
and of ***symmetric tensors***
$$s(V^{\otimes k}) = \{x \mid \pi x = x \text{ for all } \pi \in S(k)\}$$
are therefore subrepresentations of $V^{\otimes k}$. Show
$$a(V^{\otimes k}) \cong \Lambda^k V \quad \text{and} \quad s(V^{\otimes k}) \cong \mathbf{S}^k(V).$$
2. Describe the representations $\Lambda^2 V$ and $\mathbf{S}^2 V$ in matrix form.
3. Prove (3.3).
4. Prove the identities in (3.1).
5. Use an invariant inner product to show that, for compact groups G and G-modules V, there is an isomorphism $\bar{V} \cong V^*$.
6. Let V be n-dimensional. Show that $\dim \Lambda^i V = \binom{n}{i}$ and $\dim \mathbf{S}^i(V) = \binom{i+n-1}{n-1}$.

4. Characters and Orthogonality Relations

We will now begin to use invariant integration on the compact Lie group G to derive some deeper insights into the structure of representations.

Let V be a (complex) representation of G. The ***fixed point set***
$$V^G = \{v \in V \mid gv = v \text{ for all } g \in G\}$$
is a linear subspace of V. For each $v \in V$, let $p(v) = \int gv\, dg$. By invariance of integration we have $xp(v) = x\int gv\, dg = \int xgv\, dg = p(v)$ for $x \in G$, so $p(v) \in V^G$. Also, if $v \in V^G$ then $\int gv\, dg = \int v\, dg = v$. Therefore

(4.1)
$$p\colon V \to V^G, \qquad v \mapsto \int gv\, dg$$
is a projection operator onto V^G.

Recall that the group G operates on $\operatorname{Hom}(V, W)$ by $(g \cdot f)v = gf(g^{-1}v)$. With this action $\operatorname{Hom}(V, W)^G = \operatorname{Hom}_G(V, W)$, the space of G-maps $V \to W$. Therefore we obtain the projection operator
$$p\colon \operatorname{Hom}(V, W) \to \operatorname{Hom}_G(V, W), \qquad f \mapsto \int (g \cdot f)\, dg.$$
If V is irreducible, $\operatorname{Hom}_G(V, V) \cong \mathbb{C}$ by Schur's lemma.

(4.2) Proposition. *Let V be irreducible. Then for $f \in \operatorname{Hom}(V, V)$*

$$\int (g \cdot f)\, dg = \frac{1}{|V|} \operatorname{Tr}(f)\, \mathrm{id}_V,$$

where $|V| = \dim_{\mathbb{C}} V$ and $\operatorname{Tr}(f)$ is the trace of f.

PROOF. We have already remarked that by Schur's lemma the integral must be a multiple $c \cdot \mathrm{id}_V$ of the identity map id_V. We compute the constant c by applying the linear map $\operatorname{Tr}\colon \operatorname{Hom}(V, V) \to \mathbb{C}$ which, as does any linear map, commutes with integration:

$$|V| \cdot c = \operatorname{Tr}(c \cdot \mathrm{id}_V) = \int \operatorname{Tr}(g \cdot f) = \int \operatorname{Tr}(l_g \circ f \circ l_g^{-1}) = \int \operatorname{Tr}(f) = \operatorname{Tr}(f).$$

Note that we have used (3.3)(ii). □

If $v_1, \ldots, v_n$ is a basis of V, then the matrix representation $G \to \mathrm{GL}(n, \mathbb{C})$, $g \mapsto (r_{ij}(g))$ is given by $gv_j = \sum_i r_{ij}(g) v_i$. Thus if $v_1^*, \ldots, v_n^*$ is the dual basis of V^*,

$$r_{ij}(g) = v_i^*(gv_j). \tag{4.3}$$

This motivates us to consider general functions of the form

$$g \mapsto \varphi(gv)$$

with $\varphi \in V^*$, $v \in V$. These are called ***representative functions*** on G. If we apply the linearity property of the integral to (4.2) we obtain

(4.4) Proposition. *Let V be irreducible. Then for $\varphi \in V^*, v \in V, f \in \operatorname{Hom}(V, V)$*

$$\int \varphi(gf(g^{-1}v))\, dg = \frac{1}{|V|} \operatorname{Tr}(f) \varphi(v).$$

We wish to apply this result to the linear operator $B\colon V \to V$, $u \mapsto \psi(u)w$, where $\psi \in V^*$ and $w \in V$. The trace of this operator is $\psi(w)$ (see (3.2)).

We compute

$$\varphi(g^{-1}B(gv)) = \varphi(g^{-1}\psi(gv)w) = \psi(gv)\varphi(g^{-1}w).$$

Note that g acts on V^* via $g\varphi(w) = \varphi(g^{-1}w)$, and so, after identifying V^{**} with V, the function $g \mapsto \varphi(g^{-1}w) = w(g\varphi)$ is seen to be a representative function for V^*.

The following two theorems essentially express results about the operator B in terms of an Hermitian inner product $\langle -, - \rangle$ on V instead of in the language of dual spaces. For an explanation of the names of these theorems, see remarks (4.7) below.

(4.5) Theorem (*Orthogonality Relations*). *Let V be irreducible. Then:*

(i) *For any $f \in \operatorname{Hom}(V, V)$ and $v, w \in V$*

$$\int \langle gf(g^{-1}v), w\rangle \, dg = \frac{1}{|V|} \operatorname{Tr}(f)\langle v, w\rangle; \quad \textit{and}$$

(ii) *For $v, w, \alpha, \beta \in V$*

$$\int \langle g^{-1}v, \alpha\rangle \langle g\beta, w\rangle \, dg = \frac{1}{|V|} \langle \beta, \alpha\rangle \langle v, w\rangle.$$

If V happens to be a unitary representation, (ii) may be written

$$\int \langle \overline{g\alpha, v}\rangle \langle g\beta, w\rangle \, dg = \frac{1}{|V|} \langle \overline{\alpha, \beta}\rangle \langle v, w\rangle.$$

PROOF. Statement (i) follows from (4.4) and linearity of the inner product and integral. Applying (i) to the linear map f defined by $f(u) = \langle u, \alpha\rangle\beta$ yields (ii). □

(4.6) Theorem (*Orthogonality Relations*). *Let V and W be nonisomorphic irreducible representations. Then for a G-invariant inner product $\langle -, -\rangle$ on V and any $\alpha, v \in V$ and $\beta, w \in W$*

$$\int \langle \overline{g\alpha, v}\rangle \langle g\beta, w\rangle \, dg = 0.$$

PROOF. Fix α and β. From $\langle \overline{g\alpha, v}\rangle = \langle v, g\alpha\rangle$ it follows that the integral is linear in v and conjugate linear in w. Thus it defines a bilinear form $b\colon V \times \overline{W} \to \mathbb{C}$ which a calculation using the invariance of the integral and the linearity of the given inner products shows to be G-invariant. Hence it defines a G-map $b'\colon V \to \operatorname{Hom}(\overline{W}, \mathbb{C})$. But $\operatorname{Hom}(\overline{W}, \mathbb{C}) = \overline{W}^* \cong W$ (see (3.4), Ex. 5), so by Schur's lemma b' must be the zero map. □

(4.7) Remarks. There is an inner product $\langle \varphi, \psi\rangle = \int \varphi\overline{\psi}$ on the space $C^0(G, \mathbb{C})$ of continuous functions $G \to \mathbb{C}$. Theorems (4.5) and (4.6) concern the behavior of representative functions under this inner product. If $v_1, \ldots, v_m$ and $w_1, \ldots, w_n$ are orthonormal bases of the irreducible unitary representations V and W, we have the matrix representations

$$r_{ij}(g, V) = \langle gv_j, v_i\rangle,$$
$$r_{kl}(g, W) = \langle gw_l, w_k\rangle$$

and (4.5) and (4.6) yield

(4.8)
$$\int r_{ij}(g, V)\overline{r_{kl}(g, V)} = \frac{1}{|V|}\delta_{ik}\delta_{jl},$$
$$\int r_{ij}(g, V)\overline{r_{kl}(g, W)} = 0, \qquad V \not\cong W.$$

Formulas (4.8) express the fact that the matrix entries coming from irreducible unitary representations form an orthogonal system with respect to the inner product $\langle u, v\rangle = \int u\bar{v}$. In Chapter III we will show that, in the appropriate sense, this orthogonal system is complete (Theorem of Peter and Weyl). The reader is also referred to the special case of SU(2) treated in the next section. It turns out that the classical orthogonal systems of special functions all come from representations of Lie groups (see Vilenkin [1]).

We now introduce characters of representations. They are important because they are functions on G which determine representations uniquely up to isomorphism, and it is easier to handle functions than matrices.

Let V be a (complex) representation of G.

(4.9) Definition. The ***character of*** V is the function

$$\chi_V\colon G \to \mathbb{C}, \qquad g \mapsto \mathrm{Tr}(l_g)$$

where $\mathrm{Tr}(l_g)$ is the trace of the linear map $l_g\colon V \to V$, $v \mapsto gv$. The character of an irreducible representation is called an ***irreducible character***.

Properties of the trace given in (3.3) immediately yield some properties of characters:

(4.10) Proposition.

(i) χ_V *is a* C^∞*-function.*
(ii) *If* V *and* W *are isomorphic then* $\chi_V = \chi_W$.
(iii) $\chi_V(ghg^{-1}) = \chi_V(h)$.
(iv) $\chi_{V\oplus W} = \chi_V + \chi_W$.
(v) $\chi_{V\otimes W} = \chi_V \cdot \chi_W$.
(vi) $\chi_{V^*}(g) = \chi_V(g^{-1})$.
(vii) $\chi_{\bar{V}}(g) = \overline{\chi_V(g)} = \chi_V(g^{-1})$.
(viii) $\chi_V(e) = \dim_{\mathbb{C}} V$.

PROOF. For (i) write V as a matrix representation $g \mapsto (r_{ij}(g))$. Then the $r_{ij}(g)$ are C^∞ by I, (3.12) and $\chi_V(g) = \sum_i r_{ii}(g)$. For (ii), let $f\colon V \to W$ be an isomorphism. Then $l_g^W f = f l_g^V$, so

$$\chi_V(g) = \mathrm{Tr}(l_g^V) = \mathrm{Tr}(f^{-1} l_g^W f) = \mathrm{Tr}(l_g^W) = \chi_W(g).$$

Properties (iii)–(v) are clear from (3.3)(ii), (iv), and (v). To demonstrate (vi), apply (3.3)(vi) noting that the left translation by g on V^* is defined to be the map $(l_{g^{-1}})^*$. Finally (vii) follows from (ii) and the isomorphism $\bar{V} \cong V^*$, and (viii) is clear. □

By (4.10)(iii) a character is constant on conjugacy classes of G. A function $f: G \to \mathbb{C}$ with this property is called a ***class function***.

The next theorem expresses the ***orthogonality relations for characters***. In Chapter III we will show that the characters of irreducible representations form a complete orthogonal system in the space of class functions.

(4.11) Theorem.

(i) $\int \chi_V(g)\, dg = \dim V^G$.

(ii) $\langle \chi_W, \chi_V \rangle = \int \bar{\chi}_V(g) \chi_W(g)\, dg = \dim \operatorname{Hom}_G(V, W)$.

(iii) *If V and W are irreducible,* $\int \bar{\chi}_V \chi_W = \begin{cases} 1 & \textit{if } V \cong W, \\ 0 & \textit{otherwise.} \end{cases}$

PROOF. The projection operator p from (4.1) has trace $\dim V^G$. Since Tr is linear and hence commutes with integration, we get

$$\dim V^G = \operatorname{Tr}(p) = \operatorname{Tr}\left(\int l_g\, dg\right) = \int \operatorname{Tr}(l_g)\, dg = \int \chi_V(g)\, dg.$$

This proves (i). To prove (ii), recall that $\operatorname{Hom}(V, W)^G = \operatorname{Hom}_G(V, W)$, so using (i) $\dim \operatorname{Hom}_G(V, W) = \int \chi_{\operatorname{Hom}(V, W)}(g)\, dg$, but $\chi_{\operatorname{Hom}(V, W)} = \chi_{V^* \otimes W} = \bar{\chi}_V \chi_W$. The last statement (iii) follows from (ii) and Schur's lemma. □

(4.12) Theorem. *A representation is determined up to isomorphism by its character.*

PROOF. Let $V = \bigoplus_j n_j V(j)$ be a decomposition of V into irreducible representations (see notation following (1.13)). Then $\chi_V = \sum_j n_j \chi_{V(j)}$ and $n_j = \langle \chi_V, \chi_{V(j)} \rangle$. □

Note. The inner product here and in the next proposition refers to $\langle \chi_V, \chi_W \rangle = \int_g \chi_V(g) \bar{\chi}_W(g)\, dg$.

(4.13) Proposition. *Suppose that $\langle \chi_V, \chi_V \rangle = 1$. Then V is irreducible.*

PROOF. If $V = \bigoplus_j n_j V(j)$, then $\langle \chi_V, \chi_V \rangle = \sum n_j^2$. □

We may apply (4.13) to determine the irreducible representations of a product $G \times H$ of compact Lie groups G and H. Note that for representations V of G and W of H we have the tensor product $V \otimes W$ with the diagonal action $(g, h)(v \otimes w) = gv \otimes hw$.

(4.14) Proposition. *If V is an irreducible representation of G and W is an irreducible representation of H, then $V \otimes W$ is an irreducible representation of $G \times H$. Furthermore, any irreducible representation of $G \times H$ is a tensor product of this form.*

PROOF. The first half follows from (4.13) and the calculation

$$\int_{G\times H} \chi_{V\otimes W}\bar{\chi}_{V\otimes W} = \int_{G\times H} \chi_V(g)\chi_W(h)\bar{\chi}_V(g)\bar{\chi}_W(h)\, dg\, dh$$

$$= \int_G \chi_V(g)\bar{\chi}_V(g)\, dg \cdot \int_H \chi_W(h)\bar{\chi}_W(h)\, dh = 1 \cdot 1 = 1.$$

For the second half, suppose U is a $G \times H$-module. Let

$$\varphi\colon \bigoplus_j \mathrm{Hom}_H(W_j, U) \otimes W_j \to U, \qquad f \otimes w \mapsto f(w)$$

be the isotypical decomposition of U considered as an H-module. G acts on $f \in \mathrm{Hom}_H(W_j, U)$ by $(gf)(w) = g \cdot f(w)$, for since the G- and H-actions on U commute, $gf \in \mathrm{Hom}_H(W_j, U)$. Note that the isomorphism φ is not only an H- but a $G \times H$-isomorphism. Now decompose $\mathrm{Hom}_H(W_j, U) = \bigoplus_i n_{ij} V_i$ as a G-module. This gives us an isomorphism $U \cong \bigoplus_{i,j} n_{ij} V_i \otimes W_j$. □

(4.15) From the proof of (4.14) we obtain a bijection

$$\mathrm{Irr}(G, \mathbb{C}) \times \mathrm{Irr}(H, \mathbb{C}) \to \mathrm{Irr}(G \times H, \mathbb{C}),$$
$$(V, W) \mapsto V \otimes W.$$

The irreducible representations V of a compact abelian Lie group G are one-dimensional (see (1.13)). Therefore the irreducible characters are homomorphisms $G \to \mathrm{U}(1)$. The pointwise product of two such homomorphisms is again a homomorphism, and so the irreducible characters form a group, the ***character group*** $\hat{G}$ of the abelian group G.

We close this section by recording for later reference a few identities concerning the behavior of representative functions and characters with respect to convolution. All of the identities follow from the orthogonality relations.

(4.16) Proposition. *Let $g \mapsto (u_{ij}(g))$ and $g \mapsto (v_{kl}(g))$ be nonisomorphic irreducible unitary matrix representations of the compact group G with characters*

χ_U and χ_V. Let $|U|, |V|$ denote the dimensions of the representations. Then

(i) $$u_{ij} * u_{kl} = \frac{1}{|U|}\delta_{jk} u_{il}, \quad \text{and}$$

$$u_{ij} * v_{kl} = 0.$$

(ii) $$\chi_V * v_{ij} = v_{ij} * \chi_V = \frac{1}{|V|} v_{ij}.$$

$$\chi_U * v_{kl} = v_{kl} * \chi_U = 0.$$

(iii) $$\chi_V * \chi_V = \frac{1}{|V|}\chi_V.$$

$$\chi_V * \chi_U = 0.$$

PROOF. For (i) we start with the definition

$$u_{ij} * u_{kl}(g) = \int u_{ij}(gh^{-1}) u_{kl}(h)\, dh$$

of the convolution and substitute the relations

$$u_{ij}(gh^{-1}) = \sum_t u_{it}(g) u_{tj}(h^{-1})$$

and

$$u_{tj}(h^{-1}) = \bar{u}_{jt}(h).$$

The result is

$$\sum_t u_{it}(g) \int \bar{u}_{jt}(h) u_{kl}(h)\, dh.$$

Using the orthogonality relation (4.8) in this sum leads to the first identity in (i), and the second identity in (i) is proved similarly. The identities (ii) and (iii) follow from (i) using $\chi_V = \sum_t v_{tt}$. □

For a finite group G, the convolution is given by $u \cdot v = |G| u * v$, with the product in the group ring $\mathbb{C}[G]$ (see §1). In this case (4.16)(iii) says that $(|v| : |G|)\chi_V = e_V$ is an idempotent element (i.e., $e_V * e_V = e_V$) and (ii) says that e_V is contained in the center of $\mathbb{C}[G]$. Since $e_U * e_V = 0$, one often says that the e_V are central orthogonal idempotents of $\mathbb{C}[G]$.

(4.17) Exercises

1. Let G be a finite abelian group. Show that $\hat{G}$ is isomorphic to G. Let G act on the group ring $\mathbb{C}[G]$ by left translation. Show that the isotypical components of $\mathbb{C}[G]$ are one-dimensional and that $\sum_g \bar{\chi}(g) g$ generates the χ-isotypical part for $\chi \in \hat{G} = \mathrm{Irr}(G, \mathbb{C})$.

2. Let $f_t\colon G \to GL(n, \mathbb{C})$, $t \in [0, 1]$, be a family of representations f_t depending continuously on $g \in G$ and $t \in [0, 1]$, i.e., a ***homotopy of representations***. Show that, for compact G, f_0 and f_1 are isomorphic representations. Show by way of a counterexample that this conclusion is not valid in the case of a noncompact group G.

   In general, the following is true: Let $f_t\colon G \to H$ be a homotopy of homomorphisms for a compact Lie group G. Then f_0 and f_1 are conjugate; see Conner and Floyd [1], Lemma 38.1. This follows from a theorem of Montgomery and Zippin [1], p. 216, which states that each subgroup H in a sufficiently small neighborhood of a compact subgroup F is conjugate to a subgroup of F. Compare Bredon [1], II.5.6.

3. Let V be an irreducible G-module with G compact. Show that $\chi_V(x)\chi_V(y) = |V| \int \chi_V(gxg^{-1}y)\, dg$.

4. Relate the orthogonality relations for representations of S^1 to the orthogonality relations between the trigonometric polynomials of classical Fourier analysis.

5. Show that $\operatorname{Irr}(G, \mathbb{C})$ is finite for finite G.

5. Representations of SU(2), SO(3), U(2), and O(3)

Our goal in this section is to give elementary descriptions of the irreducible representations of the groups mentioned in the section heading.

We begin with SU(2). Let V_0 be the trivial representation on $\mathbb{C}$ and let V_1 be the standard representation on $\mathbb{C}^2$ (the operation being given by matrix multiplication). In the terminology of §3 the other irreducible representations will have the nth symmetric powers $\mathbf{S}^n V_1$, $n \geq 2$, as their representation spaces. More explicitly: Let V_n be the space of homogeneous polynomials of degree n in two variables z_1 and z_2. The dimension of V_n is $n + 1$. Viewing polynomials as functions on $\mathbb{C}^2$, we obtain a left action of $GL(2, \mathbb{C})$ and hence SU(2) on the polynomials by letting

$$(gP)z = P(zg),$$

where

$$P \in \mathbb{C}[z_1, z_2], \qquad g = \begin{pmatrix} a & b \\ c & d \end{pmatrix}, \qquad z = (z_1, z_2),$$

and

$$zg = (az_1 + cz_2, bz_1 + dz_2).$$

Since each g acts as a homogeneous linear transformation, the subspaces $V_n \subseteq \mathbb{C}[z_1, z_2]$ are SU(2)-invariant. Note that, in this description, $n = 0$ yields the trivial representation and V_1 may be identified with the standard representation.

A basis of the space V_n, consisting of the polynomials

$$P_k(z_1, z_2) = z_1^k z_2^{n-k}, \qquad 0 \le k \le n,$$

is used in the proof of the next proposition.

(5.1) Propbsition. *The representations V_n are irreducible.*

PROOF. It suffices to show that each SU(2)-equivariant endomorphism A of V_n is a multiple of the identity. So let A be equivariant and, for $a \in U(1)$, set

$$g_a = \begin{pmatrix} a & 0 \\ 0 & a^{-1} \end{pmatrix} \in SU(2).$$

Then $g_a P_k = a^{2k-n} P_k$ and $g_a A P_k = A g_a P_k = A a^{2k-n} P_k = a^{2k-n} A P_k$.

Now choose a such that all the powers a^{2k-n}, $0 \le k \le n$, are distinct. It is not difficult to verify that, with a so chosen, the a^{2k-n}-eigenspace of g_a in V_n is generated by P_k. Thus $AP_k = c_k P_k$ for some $c_k \in \mathbb{C}$.

We now consider the real rotations

$$r_t = \begin{pmatrix} \cos t & -\sin t \\ \sin t & \cos t \end{pmatrix} \in SU(2), \qquad t \in \mathbb{R}$$

and compute:

$$\begin{aligned} Ar_t P_n &= A(z_1 \cos t + z_2 \sin t)^n \\ &= \sum_k \binom{n}{k} \cos^k t \cdot \sin^{n-k} t \cdot AP_k \\ &= \sum_k \binom{n}{k} \cos^k t \cdot \sin^{n-k} t \cdot c_k \cdot P_k. \end{aligned}$$

Similarly,

$$r_t A P_n = \sum_k \binom{n}{k} \cos^k t \cdot \sin^{n-k} t \cdot c_n P_k.$$

Comparing coefficients shows that $c_k = c_n$, so $A = c_n \cdot \mathrm{id}$. □

Let

$$e(t) = \begin{pmatrix} \exp(it) & 0 \\ 0 & \exp(-it) \end{pmatrix}.$$

Since any element in SU(2) is conjugate to a diagonal matrix, any element is conjugate to some $e(t)$. Furthermore, $e(t)$ and $e(s)$ are conjugate if and only if $s \equiv \pm t \bmod 2\pi$. Thus if $f\colon SU(2) \to \mathbb{C}$ is a class function, $fe\colon \mathbb{R} \to \mathbb{C}$, $t \mapsto f(e(t))$, is an even 2π-periodic function. The space of continuous class functions may thus be identified with the space of even 2π-periodic continuous functions $\mathbb{R} \to \mathbb{C}$. The character χ_n of V_n has the value

$$\sum_{k=0}^{n} e^{i(n-2k)t}$$

at $e(t)$. For t not an integer multiple of π, this sum equals $\sin(n+1)t/\sin t$, which we will denote by $\kappa_n(t)$. Using the addition theorem for sin we get

$$\kappa_n(t) = \cos nt + \kappa_{n-1}(t) \cos t,$$

so $\kappa_0(t), \ldots, \kappa_n(t)$ generate the same vector space as $1, \cos t, \ldots, \cos nt$. It is well known from elementary Fourier analysis that the space generated by $\cos nt$, $n \in \mathbb{N}_0$ is uniformly dense in the space of even 2π-periodic continuous functions $\mathbb{R} \to \mathbb{C}$, so we have shown that the characters χ_n are uniformly dense in the space of class functions on SU(2).

(5.2) Proposition. *For continuous class functions f on* SU(2) *one has*

$$\int_{SU(2)} f(x)\, dx = \frac{2}{\pi} \int_0^{\pi} fe(t) \sin^2 t\, dt.$$

PROOF. Since the V_n are irreducible, we know from (4.11) that the integral $\int \chi_n = 1$ for $n = 0$ and $\int \chi_n = 0$ for $n > 0$. Since $\chi_n(e(t)) \sin^2 t = \sin(n+1)t \cdot \sin t$, it is easy to work out that the right-hand side gives the same result. And because the χ_n generate a dense subspace, the stated equation is true by continuity. □

(5.3) Proposition. *Every irreducible unitary representation of* SU(2) *is isomorphic to one of the V_n.*

PROOF. Suppose the irreducible representation W with character χ were different from all the V_n. By orthogonality (4.11)(iii) $\langle \chi, \chi_n \rangle = 0$ and $\langle \chi, \chi \rangle = 1$, but this is a contradiction because the χ_n generate a dense subspace. □

We now turn to irreducible representations of SO(3). There is an epimorphism (I, (6.18)) π: SU(2) $\to$ SO(3) with kernel the diagonal matrices $\{E, -E\}$. If W is an irreducible representation of SO(3), then the corresponding representation $\pi^* W$ of SU(2) is irreducible and $-E$ acts as the identity. Conversely, if $-E$ acts as the identity on the SU(2)-representation V, then we obtain an associated representation of SO(3). Therefore:

(5.4) The irreducible representations of SO(3) are in bijective correspondence with the irreducible representations V_n of SU(2) in which $-E$ acts as the identity.

We know that $-E$ acts as multiplication by $(-1)^n$ on V_n, so the V_{2n} yield the irreducible representations W_n of SO(3). Note that W_n has dimension

$2n + 1$. We will show later that the W_n can be realized as suitable SO(3)-invariant spaces of polynomials on S^2—the so-called spherical harmonics.

The next result is often referred to as the ***Clebsch-Gordan formula.***

(5.5) Proposition.

$$V_k \otimes V_l = \bigoplus_{j=0}^{q} V_{k+l-2j}, \quad \text{with} \quad q = \min\{k, l\}.$$

PROOF. We will prove the proposition by proving the corresponding result for characters. Since characters are determined by their values on the elements $e(t)$, it suffices to verify the purely combinatorial identity

$$\left(\sum_{\mu=0}^{k} x^{k-2\mu}\right)\left(\sum_{\nu=0}^{l} x^{l-2\nu}\right) = \sum_{j=0}^{q}\left(\sum_{\nu=0}^{k+l-2j} x^{k+l-2j-2\nu}\right)$$

and replace x by $\exp(it)$. To this end, we may assume that $l \leq k$. Now arrange the pairs of indices $(k - 2\mu, l - 2\nu)$ in a rectangular scheme—to each such pair there corresponds a summand $x^{k-2\mu}x^{l-2\nu}$ in the left-hand product. The right-hand sum then comes about by summing first over the pairs of indices on the separate lines "j = const." indicated in Figure 16 and then over j.

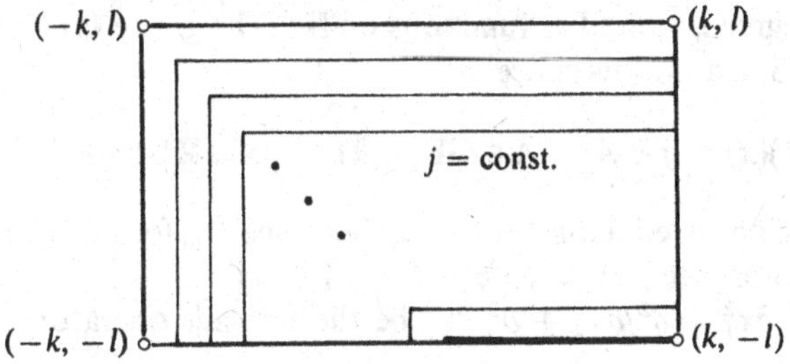

Figure 16 □

Remark. The representations V_k are sometimes enumerated by half-integers in the literature, say $V_k = V(k/2)$. The Clebsch–Gordan formula then reads

$$V(a) \otimes V(b) \cong V(|a - b|) \oplus V(|a - b| + 1) \oplus \cdots \oplus V(a + b).$$

Turning to our other groups, recall that there is an epimorphism $S^1 \times SU(2) \to U(2)$ with kernel $\{(1, E), (-1, -E)\}$ (see I, (1.16), Ex. 12). The irreducible representations of $S^1 \times SU(2)$ are given as tensor products (see (4.15)) $A_m \otimes V_n$, where A_m is the representation $S^1 \times \mathbb{C} \to \mathbb{C}, (\lambda, z) \mapsto \lambda^m z$ for $m \in \mathbb{Z}$. Of course, we have used the fact that all the irreducible representations of S^1 are of the form A_m (compare I, (4.15), Ex. 9 or §8). The element $(-1, -E)$ is contained in the kernel of $A_m \otimes V_n$ if and only if $(m + n)$ is even. Thus the representations $A_m \otimes V_n$, for $(m + n)$ even, yield the irreducible representations of U(2).

The group O(3) is isomorphic to the direct product $SO(3) \times \mathbb{Z}/2$ (see I, (1.16), Ex. 12), so an application of (4.15) to the results above gives a description of its irreducible representations.

In the rest of this section we will delve further into the previously mentioned connection between representations of SO(3) and spherical harmonic functions. The homomorphism $SU(2) \to SO(3)$ of I, (6.18) maps

$$e(t) = \begin{pmatrix} \exp(it) & 0 \\ 0 & \exp(-it) \end{pmatrix} \quad \text{to} \quad R(2t) = \begin{pmatrix} 1 & 0 & 0 \\ 0 & \cos 2t & -\sin 2t \\ 0 & \sin 2t & \cos 2t \end{pmatrix}.$$

The value of the character of W_n at $R(t)$ is the same as the value of χ_{2n} at $e(t/2)$, and this value is

$$\sum_{k=0}^{2n} e^{i(n-k)t} = \frac{e^{i(n+1)t} - e^{-i(n+1)t}}{e^{it} - e^{-it}}.$$

Since every element of SO(3) is conjugate to an element $R(t)$, the character of an SO(3)-module is determined by its restriction to the subgroup T generated by the $R(t)$.

Let P_l be the complex vector space of homogeneous polynomials in three variables of degree l, viewed as functions on $\mathbb{R}^3$. The group $GL(3, \mathbb{R})$ and its subgroup SO(3) act on this space:

$$(Af)(x) = f(xA), \quad A \in GL(3, \mathbb{R}), \qquad x \in \mathbb{R}^3, \quad f \in P_l.$$

The space P_l is not irreducible for $l \geq 2$; the space P_2, for instance, contains the invariant subspace generated by $x_1^2 + x_2^2 + x_3^2$.

Let $\Delta = \partial^2/\partial x_1^2 + \partial^2/\partial x_2^2 + \partial^2/\partial x_3^2$ be the Laplace operator on $\mathbb{R}^3$. The vector space

$$\mathfrak{H}_l = \{f \in P_l \mid \Delta f = 0\}$$

is called the space of ***harmonic polynomials*** of degree l. Restricting functions in $\mathfrak{H}_l$ to the sphere S^2 yields the ***spherical harmonics*** of degree l. Note that a homogeneous function on $\mathbb{R}^3$ is uniquely determined by its restriction to S^2.

(5.6) Lemma.

$$\dim P_l = \tfrac{1}{2}(l+1)(l+2) \quad \textit{and} \quad \dim \mathfrak{H}_l = 2l + 1.$$

PROOF. The monomials $x_1^p x_2^q x_3^r$, $p + q + r = l$, form a basis for P_l. There are $(k + 1)$ ways for the nonnegative integers p and q to add to k, so the dimension of P_l is

$$\sum_{k=0}^{l} (k + 1) = \tfrac{1}{2}(l + 1)(l + 2).$$

Now a polynomial $f \in P_l$ may be written in the form

$$f(x_1, x_2, x_3) = \sum_{k=0}^{l} \frac{x_1^k}{k!} f_k(x_2, x_3),$$

where f_k is homogeneous of degree $l - k$ in x_2, x_3. Thus we have

$$\Delta f = \sum_{k=0}^{l-2} \frac{x_1^k}{k!} f_{k+2} + \sum_{k=0}^{l} \frac{x_1^k}{k!} \left(\frac{\partial^2 f_k}{\partial x_2^2} + \frac{\partial^2 f_k}{\partial x_3^2} \right).$$

Consequently $\Delta f = 0$ if and only if

(5.7) $$f_{k+2} = -\left(\frac{\partial^2 f_k}{\partial x_2^2} + \frac{\partial^2 f_k}{\partial x_3^2} \right), \quad 0 \le k \le l - 2.$$

An element of $\mathfrak{H}_l$ is therefore uniquely determined by f_0 and f_1, since all the higher f_k may be computed from (5.7). Thus the dimension of $\mathfrak{H}_l$ is the sum of the dimensions of the spaces of homogeneous polynomials in two variables of degrees l and $l - 1$, i.e., $\dim \mathfrak{H}_l = (l + 1) + l = 2l + 1$. □

We assume the following simple lemma:

(5.8) Lemma. *The action of the Laplace-operator on the space of C^∞-functions $\mathbb{R}^3 \to \mathbb{C}$ commutes with the action of* SO(3), *i.e., Δ is* SO(3)*-equivariant.*

As a corollary to (5.8) we obtain:

(5.9) Corollary. *$\mathfrak{H}_l$ is an* SO(3)*-invariant subspace of P_l.*

Having obtained the SO(3)-module $\mathfrak{H}_l$, it is natural to ask for its decomposition into irreducible modules. Since $\dim \mathfrak{H}_l = 2l + 1 = \dim W_l$, the following result is not too surprising.

(5.10) Proposition. *The space $\mathfrak{H}_l$ of harmonic polynomials of degree l is an irreducible* SO(3)*-module.*

Proof. We will show that $\mathfrak{H}_l \cong W_l$. In fact, suppose that we have a decomposition into irreducible SO(3)-modules

$$\mathfrak{H}_l \cong \bigoplus_{\nu} W_{n_\nu}.$$

It suffices by reason of dimension to show that $n_\nu \ge l$ for some n_ν. To do this we restrict our attention to the subgroup $T \subseteq$ SO(3) of the matrices $R(t)$ introduced above. We have already computed the character value of W_n at $R(t)$, and thus we know that the character value of $\mathfrak{H}_l$ at $R(t)$ is a linear combination of $\exp(ikt)$, $|k| \le \max n_\nu$. Hence we will be done if we can find a T-invariant subspace of $\mathfrak{H}_l$ on which $R(t)$ acts via multiplication by $\exp(\pm ilt)$.

To this end we consider $f_l(x_1, x_2, x_3) = (x_2 + ix_3)^l$. Then $f_l \in \mathfrak{H}_l$, and

$$\begin{aligned} R(t)f_l(x_1, x_2, x_3) &= (x_2 \cos t + x_3 \sin t + i(x_2(-\sin t) + x_3 \cos t))^l \\ &= e^{-ilt} f_l(x_1, x_2, x_3). \end{aligned}$$

This completes the proof. □

Spherical functions are related to classical special functions (Legendre polynomials). For a representation theoretic explanation of this phenomenon, see §10.

Representations of compact groups appear in physics in many guises. For instance, SO(3) acts as the group of orientation preserving orthogonal symmetries of $\mathbb{R}^3$, and invariance under this action embodies the intuitive principle that physical reactions such as those between elementary particles should not depend on the observer's vantage point. But there are also situations in which physical quantities are not represented by single vectors like position vectors but by classes of vectors. States in quantum mechanics, for example, are given by a class λx, $\lambda \in S^1$, $|x| = 1$, of vectors, or equivalently a line through the origin, in a Hilbert space. Consequently physicists are often interested in actions of compact groups on projective spaces instead of vector spaces.

We are thus motivated to consider ***projective representations***, which are homomorphisms

$$G \to \mathrm{PGL}(n, \mathbb{C}) = \mathrm{GL}(n, \mathbb{C})/\mathbb{C}^* = \mathrm{SL}(n, \mathbb{C})/C_n,$$

where G is compact and C_n is the group of nth roots of unity viewed as a subgroup of $\mathrm{SL}(n, \mathbb{C})$ via the inclusion $\zeta \mapsto \zeta E$ for $\zeta^n = 1$. Our goal is to describe such homomorphisms, up to conjugation, for the group SO(3).

We know that linear matrix representations are similar to unitary ones. The next lemma gives an analogous statement for projective representations. The proof uses a general categorical construction called a ***fibered product***.

(5.11) Lemma. *Every homomorphism* $G \to \mathrm{SL}(n, \mathbb{C})/C_n$ *is conjugate to a homomorphism whose image lies in* $\mathrm{SU}(n)/C_n$. *(That is to say,* $\mathrm{SU}(n)/C_n$ *is a maximal compact subgroup of* $\mathrm{PGL}(n, \mathbb{C})$*.)*

PROOF. Given φ, consider the pullback diagram below, where p is projection

$$\begin{array}{ccc} H & \xrightarrow{\tilde{\varphi}} & \mathrm{SL}(n, \mathbb{C}) \\ {\scriptstyle \tilde{p}}\downarrow & & \downarrow{\scriptstyle p} \\ G & \xrightarrow[\varphi]{} & \mathrm{PGL}(n, \mathbb{C}). \end{array}$$

Thus $H = \{(g, A) \mid g \in G, A \in \mathrm{SL}(n, \mathbb{C}), pA = \varphi(g)\}$ is the fibered product of G and $\mathrm{SL}(n, \mathbb{C})$ over $\mathrm{PGL}(n, \mathbb{C})$. The maps $\tilde{\varphi}$ and $\tilde{p}$ are simply given by $\tilde{\varphi}(g, A) = A$ and $\tilde{p}(g, A) = g$. Then H is a group and is compact because G

is compact and $\tilde{p}$ has finite kernel C_n. Thus by II, (1.7) the matrix representation $\tilde{\varphi}$ is similar to a homomorphism $H \to \mathrm{SU}(n)$, and the lemma follows. □

We are now reduced to considering homomorphisms

$$\varphi\colon \mathrm{SO}(3) \to \mathrm{SU}(n)/C_n.$$

Since $\mathrm{SU}(n)$ is simply connected, the projections $\mathrm{SU}(n) \to \mathrm{SU}(n)/C_n$ in general, and the projection $\mathrm{SU}(2) \to \mathrm{SU}(2)/C_2 = \mathrm{SO}(3)$ in particular, are universal coverings. The theory of covering spaces tells us that there is a lifting $\tilde{\varphi}$ of φ:

$$\begin{array}{ccc} \mathrm{SU}(2) & \xrightarrow{\tilde{\varphi}} & \mathrm{SU}(n) \\ \downarrow & & \downarrow \\ \mathrm{SO}(3) & \xrightarrow[\varphi]{} & \mathrm{SU}(n)/C_n \end{array}$$

It is for this reason that physicists are often interested in unitary representations of the quaternion group $\mathrm{SU}(2) = \mathrm{Spin}(3)$ rather than in those of the more natural symmetry group $\mathrm{SO}(3)$. In fact, recovering the projective representations of $\mathrm{SO}(3)$ from the unitary representations of $\mathrm{SU}(2)$ is not difficult.

First note that a representation $\mathrm{SU}(2) \to \mathrm{U}(n)$ automatically lands in $\mathrm{SU}(n)$ since (5.3) shows that there are no nontrivial homomorphisms $\mathrm{SU}(2) \to \mathrm{U}(1)$ (also see (5.13), Ex. 7). So we only need to ask ourselves which homomorphisms $\tilde{\varphi}\colon \mathrm{SU}(2) \to \mathrm{SU}(n)$ may be pushed down to homomorphisms $\mathrm{SO}(3) \to \mathrm{SU}(n)/C_n$. Recalling that $\mathrm{SO}(3) = \mathrm{SU}(2)/C_2$, the necessary and sufficient condition needed on $\tilde{\varphi}$ is clearly that $\tilde{\varphi}(-E)$ be a multiple of the identity. Hence we find

(5.12) Proposition. *The projective representations of* SO(3) *are given up to conjugation by the representations of* SU(2) *which have the form either*

$$\bigoplus_n k_n V_{2n} \quad \text{or} \quad \bigoplus_n k_n V_{2n+1}, \qquad k_n \in \mathbb{N}_0.$$

The first kind are called ***even*** *and the second kind are called* ***odd***.

Proof. $(-E)$ operates by id on the first space and $(-\mathrm{id})$ on the second space. □

For an elementary treatment of the representation theory of SU(2) and SO(3) see Sugiura [1] and Gelfand, Minlos, and Shapiro [1]. For an extensive bibliography on projective representations see Beyl and Tappe [1].

(5.13) Exercises

1. Show that the space of conjugacy classes of $G = \mathrm{SU}(2)$—i.e., the orbit space of the action $G \times G \to G, (g, x) \mapsto gxg^{-1}$—is homeomorphic to a compact interval.

2. Show that

$$\langle \sum a_k P_k, \sum b_k P_k \rangle = \sum k!\,(n-k)!\, a_k \bar{b}_k$$

defines an SU(2)-invariant inner product on V_n. *Hint*: Use the fact that an invariant inner product $\langle\langle\ ,\ \rangle\rangle$ exists, and restrict to the action of $\{g_a | a \in \mathrm{U}(1)\}$. The P_k generate different irreducible summands for this action and hence must be pairwise orthogonal. Thus $\langle\langle \sum a_k P_k, \sum b_k P_k \rangle\rangle = \sum c_k \cdot a_k \bar{b}_k$. To compute c_k, use $\langle\langle r_t P_n, r_t P_n \rangle\rangle = \langle\langle P_n, P_n \rangle\rangle$ and show $c_n/c_k = \binom{n}{k}$.

3. Describe a method to compute the multiplicity of V_i in $\Lambda^k(V_n)$.

4. The representation of U(2) on the space V_n of homogeneous polynomials is irreducible. Which representation is it in terms of the classification $A_m \otimes V_n$ given above?

5. Verify the Clebsch–Gordan formula for representations of SO(3):

$$W_k \otimes W_l \cong W_{|k-l|} \oplus W_{|k-l|+1} \oplus \cdots \oplus W_{k+l}.$$

For $(m + p)$ even, let $T^{(m,p)}$ be the irreducible representation of U(2) derived from $A_m \otimes V_p$ as in the text. Show that

$$T^{(m,p)} \otimes T^{(n,q)} \cong \bigoplus_j T^{(m+n,\, p+q-2j)},$$

where the summation is over integers j such that $0 \le j \le \min\{p, q\}$.

6. Show lemma (5.8). An easy argument is $\Delta = \operatorname{div} \operatorname{grad}$.

7. Let $p\colon \tilde{G} \to G$ be the universal cover of a connected group. Show that every homomorphism $\varphi\colon G \to \mathrm{SU}(n)/C_n$ has a unique lifting $\tilde{\varphi}$ which makes the diagram

$$\begin{array}{ccc} \tilde{G} & \xrightarrow{\ \tilde{\varphi}\ } & \mathrm{SU}(n) \\ {\scriptstyle p}\big\downarrow & & \big\downarrow \\ G & \xrightarrow[\ \varphi\ ]{} & \mathrm{SU}(n)/C_n \end{array}$$

commute—really, show that the universal cover is a functor.

If $\tilde{G}$ is compact, show that every homomorphism $\tilde{\varphi}\colon \tilde{G} \to \mathrm{U}(n)$ has image in $\mathrm{SU}(n)$, and if the representation defined by $\tilde{\varphi}$ is irreducible, then from $\tilde{\varphi}$ one can produce a φ such that the diagram commutes. Thus in the latter case the irreducible projective representations of G (i.e., those without G-invariant projective subspace) correspond uniquely to the irreducible representations of $\tilde{G}$.

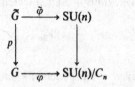

Remark. Note that φ gives rise to a pullback diagram

$$\begin{array}{ccc} H & \longrightarrow & \mathrm{SU}(n) \\ \downarrow & & \downarrow \\ G & \longrightarrow & \mathrm{SU}(n)/C_n \end{array}$$

This provides a covering $H \to G$, which is classified by the homomorphism $\pi_1(G) \to \pi_1(\mathrm{SU}(n)/C_n) \cong C_n$. Thus if this homomorphism is trivial, φ may be lifted to a homomorphism $\tilde{G} \to \mathrm{SU}(n)$. Apply this to $\tilde{\varphi}$ instead of φ. For the second part, show that there is no nontrivial homomorphism $\tilde{G} \to S^1$ and apply Schur's lemma.

6. Real and Quaternionic Representations

If U is a finite-dimensional left vector space over the real numbers $\mathbb{R}$ (resp. the quaternions $\mathbb{H}$), then a ***real*** (resp. ***quaternionic***) ***representation of*** G ***on*** U is a continuous action $\rho: G \times U \to U$ such that the left translations are $\mathbb{R}$-linear (resp. $\mathbb{H}$-linear). The pair (U, ρ) is called a real (resp. quaternionic) representation of G, or a real (resp. quaternionic) G-module. It will usually be denoted simply by U.

Choosing a basis for U yields corresponding matrix representations $G \to \mathrm{GL}(n, \mathbb{R})$ (resp. $G \to \mathrm{GL}(n, \mathbb{H})$).

If the representation space U carries a G-invariant inner product, we consider ***orthogonal*** (resp. ***symplectic***) representations or G-modules. Matrix forms are, in this case, homomorphisms $G \to \mathrm{O}(n)$ (resp. $G \to \mathrm{Sp}(n)$).

As in the case of complex representations, one can show that invariant inner products exist for compact G. This leads to the conclusion that every real (resp. quaternionic) G-module is a direct sum of irreducible submodules: real and quaternionic representations of compact groups are semisimple (cf. §2).

For the remainder of this section we assume that G is compact.

We want to analyze the interaction between real, complex, and quaternionic representations. This requires a certain amount of bookkeeping, but is not mathematically deep. In order to start the bookkeeping, we do two things: We introduce notation for various categories of representations and functors between them, and we treat complex representations as basic objects, viewing real and quaternionic G-modules as complex G-modules with additional structure.

We begin by defining the additional structure. Let V be a complex G-module. A ***real structure*** on V is a conjugate-linear G-map $\mathcal{J}: V \to V$ such that $\mathcal{J}^2 = \mathrm{id}$. A ***quaternionic structure*** on V is a conjugate-linear G-map $\mathcal{J}: V \to V$ such that $\mathcal{J}^2 = -\mathrm{id}$. In both cases $\mathcal{J}$ is called a ***structure map***. A complex representation is said to be of ***real*** (resp. ***quaternionic***) type if it admits a real (resp. quaternionic) structure.

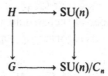

At this point the reader should notice that a representation can be simultaneously of real and quaternionic type (example?). Furthermore, a complex vector space has a large number of real structures ((6.10), Ex. 4).

Next we introduce some categories of representations and functors between them. We let

$$K = \mathbb{R}, \mathbb{C}, \text{ or } \mathbb{H}$$

and let

$$\mathrm{Rep}(G, K)$$

be the category whose objects are G-representations on K-vector spaces and whose morphisms are K-linear G-equivariant maps. Moreover, we let

$$\mathrm{Rep}_+(G, \mathbb{C}) \quad \text{and} \quad \mathrm{Rep}_-(G, \mathbb{C})$$

be the categories of complex G-modules with real and quaternionic structures. Morphisms in these categories are $\mathbb{C}$-linear G-equivariant maps which commute with the structure maps.

We continue by showing that $\mathrm{Rep}(G, \mathbb{R})$ and $\mathrm{Rep}_+(G, \mathbb{C})$ are equivalent categories. This is the formalization of the statement that real representations are complex representations with additional structure.

Given a real G-module U, let $e_+(U) = \mathbb{C} \otimes_{\mathbb{R}} U$ with the structure map $\mathscr{J}(z \otimes u) = \bar{z} \otimes u$. Conversely, given $(V, \mathscr{J})$, let V_+ and V_- be the $(+1)$ and (-1) eigenspaces of $\mathscr{J}$. The equality $2v = (v + \mathscr{J}v) + (v - \mathscr{J}v)$ implies that $V = V_+ \oplus V_-$. Also, note that multiplication by $i \in \mathbb{C}$ induces isomorphisms $V_+ \to V_-$ and $V_- \to V_+$. Let $s_+(V, \mathscr{J}) = V_+$, considered as a real G-module. These constructions supply functors

$$e_+ : \mathrm{Rep}(G, \mathbb{R}) \to \mathrm{Rep}_+(G, \mathbb{C}),$$

$$s_+ : \mathrm{Rep}_+(G, \mathbb{C}) \to \mathrm{Rep}(G, \mathbb{R}).$$

Furthermore, the compositions $e_+ s_+$ and $s_+ e_+$ are naturally equivalent to the identity: The reader may check that a natural isomorphism $e_+ s_+(V, \mathscr{J}) \cong (V, \mathscr{J})$ is given by $\mathbb{C} \otimes_{\mathbb{R}} V_+ \to V$, $z \otimes v \mapsto zv$ and a natural isomorphism $s_+ e_+ U \cong U$ is given by $U \to (\mathbb{C} \otimes_{\mathbb{R}} U)_+$, $u \mapsto 1 \otimes u$.

In the case of quaternionic representations we have functors

$$e_- : \mathrm{Rep}(G, \mathbb{H}) \to \mathrm{Rep}_-(G, \mathbb{C}),$$

$$s_- : \mathrm{Rep}_-(G, \mathbb{C}) \to \mathrm{Rep}(G, \mathbb{H}).$$

Here $e_-(W)$ is $(W, \mathscr{J})$, where W is considered as a $\mathbb{C}$-vector space by restricting scalars from $\mathbb{H}$ to $\mathbb{C}$. The G-action on W remains the same, and $\mathscr{J}$ is defined as multiplication by $j \in \mathbb{H}$. We define $s_-(V, \mathscr{J})$ to be V viewed as an $\mathbb{H}$-module, where the action of $j \in \mathbb{H}$ is defined as $\mathscr{J}$. A straightforward calculation similar to that above shows that $e_- s_-$ and $s_- e_-$ are naturally isomorphic to the identity.

We will also consider relations (i.e., functors) between different types of representations coming from restriction, extension, and conjugation.

First, we may view a complex representation as real, forget about the structure maps $\mathscr{J}$, and so on. These processes lead to the following maps called ***restriction maps***. Thus, for example, $r_{\mathbb{R}}^{\mathbb{C}}$ of the complex G-module V is V viewed as an $\mathbb{R}$-vector space with the same G-action, while $r_+(V, \mathscr{J}) = V$ as a complex G-module with no additional structure.

$$r_{\mathbb{R}}^{\mathbb{C}}\colon \operatorname{Rep}(G, \mathbb{C}) \to \operatorname{Rep}(G, \mathbb{R}),$$
$$r_{\mathbb{C}}^{\mathbb{H}}\colon \operatorname{Rep}(G, \mathbb{H}) \to \operatorname{Rep}(G, \mathbb{C}),$$
$$r_{\mathbb{R}}^{\mathbb{H}}\colon \operatorname{Rep}(G, \mathbb{H}) \to \operatorname{Rep}(G, \mathbb{R}), \qquad r_{\mathbb{R}}^{\mathbb{H}} = r_{\mathbb{R}}^{\mathbb{C}} \circ r_{\mathbb{C}}^{\mathbb{H}},$$
$$r_+\colon \operatorname{Rep}_+(G, \mathbb{C}) \to \operatorname{Rep}(G, \mathbb{C}),$$
$$r_-\colon \operatorname{Rep}_-(G, \mathbb{C}) \to \operatorname{Rep}(G, \mathbb{C}).$$

Second, we have ***extension maps***,

$$e_{\mathbb{R}}^{\mathbb{C}}\colon \operatorname{Rep}(G, \mathbb{R}) \to \operatorname{Rep}(G, \mathbb{C}),$$
$$e_{\mathbb{C}}^{\mathbb{H}}\colon \operatorname{Rep}(G, \mathbb{C}) \to \operatorname{Rep}(G, \mathbb{H}),$$
$$e_{\mathbb{R}}^{\mathbb{H}}\colon \operatorname{Rep}(G, \mathbb{R}) \to \operatorname{Rep}(G, \mathbb{H}).$$

The map $e_{\mathbb{R}}^{\mathbb{C}}$ is defined by $e_{\mathbb{R}}^{\mathbb{C}}(U) = \mathbb{C} \otimes_{\mathbb{R}} U$, and the map $e_{\mathbb{C}}^{\mathbb{H}}$ is defined by $e_{\mathbb{C}}^{\mathbb{H}}(V) = \mathbb{H} \otimes_{\mathbb{C}} V$ where we view $\mathbb{H}$ as a right $\mathbb{C}$-module via right multiplication: $\mathbb{H} \times \mathbb{C} \to \mathbb{H}$, $(w, v) \mapsto wv$. Finally, we let $e_{\mathbb{R}}^{\mathbb{H}} = e_{\mathbb{C}}^{\mathbb{H}} e_{\mathbb{R}}^{\mathbb{C}}$, which is the same thing as saying that $e_{\mathbb{R}}^{\mathbb{H}}(V) = \mathbb{H} \otimes_{\mathbb{R}} V$.

Third, we have ***conjugation***

$$c\colon \operatorname{Rep}(G, \mathbb{C}) \to \operatorname{Rep}(G, \mathbb{C})$$

mapping V to the conjugate module $\overline{V}$.

(6.1) Proposition. *The maps defined above satisfy the following relations*:

$$r_{\mathbb{R}}^{\mathbb{C}} e_{\mathbb{R}}^{\mathbb{C}} = 2, \qquad c e_{\mathbb{R}}^{\mathbb{C}} = e_{\mathbb{R}}^{\mathbb{C}},$$
$$e_{\mathbb{R}}^{\mathbb{C}} r_{\mathbb{R}}^{\mathbb{C}} = 1 + c, \qquad r_{\mathbb{R}}^{\mathbb{C}} c = r_{\mathbb{R}}^{\mathbb{C}},$$
$$e_{\mathbb{C}}^{\mathbb{H}} r_{\mathbb{C}}^{\mathbb{H}} = 2, \qquad c r_{\mathbb{C}}^{\mathbb{H}} = r_{\mathbb{C}}^{\mathbb{H}},$$
$$r_{\mathbb{C}}^{\mathbb{H}} e_{\mathbb{C}}^{\mathbb{H}} = 1 + c, \qquad e_{\mathbb{C}}^{\mathbb{H}} c = e_{\mathbb{C}}^{\mathbb{H}},$$
$$e_{\mathbb{R}}^{\mathbb{C}} = r_+ e_+, \qquad r_{\mathbb{C}}^{\mathbb{H}} = r_- e_-,$$
$$c^2 = \mathrm{id}.$$

Proof. The magnanimous reader will not misinterpret these statements but will understand that the first claims the existence of a natural isomorphism $r_{\mathbb{R}}^{\mathbb{C}} e_{\mathbb{R}}^{\mathbb{C}}(U) \cong U \oplus U$, for real G-modules U, the second maintains that $e_{\mathbb{R}}^{\mathbb{C}} r_{\mathbb{R}}^{\mathbb{C}}(V) \cong V \oplus \overline{V}$ for complex G-modules V, and so on. The verification of these relations is chiefly a matter of patience. As an example, we will prove that $e^{\mathbb{C}}_{\mathbb{R}} r^{\mathbb{C}}_{\mathbb{R}} = 1 + c$. Define a map $\alpha\colon V \oplus V \to \mathbb{C} \otimes_{\mathbb{R}} V$ by

$$\alpha(v, w) = \tfrac{1}{2}(1 \otimes v - i \otimes iv) + \tfrac{1}{2}(1 \otimes w + i \otimes iw).$$

Then α is $\mathbb{R}$-linear by construction and satisfies the relation $\alpha(iv, -iw) = i\alpha(v, w)$. Thus α may be considered as a $\mathbb{C}$-linear map $V \oplus \bar{V} \to \mathbb{C} \otimes_{\mathbb{R}} V$. An inverse to α is given by the map $\beta\colon \mathbb{C} \otimes_{\mathbb{R}} V \to V \oplus \bar{V}$, $z \otimes v \mapsto (zv, \bar{z}v)$.

As another example, the relation $r^{\mathbb{H}}_{\mathbb{C}} e^{\mathbb{H}}_{\mathbb{C}} = 1 + c$ comes from the $\mathbb{C}$-isomorphism

$$\mathbb{H} \otimes_{\mathbb{C}} V \to V \oplus \bar{V}, \qquad (z_1 + jz_2) \otimes v \mapsto (z_1 v, z_2 v).$$

Here we view $\{1, j\}$ as a $\mathbb{C}$-basis of $\mathbb{H}$ with respect to right multiplication, and $z_1, z_2 \in \mathbb{C}$. The rest of the proof is no more difficult than the examples given. □

We now come to the main goal of this section: To relate the irreducible real and quaternionic representations to the irreducible complex representations. We do this by considering the sets

$$\mathrm{Irr}(G, K), \qquad K = \mathbb{R}, \mathbb{C}, \mathbb{H}$$

of irreducible G-modules over K and partitioning each of these into three disjoint (but possibly empty) parts. Thus we end up with nine sets—admittedly a little rough on the reader who rebels against bookkeeping.

In order to simplify the exposition, we will adhere to the following ***notational convention***: The letters U, V, and W will, respectively, denote elements of $\mathrm{Irr}(G, \mathbb{R})$, $\mathrm{Irr}(G, \mathbb{C})$, and $\mathrm{Irr}(G, \mathbb{H})$, regardless of the presence of indices or conjugation signs.

(6.2) Table and Definitions. The following table gives the definitions of the various subsets $\mathrm{Irr}(G, L)_K$ of irreducible representations.

$K =$	$\mathbb{R}$	$\mathbb{C}$	$\mathbb{H}$
$U \in \mathrm{Irr}(G, \mathbb{R})_K$	$e^{\mathbb{C}}_{\mathbb{R}} U = V$ V of real type	$U = r^{\mathbb{C}}_{\mathbb{R}} V$ $V \not\cong \bar{V}$	$U = r^{\mathbb{C}}_{\mathbb{R}} V$ V of quaternionic type
$V \in \mathrm{Irr}(G, \mathbb{C})_K$	V of real type	$V \not\cong \bar{V}$	V of quaternionic type
$W \in \mathrm{Irr}(G, \mathbb{H})_K$	$W = e^{\mathbb{H}}_{\mathbb{C}} V$ V of real type	$W = e^{\mathbb{H}}_{\mathbb{C}} V$ $V \not\cong \bar{V}$	$r^{\mathbb{H}}_{\mathbb{C}} W = V$ V of quaternionic type

In words this means: $U \in \mathrm{Irr}(G, \mathbb{R})_{\mathbb{R}}$ if and only if $e^{\mathbb{C}}_{\mathbb{R}} U$ is an irreducible complex G-module V and V possesses a real structure. Also, $U \in \mathrm{Irr}(G, \mathbb{R})_{\mathbb{C}}$ precisely if there is an irreducible G-module V, not isomorphic to its conjugate

$\bar{V}$, such that U is isomorphic to $r_{\mathbb{R}}^{\mathbb{C}} V$. The rest of the definitions should be apparent from these examples.

The reader may find it comfortable at this point to think of the set $\mathrm{Irr}(G, \mathbb{C})$ as the set of isomorphism classes of irreducible representations, whereas earlier we had to insist on choosing particular representatives.

We will call an ***irreducible*** representation in a set with index $\mathbb{R}$, $\mathbb{C}$, or $\mathbb{H}$, respectively, of ***real type***, ***complex type***, or ***quaternionic type***. Thus the columns of table (6.2) contain irreducible representations of like type, and a glance at the second row shows that our more general usage of type is consistent with usage for complex representations.

The next theorem constitutes the obvious next step in our plan. Despite the simplicity of its statement given here, its proof must be delayed for the sake of developing some more background.

(6.3) Theorem. *For* $L = \mathbb{R}, \mathbb{C}$, *and* $\mathbb{H}$, *the set* $\mathrm{Irr}(G, L)$ *is the disjoint union of its subsets* $\mathrm{Irr}(G, L)_{\mathbb{R}}$, $\mathrm{Irr}(G, L)_{\mathbb{C}}$, *and* $\mathrm{Irr}(G, L)_{\mathbb{H}}$.

The next proposition gives us a criterion which will be used in the proof of (6.3).

(6.4) Proposition. *A complex representation V is of real (resp. quaternionic) type if and only if there exists a nonsingular symmetric (resp. skew-symmetric) G-invariant bilinear form $B\colon V \times V \to \mathbb{C}$.*

PROOF. Suppose such a form B is given. Then $B(v, w) = \varepsilon B(w, v)$, where $\varepsilon = \pm 1$. Choose a G-invariant Hermitian inner product $\langle\ ,\ \rangle$ on V and define a map $f\colon V \to V$ by requiring that $B(v, w) = \langle v, fw\rangle$ for all $v \in V$. Then f is conjugate-linear, G-equivariant, and an isomorphism since B is nonsingular. From the relation $\langle v, fw\rangle = B(v, w) = \varepsilon B(w, v) = \varepsilon\langle w, fv\rangle = \varepsilon\overline{\langle fv, w\rangle}$ applied twice, we find that

$$\langle v, f^2 w\rangle = \varepsilon\overline{\langle fv, fw\rangle} = \langle f^2 v, w\rangle.$$

Thus εf^2 is Hermitian and positive-definite (i.e., $\langle v, \varepsilon f^2 v\rangle \geq 0$). Hence εf^2 has positive real eigenvalues and we may decompose V into the orthogonal sum of eigenspaces of εf^2. It is easy to check that each such eigenspace is G-invariant and invariant under f, since f commutes with real scalars. Hence we may define a G-automorphism $h\colon V \to V$ by setting $h = \sqrt{\lambda}\cdot \mathrm{id}$ on the eigenspace of εf^2 with associated eigenvalue $\lambda \in \mathbb{R}_+$. Then $hf = fh$ since the eigenspaces of εf^2 are invariant under f, and $\varepsilon f^2 = h^2$. The conjugate-linear G-map $\mathscr{J} = hf^{-1}$ then satisfies $\mathscr{J}^2 = \varepsilon \cdot \mathrm{id}$.

Conversely, suppose that V has a structure map $\mathscr{J}$ such that $\mathscr{J}^2 = \varepsilon \cdot \mathrm{id}$. If $\varepsilon = 1$, then $V \cong \mathbb{C} \otimes_{\mathbb{R}} V_+$, and any nonsingular G-invariant symmetric $\mathbb{R}$-bilinear form on V_+ may be extended to a $\mathbb{C}$-bilinear form B on V, which again is nonsingular, G-invariant and symmetric. If $\varepsilon = -1$, we consider

V to be an $\mathbb{H}$-module, the action of $j \in \mathbb{H}$ being that of $\mathscr{J}$. Thus V carries a G-invariant symplectic inner product $\langle\ ,\ \rangle$. We write

$$\langle u, v\rangle = H(u, v) + B(u, v)j$$

with $H(u, v)$ and $B(u, v)$ in $\mathbb{C}$. Using the relations $\langle \lambda u, v\rangle = \lambda\langle u, v\rangle$ and $\langle u, \lambda v\rangle = \langle u, v\rangle\bar{\lambda}$ for $u, v \in V$ and $\lambda \in \mathbb{H}$ is easy to verify that B is $\mathbb{C}$-bilinear. Moreover, since $\langle u, v\rangle = \langle \overline{v, u}\rangle$, we have $H(u, v) + B(u, v)j = (H(v, u) + B(v, u)j)^{-} = \bar{H}(v, u) - j\bar{B}(v, u) = \bar{H}(v, u) - B(v, u)j$. This shows that B is skew-symmetric. To see that B is nonsingular, suppose $B(u, v_0) = 0$ for all $u \in V$. This means $\langle u, v_0\rangle \in \mathbb{C}$ for all $u \in V$, and, since $\langle ju, v_0\rangle = j\langle u, v_0\rangle$, this implies $\langle u, v_0\rangle = 0$ for all $u \in \mathbb{C}$. Since symplectic inner products are nonsingular, $v_0 = 0$. □

A complex representation V is called ***self-conjugate*** if $V \cong \bar{V}$. The next proposition is essentially a corollary of (6.4).

(6.5) Proposition. *Let V be an irreducible self-conjugate representation. Then V is of real or quaternionic type and not of both.*

PROOF. We intend to use (6.4), so we will examine the space $\mathrm{Hom}(V \otimes V, \mathbb{C}) \cong V^* \otimes V^*$ of bilinear forms on V. This space is the direct sum $S \oplus A$ of the space S of symmetric and the space A of skew-symmetric bilinear forms. The G-invariant forms are elements of the fixed-point set $(S \oplus A)^G = S^G \oplus A^G$. Since V is self-conjugate we have $V \cong \bar{V} \cong V^*$, and since V is irreducible $\mathrm{Hom}(V \otimes V, \mathbb{C})^G \cong \mathrm{Hom}(V, V^*)^G$ is one-dimensional by Schur's lemma. Thus there are just two cases: If $\dim S^G = 1$ and $\dim A^G = 0$, then V admits a symmetric G-invariant form which corresponds to an isomorphism $V \cong V^*$ and hence is nonsingular, and V admits no nonsingular skew-symmetric G-invariant forms at all. The other case is that $\dim S^G = 0$ and $\dim A^G = 1$, and here the situation is reversed. □

We now present a

PROOF OF THEOREM (6.3). The proof is divided into three parts corresponding to $L = \mathbb{C}$, $\mathbb{R}$, and $\mathbb{H}$. Letters U, V, W with or without indices denote irreducible representations over $\mathbb{R}$, $\mathbb{C}$, and $\mathbb{H}$, respectively.

First part: Let $V \in \mathrm{Irr}(G, \mathbb{C})$. If V is of either real or quaternionic type, the structure map $\mathscr{J}$ provides an isomorphism $V \cong \bar{V}$. This observation, together with proposition (6.5) and a glance at the table (6.2) takes care of the case $L = \mathbb{C}$.

Second part: Let $U \in \mathrm{Irr}(G, \mathbb{R})$. Writing simply e and r for $e_{\mathbb{R}}^{\mathbb{C}}$ and $r_{\mathbb{R}}^{\mathbb{C}}$, the argument given in the proof of (6.4) shows that if eU is irreducible, then $eU = V$ carries a symmetric bilinear form. Hence by (6.5) V is of real type. If eU is not irreducible, let $eU = V_1 \oplus \cdots \oplus V_t$ be a decomposition into irreducible complex G-modules. The relation $reU = U \oplus U$ from (6.1) shows that $U \oplus U = rV_1 \oplus \cdots \oplus rV_t$. Hence $t = 2$ and $rV_1 = rV_2 = U$.

Applying another relation from (6.1), $eU = erV_1 = V_1 \oplus \bar{V}_1$, so we must have $V_2 = \bar{V}_1$. There are now two possibilities: If $V_1 \ncong \bar{V}_1$, then $U \in \mathrm{Irr}(G, \mathbb{R})_{\mathbb{C}}$. And if $V_1 \cong \bar{V}_1$ we can show that V_1 is not of real type, for if it were, we would have $rV_1 = (V_1)_+ \oplus (V_1)_-$ contradicting $rV_1 = U$. Thus if $V_1 \cong \bar{V}_1$, then $V_1 \in \mathrm{Irr}(G, \mathbb{C})_{\mathbb{H}}$ and we have completed the case $L = \mathbb{R}$.

Third part: Let $W \in \mathrm{Irr}(G, \mathbb{H})$ and let r and e stand for $r_{\mathbb{C}}^{\mathbb{H}}$ and $e_{\mathbb{C}}^{\mathbb{H}}$. If $rW = V$ is irreducible, it carries a quaternionic structure, so $W \in \mathrm{Irr}(G, \mathbb{H})_{\mathbb{H}}$ in this case. If rW is not irreducible, let $rW = V_1 \oplus \cdots \oplus V_t$. As above the relations $er = 2$ and $re = 1 + c$ from (6.1) imply that $t = 2$ and $V_2 = \bar{V}_1$. This time we have $eV_1 = W$. If $V_1 \ncong \bar{V}_1$, then $W \in \mathrm{Irr}(G, \mathbb{H})_{\mathbb{C}}$. It only remains to demonstrate that if $V_1 \cong \bar{V}_1$, then V_1 is not of quaternionic type. But if it were, we could write $V_1 = rX$ and $W = eV_1 = erX = X \oplus X$ would contradict the irreducibility of W. This finishes the case $L = \mathbb{H}$ and the theorem is proved. □

In the course of proving (6.3) we have established the first six of the following implications. Analogous reasoning leads to the remaining ones, so we leave the remainder of the proof of the next proposition as an exercise.

(6.6) Proposition.

(i) $U \in \mathrm{Irr}(G, \mathbb{R})_{\mathbb{R}} \Rightarrow e_{\mathbb{R}}^{\mathbb{C}}U = V$, $\quad V \in \mathrm{Irr}(G, \mathbb{C})_{\mathbb{R}}$.

(ii) $U \in \mathrm{Irr}(G, \mathbb{R})_{\mathbb{C}} \Rightarrow e_{\mathbb{R}}^{\mathbb{C}}U = V \oplus \bar{V}$, $\quad V \in \mathrm{Irr}(G, \mathbb{C})_{\mathbb{C}}$.

(iii) $U \in \mathrm{Irr}(G, \mathbb{R})_{\mathbb{H}} \Rightarrow e_{\mathbb{R}}^{\mathbb{C}}U = V \oplus V$, $\quad V \in \mathrm{Irr}(G, \mathbb{C})_{\mathbb{H}}$.

(iv) $W \in \mathrm{Irr}(G, \mathbb{H})_{\mathbb{R}} \Rightarrow r_{\mathbb{C}}^{\mathbb{H}}W = V \oplus V$, $\quad V \in \mathrm{Irr}(G, \mathbb{C})_{\mathbb{R}}$.

(v) $W \in \mathrm{Irr}(G, \mathbb{H})_{\mathbb{C}} \Rightarrow r_{\mathbb{C}}^{\mathbb{H}}W = V \oplus \bar{V}$, $\quad V \in \mathrm{Irr}(G, \mathbb{C})_{\mathbb{C}}$.

(vi) $W \in \mathrm{Irr}(G, \mathbb{H})_{\mathbb{H}} \Rightarrow r_{\mathbb{C}}^{\mathbb{H}}W = V$, $\quad V \in \mathrm{Irr}(G, \mathbb{C})_{\mathbb{H}}$.

(vii) $V \in \mathrm{Irr}(G, \mathbb{C})_{\mathbb{R}} \Rightarrow r_{\mathbb{R}}^{\mathbb{C}}V = U \oplus U$, $\quad U \in \mathrm{Irr}(G, \mathbb{R})_{\mathbb{R}}$,
$e_{\mathbb{C}}^{\mathbb{H}}V = W$, $\quad W \in \mathrm{Irr}(G, \mathbb{H})_{\mathbb{R}}$.

(viii) $V \in \mathrm{Irr}(G, \mathbb{C})_{\mathbb{C}} \Rightarrow r_{\mathbb{R}}^{\mathbb{C}}V = U = r_{\mathbb{R}}^{\mathbb{C}}\bar{V}$, $\quad U \in \mathrm{Irr}(G, \mathbb{R})_{\mathbb{C}}$,
$e_{\mathbb{C}}^{\mathbb{H}}V = W = e_{\mathbb{C}}^{\mathbb{H}}\bar{V}$, $\quad W \in \mathrm{Irr}(G, \mathbb{H})_{\mathbb{C}}$.

(ix) $V \in \mathrm{Irr}(G, \mathbb{C})_{\mathbb{H}} \Rightarrow r_{\mathbb{R}}^{\mathbb{C}}V = U$, $\quad U \in \mathrm{Irr}(G, \mathbb{R})_{\mathbb{H}}$,
$e_{\mathbb{C}}^{\mathbb{H}}V = W \oplus W$, $\quad W \in \mathrm{Irr}(G, \mathbb{H})_{\mathbb{H}}$.

The next result is an important and easily remembered characterization of the sets $\mathrm{Irr}(G, \mathbb{R})_K$.

(6.7) Theorem. *The endomorphism algebra* $\mathrm{Hom}_G(U, U)$ *of* $U \in \mathrm{Irr}(G, \mathbb{R})$ *is isomorphic to* K *if and only if* $U \in \mathrm{Irr}(G, \mathbb{R})_K$.

PROOF. By Schur's lemma every nonzero $\varphi \in \mathrm{Hom}_G(U, U)$ is invertible and hence $\mathrm{Hom}_G(U, U)$ is a division algebra over $\mathbb{R}$. There are only three possibilities, namely $\mathbb{R}$, $\mathbb{C}$, and $\mathbb{H}$ (Frobenius' theorem, see Jacobson [2], 7.7, p. 430). Suppose $\mathrm{Hom}_G(U, U) = \mathbb{C}$. This means $\mathbb{C}$ acts on U, so $U = r_{\mathbb{R}}^{\mathbb{C}}V$ for some irreducible V. If V were of real type, then $r_{\mathbb{R}}^{\mathbb{C}}V = V_+ \oplus V_-$ would not be irreducible, and if V were of quaternionic type, then $\mathbb{H}$ would be contained in $\mathrm{Hom}_G(U, U)$. Thus $V \in \mathrm{Irr}(G, \mathbb{C})_{\mathbb{C}}$ and $U \in \mathrm{Irr}(G, \mathbb{R})_{\mathbb{C}}$. Next suppose $\mathrm{Hom}_G(U, U) = \mathbb{H}$. Then $U = r_{\mathbb{R}}^{\mathbb{H}} W$ and W is irreducible. From (6.6) we see that $W \in \mathrm{Irr}(G, \mathbb{H})_{\mathbb{H}}$, so $U \in \mathrm{Irr}(G, \mathbb{R})_{\mathbb{H}}$. Finally, suppose $\mathrm{Hom}_G(U, U) = \mathbb{R}$. Then U cannot be of the form $r_{\mathbb{R}}^{\mathbb{C}} V$ for any irreducible V of complex or quaternionic type, for in either case $\mathbb{C} \subset \mathrm{Hom}_G(U, U)$. This completes the proof. □

We have just seen that the type of an irreducible real representation is determined by its endomorphism algebra. The next proposition shows how the type of an irreducible complex representation is determined by its character.

(6.8) Proposition. *Let V be an irreducible complex representation of G with character $\chi_V \colon G \to \mathbb{C}$. Then*

$$\int \chi_V(g^2)\, dg = \begin{cases} 1 \Leftrightarrow V & \text{is of real type,} \\ 0 \Leftrightarrow V & \text{is of complex type,} \\ -1 \Leftrightarrow V & \text{is of quaternionic type.} \end{cases}$$

Proof. The representation $V \otimes V$ splits into $S \oplus A$ where S is the space of symmetric and A is the space of antisymmetric tensors. Both are G-modules. We claim that $\chi_V(g^2) = \chi_S(g) - \chi_A(g)$ for each $g \in G$. For this we may view V, S, and A as representations of the closed abelian subgroup H of G generated by the element $g \in G$. As such $V = \bigoplus_i V(i)$ where all the $V(i)$ are irreducible H-modules and hence one-dimensional. Now $A = \Lambda^2(V) = \Lambda^2(\bigoplus_i V(i))$. Combining the general fact that $\Lambda^k(B \oplus C) = \bigoplus_j \Lambda^j(B) \otimes \Lambda^{k-j}(C)$ and the observation that $\Lambda^2(V(i)) = 0$ because $V(i)$ is one-dimensional shows that $A = \bigoplus_{i<j} V(i) \otimes V(j)$. Thus

$$\begin{aligned} \chi_V(g^2) &= \sum_i \chi_{V(i)}(g^2) = \sum_i (\chi_{V(i)}(g))^2 \\ &= \Big(\sum_i \chi_{V(i)}(g)\Big)^2 - 2 \sum_{i<j} \chi_{V(i)}(g)\chi_{V(j)}(g) = \chi_{V\otimes V}(g) - 2\chi_A(g) \\ &= (\chi_{V\otimes V}(g) - \chi_A(g)) - \chi_A(g) = \chi_S(g) - \chi_A(g), \end{aligned}$$

establishing the claim. Integrating yields

$$\int \chi_V(g^2) = \int \chi_S(g) - \int \chi_A(g) = \dim S^G - \dim A^G.$$

But $\dim S^G = \dim(S^*)^G$, which is the dimension of the symmetric G-invariant bilinear forms on V. The same is true for A^G with respect to antisymmetric forms, so the current proposition follows directly from (6.4). □

There is a ***canonical decomposition*** of real G-modules X which is analogous to (1.14). For $U \in \operatorname{Irr}(G, \mathbb{R})$, let $D(U) = \operatorname{Hom}_G(U, U)$ be the endomorphism ring of U. Then $\operatorname{Hom}_G(U, X)$ is a right $D(U)$-module via composition, and U is a left $D(U)$-module via evaluation. Therefore the tensor product $\operatorname{Hom}_G(U, X) \otimes_{D(U)} U$ is defined and the map $\operatorname{Hom}_G(U, X) \otimes U \to X$, $\varphi \otimes u \to \varphi(u)$ factors through this tensor product. Thus we obtain a homomorphism

$$d\colon \bigoplus_{U \in \operatorname{Irr}(G, \mathbb{R})} \operatorname{Hom}_G(U, X) \otimes_{D(U)} U \to X.$$

(6.9) Proposition. *The map d is an isomorphism.*

PROOF. See the proof of (1.14). □

The image of $\operatorname{Hom}_G(U, X) \otimes_{D(U)} U$ is the U-isotypical part of X.
An original source for this section is Malcev [1], §2.

(6.10) Exercises

1. Let G be a finite group of odd order. Using (6.8) show that every nontrivial irreducible complex representation is of complex type.

2. Let χ be the character of an irreducible complex representation V of a compact group. Show that if χ is real-valued, then V is of real or quaternionic type.

3. Let U be a real G-module, G compact. The real character $\chi_U^{\mathbb{R}}\colon G \to \mathbb{R}$ is defined as $\chi_U^{\mathbb{R}}(g) = \operatorname{Tr}(l_g)$. For $U, U' \in \operatorname{Irr}(G, \mathbb{R})$, show that

$$\int_G \chi_U^{\mathbb{R}} \chi_U^{\mathbb{R}} = \dim_{\mathbb{R}} \operatorname{Hom}_G(U, U), \quad \text{and}$$

$$\int_G \chi_U^{\mathbb{R}} \chi_{U'}^{\mathbb{R}} = 0 \quad \text{if } U' \not\cong U.$$

Let (a_{ij}) and (b_{kl}) be inequivalent irreducible real matrix representations. Show that $\int_G a_{ij} b_{kl} = 0$.

4. Let X be the set of real structures on an n-dimensional complex vector space $\mathbb{C}^n$. Then $\operatorname{GL}(n, \mathbb{C})$ acts on X by $(A, \mathscr{J}) \mapsto A \mathscr{J} A^{-1}$. Show that X is isomorphic to $\operatorname{GL}(n, \mathbb{C})/\operatorname{GL}(n, \mathbb{R})$ as a $\operatorname{GL}(n, \mathbb{C})$-space.

5. Let G be compact. Show that two homomorphisms $\alpha, \beta\colon G \to \mathrm{O}(n)$ are similar (conjugate) if and only if the corresponding homomorphisms $G \to \operatorname{GL}(n, \mathbb{R})$ are similar.

6. Verify the assertions of (6.6).

7. The representations in $\mathrm{Irr}(G, \mathbb{C})_{\mathbb{C}}$ occur in pairs $(V, \bar{V})$. Let $\frac{1}{2}\mathrm{Irr}(G, \mathbb{C})_{\mathbb{C}} \subset \mathrm{Irr}(G, \mathbb{C})_{\mathbb{C}}$ be a subset containing exactly one from each pair. Show that the following maps are defined and are bijections.
$$\mathrm{Irr}(G, \mathbb{R})_{\mathbb{R}} \xrightarrow{e_{\mathbb{R}}^{\mathbb{C}}} \mathrm{Irr}(G, \mathbb{C})_{\mathbb{R}} \xrightarrow{e_{\mathbb{C}}^{\mathbb{H}}} \mathrm{Irr}(G, \mathbb{H})_{\mathbb{R}},$$
$$\mathrm{Irr}(G, \mathbb{R})_{\mathbb{C}} \xleftarrow{r_{\mathbb{R}}^{\mathbb{C}}} \tfrac{1}{2}\mathrm{Irr}(G, \mathbb{C})_{\mathbb{C}} \xrightarrow{e_{\mathbb{C}}^{\mathbb{H}}} \mathrm{Irr}(G, \mathbb{H})_{\mathbb{C}},$$
$$\mathrm{Irr}(G, \mathbb{R})_{\mathbb{H}} \xleftarrow{r_{\mathbb{R}}^{\mathbb{C}}} \mathrm{Irr}(G, \mathbb{C})_{\mathbb{H}} \xleftarrow{r_{\mathbb{C}}^{\mathbb{H}}} \mathrm{Irr}(G, \mathbb{H})_{\mathbb{H}}.$$
For example, to say that $e_{\mathbb{R}}^{\mathbb{C}}U$ is defined means that for $U \in \mathrm{Irr}(G, \mathbb{R})_{\mathbb{C}}$ the representation $e_{\mathbb{R}}^{\mathbb{C}}U$ is irreducible and of real type.

8. Let X and Y be real G-modules. Show that
$$e_+(X \otimes_{\mathbb{R}} Y) \cong e_+(X) \otimes_{\mathbb{C}} e_+(Y).$$
If $X \in \mathrm{Rep}(G, \mathbb{R})$ and $Y \in \mathrm{Rep}(G, \mathbb{H})$, then $X \otimes_{\mathbb{R}} Y \in \mathrm{Rep}(G, \mathbb{H})$, with $h \in \mathbb{H}$ acting via $h(x \otimes y) = x \otimes hy$. Show that
$$e_+(X) \otimes_{\mathbb{C}} e_-(Y) \cong e_-(X \otimes_{\mathbb{R}} Y).$$
If $X, Y \in \mathrm{Rep}(G, \mathbb{H})$, view X as a right $\mathbb{H}$-module X' via $xh = \bar{h}x$. Then $X' \otimes_{\mathbb{H}} Y$ is a real G-module. Show
$$X' \otimes_{\mathbb{H}} Y \cong s_+(e_- X \otimes_{\mathbb{C}} e_- Y).$$
Describe the behavior of tensor products of complex representations possessing structure maps.

9. Use (4.14) and the previous exercise to give a detailed description of $\mathrm{Irr}(G \times H, \mathbb{R})$ in terms of $\mathrm{Irr}(G, K)_L$ and $\mathrm{Irr}(H, K)_L$. In particular, show that for $U \in \mathrm{Irr}(G, \mathbb{R})$ the representation $U^* \otimes_{D(U)} U$ is in $\mathrm{Irr}(G \times G, \mathbb{R})$. Here $D(U)$ is the endomorphism algebra of U.

10. Show that a representation of G has a real (resp. quaternionic) structure if and only if it is self-conjugate and its quaternionic (resp. real) irreducible components have even multiplicity. *Hint*: Use (6.1) and first show that if U and $U \oplus V$ are real (resp. quaternionic), then so is V. Or use Exercise 7.

7. The Character Ring and the Representation Ring

Let G be a compact Lie group. The relations $\chi_{V \oplus W} = \chi_V + \chi_W$ and $\chi_{V \otimes W} = \chi_V \chi_W$ suggest that characters might generate an interesting ring of functions on G. Therefore we let
$$\mathrm{R}(G) = \mathrm{R}(G, \mathbb{C})$$
be the additive group of functions $G \to \mathbb{C}$ generated by characters of complex representations. Using orthogonality of irreducible characters and semisimplicity we note that the characters of irreducible representations are linearly independent and generate $\mathrm{R}(G)$. Thus $\mathrm{R}(G)$ is the free abelian group

generated by these characters. Since $\chi_V \cdot \chi_W = \chi_{V \otimes W}$, the group R($G$) is closed under multiplication. In fact, R(G) is a commutative ring with unit and a subring of the ring of C^∞-functions on G. We call R(G) the ***character ring*** of (complex representations of) G. Elements of R(G) are called ***virtual characters***.

The character ring is a weaker invariant of G than the set of representations itself. This is because we cannot distinguish which elements of R(G) are ***positive***, i.e., are actually characters. Moreover we cannot necessarily recover the irreducible characters from R(G).

The character ring has an alternate description which exhibits a useful universal property. Let $\mathrm{R}^+(G)$ be the set of isomorphism classes of complex G-modules together with the two composition laws $\oplus$ and $\otimes$. This object satisfies the axioms of a commutative ring, except that there are no additive inverses. Mapping representations to their characters gives a map $\chi\colon \mathrm{R}^+(G) \to \mathrm{R}(G)$ satisfying the following:

Universal Property. Given any map $\varphi\colon \mathrm{R}^+(G) \to R$ into a commutative ring such that $\varphi(V \oplus W) = \varphi(V) + \varphi(W)$, $\varphi(V \otimes W) = \varphi(V)\varphi(W)$, and $\varphi(1) = 1$, there is a unique homomorphism $\Phi\colon \mathrm{R}(G) \to R$ such that $\Phi \circ \chi = \varphi$.

Similarly, if φ is just a homomorphism into an abelian group R, there is a unique homomorphism of abelian groups $\Phi\colon \mathrm{R}(G) \to R$ such that $\Phi \circ \chi = \varphi$.

The proof of these assertions is straightforward using the fact that the additive structure of R(G) is that of a free abelian group on the irreducible characters. Naturally, there is a purely formal construction of this universal ring which doesn't refer to characters, often called the Grothendieck construction. In this context R(G) is also called the (complex) ***representation ring*** of G.

We use the universal property to give R(G) more structure. Recall that a fundamental property of the kth exterior power $\Lambda^k V$ of a representation V is that

(7.1)
$$\Lambda^k(V \oplus W) = \sum_{i+j=k} \Lambda^i(V) \otimes \Lambda^j(W).$$

$\Lambda^0(V) = \mathbb{C}$ with the trivial action corresponds to the unit of R(G). By using the formal power series with coefficients in R(G)

$$\lambda_t(V) = 1 + \Lambda^1(V)t + \Lambda^2(V)t^2 + \cdots,$$

we can express all the relations (7.1) with the single relation

(7.2)
$$\lambda_t(V \oplus W) = \lambda_t(V) \cdot \lambda_t(W).$$

Thus $V \mapsto \lambda_t(V)$ maps the additive semigroup $\mathrm{R}^+(G)$ homomorphically into the multiplicative group

$$1 + t\mathrm{R}(G)[[t]]$$

of formal power series over R(G) with constant term 1.

The universal property yields an induced homomorphism

$$\lambda_t\colon \mathrm{R}(G) \to 1 + t\mathrm{R}(G)[[t]].$$

For $x \in \mathrm{R}(G)$ we set

$$\lambda_t(x) = \lambda^0(x) + \lambda^1(x)t + \lambda^2(x)t^2 + \cdots,$$

thereby obtaining maps $\lambda^i\colon \mathrm{R}(G) \to \mathrm{R}(G)$ such that

$$\text{(7.3)}\qquad \lambda^0(x) = 1, \qquad \lambda^1(x) = x, \qquad \lambda^k(x + y) = \sum_i \lambda^i(x)\cdot\lambda^{k-i}(y).$$

A ring R together with maps $\lambda^i\colon R \to R$ satisfying (7.3) is called a ***λ-ring***.

The behavior of the λ^i with respect to composition and products is quite complicated and is given by certain universal polynomials with integer coefficients (independent of G)

$$\lambda^i\lambda^j(x) = P_{i,j}(\lambda^1 x, \ldots, \lambda^{ij}x),$$
$$\lambda^k(xy) = P_k(\lambda^1 x, \ldots, \lambda^k x; \lambda^1 y, \ldots, \lambda^k y),$$

making $\mathrm{R}(G)$ into a so-called ***special*** λ-ring. Definitions and details may be found in Atiyah and Tall [1].

We merely exhibit the definition of the polynomials $P_{i,j}$ and P_k. Let $x_1, \ldots, x_p, y_1, \ldots, y_q$ be indeterminates and let u_i, v_i be the ith elementary symmetric functions in the $x_1, \ldots, x_p$ and $y_1, \ldots, y_q$, respectively. We define polynomials with integer coefficients by requiring $P_n(u_1, \ldots, u_n; v_1, \ldots, v_n)$ to be the coefficient of t^n in $\prod_{i,j}(1 + x_i y_j t)$ and $P_{n,m}(u_1, \ldots, u_{mn})$ to be the coefficient of t^n in $\prod_{i_1 < \cdots < i_m}(1 + x_{i_1} \cdot \ldots \cdot x_{i_m} t)$.

Then P_n is a polynomial of weight n in the u_i and v_i, and $P_{n,m}$ is of weight nm in the u_i. If we assume that $p \geq m$, $q \geq n$ (resp. $p \geq mn$), then none of the variables u_i, v_i are zero and the resulting polynomials are independent of p and q.

We also mention the important ***Adams operations***

$$\Psi^k\colon \mathrm{R}(G) \to \mathrm{R}(G), \qquad k \in \mathbb{N},$$

which are ring homomorphisms satisfying $\Psi^k\Psi^l = \Psi^{kl}$. They, too, are certain polynomials in the λ^i-maps defined as follows:

Consider the sum

$$x_1^k + \cdots + x_m^k$$

in $m \geq k$ variables x_i and express it as a polynomial $Q_k(\sigma_1, \ldots, \sigma_k)$ in the elementary symmetric polynomials σ_i of $x_1, \ldots, x_m$. Now the polynomial Q_k is independent of m for $m \geq k$, as may be seen by setting $x_{m+1} = 0$ and noting that $\sigma_k(x_1, \ldots, x_m, 0) = \sigma_k(x_1, \ldots, x_m)$. We define

$$\Psi^k(x) = Q_k(\lambda^1(x), \ldots, \lambda^k(x)) \in \mathrm{R}(G).$$

(7.4) Proposition. *If we view x and $\Psi^k x$ as virtual characters, then*

$$\Psi^k(x)(g) = x(g^k)$$

for all $g \in G$.

PROOF. Let $x = V$ be a G-module. We view V as an H-module, where H is the closed abelian subgroup of G generated by g. Expressing V as the direct sum of one-dimensional H-modules gives a basis $v_1, \dots, v_n$ of eigenvectors for the action of g, say $gv_i = \lambda_i v_i$. The $v_{i_1} \wedge \cdots \wedge v_{i_r}$, $i_1 < \cdots < i_r$ form a basis for $\Lambda^r V$, so g acts on $\Lambda^r V$ with trace $\sigma_r(\lambda_1, \dots, \lambda_r)$, σ_r being the rth elementary symmetric function in the λ_i. Hence $\Psi^k(V)(g) = Q_k(\sigma_1, \dots, \sigma_k) = \lambda_1^k + \cdots + \lambda_r^k$. Since the trace of g^k on V is $\lambda_1^k + \cdots + \lambda_r^k$, this proves the proposition for x actually a G-module.

Now if x is any virtual character, we define a formal power series $\Psi_t(x) = \sum_{n \geq 1} \Psi(x, n)t^n$ by

$$\text{(7.5)} \qquad \Psi_{-t}(x) = -t\left(\frac{d}{dt}\lambda_t(x)\right)/\lambda_t(x) = -t\frac{d}{dt}\log \lambda_t(x).$$

It can be shown ((7.10), Ex. 2) that $\Psi(x, n) = Q_n(\lambda^1(x), \dots, \lambda^n(x)) = \Psi^n(x)$. From the definition of the $\Psi(x, n)$ it is clear that they are additive in x. This, together with the validity of the proposition for positive virtual characters, is enough to finish the proof. □

(7.6) Corollary. *The Adams operation Ψ^k is a ring homomorphism and $\Psi^k\Psi^l = \Psi^{kl}$.*

(7.7) Proposition. *If G and H are compact Lie groups there is a canonical isomorphism of rings*

$$\mathrm{R}(G \times H) \cong \mathrm{R}(G) \otimes \mathrm{R}(H).$$

PROOF. The isomorphism is induced by mapping the element

$$V \otimes W \in \mathrm{R}(G) \otimes \mathrm{R}(H)$$

to the $G \times H$-module $V \otimes W$. This is an isomorphism due to (4.15). □

There is also a character ring $\mathrm{R}(G, \mathbb{R})$ of real characters. Other notations for this ring include $\mathrm{RO}(G)$ and $\mathrm{K}_{\mathbb{R}}(G)$. It enjoys a similar universal property and λ^i and Ψ^i operations. But quaternionic representations only give an additive group $\mathrm{R}(G, \mathbb{H})$, since the tensor product of quaternionic representations is real. The maps appearing in proposition (6.1) are compatible with direct sums and therefore yield additive maps between corresponding representation rings (groups). The identities of (6.1) remain valid, and from $r_{\mathbb{R}}^{\mathbb{C}} e_{\mathbb{R}}^{\mathbb{C}} = 2$ and $e_{\mathbb{C}}^{\mathbb{H}} r_{\mathbb{C}}^{\mathbb{H}} = 2$ we conclude:

(7.8) *The maps*

$$e_{\mathbb{R}}^{\mathbb{C}}: \mathrm{R}(G, \mathbb{R}) \to \mathrm{R}(G, \mathbb{C}), \quad \text{and}$$
$$r_{\mathbb{C}}^{\mathbb{H}}: \mathrm{R}(G, \mathbb{H}) \to \mathrm{R}(G, \mathbb{C})$$

are injective.

As a corollary:

(7.9) *If U_1 and U_2 are real G-modules such that*

$$U_1 \otimes_{\mathbb{R}} \mathbb{C} \cong U_2 \otimes_{\mathbb{R}} \mathbb{C}, \quad \text{then} \quad U_1 \cong U_2.$$

R(G) and RO(G) may be considered as functors: If $\rho: G \to H$ is a homomorphism, then the assignment $V \mapsto \rho^*V$ (see end of §1) induces ring homomorphisms

$$\rho^*: \mathrm{R}(H) \to \mathrm{R}(G), \qquad \rho^*: \mathrm{RO}(H) \to \mathrm{RO}(G).$$

(7.10) Exercises

1. ***Splitting principle.*** Note that the homomorphism
$$(\mathrm{res}_S^G): \mathrm{R}(G) \to \prod_S \mathrm{R}(S)$$
is injective, where S ranges over the family of closed topologically cyclic subgroups of G. Consequently, if $\varphi, \psi: \mathrm{R}(G) \to \mathrm{R}(G)$ are natural transformations of functors (or at least transformations compatible with restrictions), which agree on all integral linear combinations of one-dimensional characters, then φ and ψ are equal.

2. Verify the identity $\Psi(x, n) = \Psi^n(x)$ as follows: First observe that if x is one-dimensional (i.e., x or $-x$ is a one-dimensional character), then (7.5) yields $\Psi(x, n) = x^n = \Psi^n(x)$. Next show that if $x = x_1 + \cdots + x_n$ is a sum of one-dimensional virtual characters, then $\Psi(x, n) = \sum_{i=1}^n x_i^n = \Psi^n(x)$. Finally, apply Exercise 1 to get the complete result.

3. Use Exercise 1 as in the previous exercise to verify the equations
$$\lambda^i\lambda^j(x) = P_{i,j}(\lambda^1 x, \ldots, \lambda^{ij}x),$$
$$\lambda^k(xy) = P_k(\lambda^1 x, \ldots, \lambda^k x; \lambda^1 y, \ldots, \lambda^k y).$$

4. Show that $\Psi^k\lambda^i = \lambda^i\Psi^k$ in R(G).

5. Let G be finite, k prime to the order of G, and V an irreducible G-module. Prove that $\Psi^k V$ is also irreducible (and not just virtual!) as follows: First notice that V may be obtained as a matrix representation with entries in some cyclotomic field $\mathbb{Q}(\omega)$, ω an nth-root of unity. Next use (7.4) to see that the character of $\Psi^k V$ is obtained by applying a suitable Galois automorphism of $\mathbb{Q}(\omega)$. This automorphism applied to the matrix entries gives the irreducible G-module $\Psi^k V$.
   Show by example that $\Psi^k V$ may be "virtual" if k is not prime to $|G|$. *Hint*: Try the symmetric group $G = S(3)$.

6. Let $0 \to V_0 \to V_1 \to \cdots \to V_k \to 0$ be an exact sequence of representations of G. Show $\sum_{\nu=0}^k (-1)^\nu V_\nu = 0$ in R(G).

7. Form the power series $\mathrm{S}_t(V) = 1 + \mathrm{S}^1(V)t + \mathrm{S}^2(V)t^2 + \cdots$, where $\mathrm{S}^i(V)$ is the ith symmetric power of V. Show that

$$\lambda_t(V - W) = \lambda_t(V)\mathrm{S}_{-t}(W)$$

for complex (or real) G-modules V and W. *Hint*: First show that for every n there is an exact sequence

$$0 \to \Lambda^n(W) \to \mathrm{S}^1(W) \otimes \Lambda^{n-1}(W) \to \cdots \to \mathrm{S}^{n-1}(W) \otimes \Lambda^1(W) \to \mathrm{S}^n(W) \to 0$$

whose morphisms are given by

$$\alpha \otimes (u_1 \wedge \cdots \wedge u_k) \mapsto \sum_{\nu=1}^{k} (-1)^\nu \alpha \cdot u_\nu \otimes (u_1 \wedge \cdots \wedge \hat{u}_\nu \wedge \cdots \wedge u_k),$$

where $\alpha \in \mathrm{S}^{n-k}(W)$, $u_\nu \in W$, and ^ means "delete this factor." Then use Exercise 6.

8. Let $\mathrm{RC}(G)$ be the subring of elements of $\mathrm{R}(G)$ fixed under complex conjugation. Let $\varphi\colon \mathrm{R}(G) \to \mathrm{R}(G)$ be defined by $\varphi(x) = x - \bar{x}$. Show that the sequence

$$0 \to \mathrm{RC}(G) \to \mathrm{R}(G) \xrightarrow{\varphi} \mathrm{R}(G) \xrightarrow{e_{\mathbb{R}}^{\mathbb{C}} r_{\mathbb{R}}^{\mathbb{C}}} \mathrm{R}(G)$$

is exact.

8. Representations of Abelian Groups

Let G be a compact abelian group. Then G is isomorphic to a product

$$G \cong S^1 \times \cdots \times S^1 \times \mathbb{Z}/m_1 \times \cdots \times \mathbb{Z}/m_k$$

of circle groups and finite cyclic groups (see I, (3.7)). The irreducible complex G-modules are one-dimensional (1.13) and given by homomorphisms $G \to S^1$. Such homomorphisms are called characters of G. In order to determine all representations of G, it suffices by (4.15) to determine the irreducible representations of S^1 and $\mathbb{Z}/m$.

(8.1) Proposition. *The irreducible complex representations of S^1 are given by the characters $z \mapsto z^m$, $m \in \mathbb{Z}$. The characters of the torus $T^n = \mathbb{R}^n/\mathbb{Z}^n$ all have the form*

$$\vartheta\colon [x] \mapsto \exp(2\pi i\alpha(x)), \qquad x \in \mathbb{R}^n,$$

with $\alpha(x) = \langle a, x\rangle = \sum_\nu a_\nu x_\nu$, $a = (a_1, \ldots, a_n) \in \mathbb{Z}^n$.

Proof. Given a character ϑ of T^n, there is a commutative diagram

$$\begin{array}{ccccccccc}
0 & \longrightarrow & \mathbb{Z}^n & \longrightarrow & LT^n & \longrightarrow & T^n & \longrightarrow & 0 \\
 & & \vdots\,\alpha & & \downarrow\,\alpha & & \downarrow\,\vartheta & & \\
0 & \longrightarrow & \mathbb{Z} & \longrightarrow & \mathbb{R} & \xrightarrow[\exp]{} & S^1 & \longrightarrow & 1
\end{array}$$

where $\alpha = \mathrm{L}\vartheta$, $\exp(t) = e^{2\pi i t}$ for $t \in \mathbb{R}$, and $\mathrm{L}T^n = \mathbb{R}^n$. The result on characters of T^n follows from this diagram, and the statement about S^1 is the case $n = 1$ expressed in multiplicative notation. □

(8.2) Definition. Let T be a torus and V a complex T-module. An irreducible character $\vartheta: T \to S^1$ is called a ***weight*** of V if the corresponding isotypical summand

$$V(\vartheta) = \{v \in V \mid xv = \vartheta(x) \cdot v \text{ for all } x \in T\}$$

is nonzero. In this case $V(\vartheta)$ is called the ***weight space*** of ϑ.

It follows from (1.14) that every complex T-module is the direct sum of its weight spaces.

Now the characters of an abelian group themselves form a group under multiplication—the character group $\hat{G}$ of G. The connection between $\hat{G}$ and the character ring $\mathrm{R}(G)$ is given by the next proposition. Note that for any discrete group S, the integral group ring $\mathbb{Z}[S]$ is additively the free abelian group on S with multiplication defined by

$$\left(\sum \lambda_g g\right) * \left(\sum \mu_h h\right) = \sum_{g,h} \lambda_g \mu_h gh$$

as in §1.

(8.3) Proposition. *If G is abelian there is a canonical isomorphism of rings*

$$\mathrm{R}(G) \cong \mathbb{Z}[\hat{G}].$$

PROOF. It is clear that the irreducible characters freely generate both $\mathrm{R}(G)$ and $\mathbb{Z}[\hat{G}]$, and multiplication is the same in both rings. This proves the proposition. □

From (8.1) we see that the character group of the torus $(S^1)^n \cong \mathbb{R}^n/\mathbb{Z}^n$ is isomorphic to $\mathbb{Z}^n$. Furthermore, the group ring $\mathbb{Z}[\mathbb{Z}^n]$ is isomorphic to the ring

$$\mathbb{Z}[a_1, a_1^{-1}, \ldots, a_n, a_n^{-1}]$$

of Laurent polynomials with coefficients in $\mathbb{Z}$. We deal quickly with $\mathbb{Z}/m$.

(8.4) Proposition. *The irreducible characters of $\mathbb{Z}/m$ are given by*

$$x \bmod m \mapsto \exp(2\pi i x j/m) \quad \textit{for } j = 0, 1, \ldots, m-1.$$

PROOF. $\vartheta: \mathbb{Z}/m \to S^1$ is determined by $\vartheta(1)$ which must be an mth root of unity. □

The nontrivial irreducible complex G-modules of a torus are not self-conjugate. Combining this observation with (6.6) produces the following proposition:

(8.5) Proposition. *The real irreducible G-modules of a torus $T^n = \mathbb{R}^n/\mathbb{Z}^n$ are*

(i) *the one-dimensional trivial G-module (which is of real type) and*
(ii) *the two-dimensional real G-modules $r_{\mathbb{R}}^{\mathbb{C}} V$, with nontrivial V as in (8.1). These are of complex type, and given two irreducible complex T^n-modules V_1 and V_2, $r_{\mathbb{R}}^{\mathbb{C}} V_1 \cong r_{\mathbb{R}}^{\mathbb{C}} V_2$ if and only if $V_1 \cong V_2$ or $V_1 \cong \overline{V}_2$.*

The representations in (8.5)(ii) may be given more explicitly by

$$[x_1, \ldots, x_n] \mapsto \begin{pmatrix} \cos 2\pi\langle a, x\rangle & \sin 2\pi\langle a, x\rangle \\ -\sin 2\pi\langle a, x\rangle & \cos 2\pi\langle a, x\rangle \end{pmatrix}$$

with $a = (a_1, \ldots, a_n) \in \mathbb{Z}^n\backslash\{0\}$ and $\langle a, x\rangle = \sum_j a_j x_j$. Moreover, a and $-a$ give equivalent (i.e., isomorphic) representations.

(8.6) Definition. Let $T^n = \mathbb{R}^n/\mathbb{Z}^n$ be a torus and U a real T^n-module. By a ***weight*** of U we mean a weight of its complexification

$$V = e_{\mathbb{R}}^{\mathbb{C}} U = \mathbb{C} \otimes_{\mathbb{R}} U.$$

We may decompose the real T^n-module U into irreducible real modules

$$U = n_0 U_0 \oplus n_1 U_1 \oplus \cdots \oplus n_k U_k$$

with U_0 trivial, $U_\nu = r_{\mathbb{R}}^{\mathbb{C}} V_\nu$ for some nontrivial complex T^n-module V_ν, and V_ν not isomorphic to V_μ or $\overline{V}_\mu$ if $\nu \neq \mu$. By (6.1) we have

$$V = e_{\mathbb{R}}^{\mathbb{C}} U = n_0 V_0 \oplus \bigoplus_{\nu>0} n_\nu e_{\mathbb{R}}^{\mathbb{C}} r_{\mathbb{R}}^{\mathbb{C}} V_\nu = n_0 V_0 \oplus \bigoplus_{\nu>0} n_\nu (V_\nu \oplus \overline{V}_\nu),$$

with trivial V_0.

If ϑ is the character of V_ν, then $V(\vartheta) = n_\nu V_\nu$, and we denote the corresponding real isotypical summand by $U(\vartheta) = U(\overline{\vartheta}) = n_\nu U_\nu$. This space is called the ***real weight space*** corresponding to ϑ. The space U may be thought of as the set of fixed points of V under conjugation (see §6). Thus $U = V^{\mathscr{J}}$ and

$$U(\vartheta) = (V(\vartheta) \oplus V(\overline{\vartheta}))^{\mathscr{J}} = (V(\vartheta) \oplus V(\overline{\vartheta})) \cap U.$$

For later use we note:

(8.7) Proposition. *Let T be a torus and U a real T-module.*

(i) *If ϑ is a weight of U, then $\overline{\vartheta}$ is also a weight of U.*
(ii) *Let $V = e_{\mathbb{R}}^{\mathbb{C}} U$. If ϑ is constant, then $V(\vartheta) = e_{\mathbb{R}}^{\mathbb{C}} U(\vartheta)$. If ϑ is not constant, then $V(\vartheta) \cap V(\overline{\vartheta}) = 0$ and*

$$U(\vartheta) = r_{\mathbb{R}}^{\mathbb{C}} V(\vartheta) = r_{\mathbb{R}}^{\mathbb{C}} V(\overline{\vartheta}) = (V(\vartheta) \oplus V(\overline{\vartheta})) \cap U,$$

$$e_{\mathbb{R}}^{\mathbb{C}} U(\vartheta) = V(\vartheta) \oplus V(\overline{\vartheta}).$$

Hence $\dim_{\mathbb{R}} U(\vartheta) = 2 \dim_{\mathbb{C}} V(\vartheta) = 2 \dim_{\mathbb{C}} V(\overline{\vartheta})$.

(iii) $U = \bigoplus_{\vartheta} U(\vartheta)$, *where* ϑ *ranges over a system of weights of* U *containing exactly one element of each pair* $\vartheta, \bar{\vartheta}$ *of conjugate weights.* □

We end this section by noting that for any compact abelian G we can say:

(8.8) Proposition. *An irreducible real G-module of the compact abelian group G is either one-dimensional and of real type or two-dimensional and of complex type.*

Proof. The proposition follows from (6.6) and the fact that irreducible complex G-modules are one-dimensional. □

Finally, note that if U is one-dimensional it is given by a homomorphism $G \to \mathrm{O}(1) \cong \mathbb{Z}/2$.

(8.9) Exercises

In the following exercises we assume that groups are compact abelian Lie groups.

1. Let V and W be real representations of a torus T such that $\dim V^H = \dim W^H$ for all subgroups H of T. Show that $V \cong W$.

2. If

$$\begin{array}{ccc} S_1 \times_S S_2 & \longrightarrow & S_1 \\ \downarrow & & \downarrow \\ S_2 & \longrightarrow & S \end{array}$$

is a pullback diagram of groups, show that the induced map $\mathrm{R}(S_1) \otimes_{\mathrm{R}(S)} \mathrm{R}(S_2) \to \mathrm{R}(S_1 \times_S S_2)$ is an isomorphism.

3. For abelian Lie groups G and H, let $\hat{G}$ and $\hat{H}$ denote the groups of irreducible characters (see §4). Let $f: G \to H$ be a homomorphism. Show that if f is injective then $\hat{f}: \hat{H} \to \hat{G}$ is surjective, and if f is surjective then $\hat{f}$ is injective. Show that

$$\prod_{\lambda=1}^{k} (\hat{G}_\lambda) = \left(\bigoplus_{\lambda=1}^{k} G_\lambda\right)^{\wedge}$$

for any family of Lie groups G_λ.

4. Show that every projective representation of a torus factors through the unitary group as in the following diagram:

$$\begin{array}{ccc} & & \mathrm{U}(n) \\ & \nearrow & \downarrow \\ T^k & \longrightarrow & \mathrm{U}(n)/S^1. \end{array}$$

Hint: Start with a lifting $\mathbb{R}^k \to \mathrm{SU}(n)$ (see (5.11), Ex. 7), or use IV, (1.12), Ex. 5 with G a suitable fiber product.

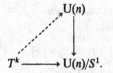

9. Representations and Lie Algebras

The purpose of this section is to give a brief introduction to the infinitesimal form of a representation. It also contains a detailed discussion of torus representations and infinitesimal weights.

Let $G \times M \to M$ be a differentiable G-action on a manifold M. Each $X \in \mathrm{L}G$ determines an action $\Phi^X: \mathbb{R} \times M \to M$, $(t, p) \mapsto \exp(tX) \cdot p$ of the one-parameter group corresponding to X on M. The map Φ^X is the global flow of the vector field $\tilde{X}$ whose value at p is the tangent vector at $t = 0$ of the map $t \mapsto \exp(tX) \cdot p$. The ***Lie derivative*** of a function $f: M \to \mathbb{R}$ with respect to the vector field $\tilde{X}$ is

$$\mathrm{L}_X f(p) = \left.\frac{d}{dt}\right|_{t=0} f(\exp(tX) \cdot p),$$

which is just the ordinary directional derivative with respect to the same vector field.

If $G \times V \to V$ is a representation, we may form the Lie derivative for $v \in V$ and $X \in \mathrm{L}G$

$$\mathrm{L}_X v := \lim_{t \to 0} \frac{1}{t} (\exp(tX) \cdot v - v).$$

The map $(X, v) \mapsto \mathrm{L}_X v$ is linear in both X and v. Moreover, as we will see in a moment,

(9.1) $$\mathrm{L}_X \mathrm{L}_Y v - \mathrm{L}_Y \mathrm{L}_X v = \mathrm{L}_{[X, Y]} v.$$

Let $\rho: G \to \mathrm{Aut}(V)$ be the homomorphism corresponding to the representation $G \times V \to V$, and let $\mathrm{L}\rho: \mathrm{L}G \to \mathrm{LAut}(V)$ be its differential. Since $\mathrm{LAut}(V)$ is canonically isomorphic to $\mathrm{End}(V)$ (see I, (2.8), (2.13)), we may interpret $\mathrm{L}\rho(X)$ in $\mathrm{End}(V)$. As such $\mathrm{L}\rho(X)$ is none other than the map $v \mapsto \mathrm{L}_X v$. This observation also proves (9.1). Note that the diagram

(9.2) $$\begin{array}{ccc} \mathrm{L}(G) & \xrightarrow{\mathrm{L}\rho} & \mathrm{End}(V) \cong \mathrm{LAut}(V) \\ \Big\downarrow{\scriptstyle \exp} & & \Big\downarrow{\scriptstyle \exp} \\ G & \xrightarrow[\rho]{} & \mathrm{Aut}(V) \end{array}$$

commutes (see I, (3.2)). The right-hand exponential map is given by $\exp(A) = \sum_\nu (1/\nu!) A^\nu = e^A$, see I, (3.3). This says that for $X \in \mathrm{L}(G)$ the left translation by $\exp X$ on V is given by the automorphism e^{L_X}.

The map $\mathrm{L}G \times V \to V$, $(X, v) \mapsto \mathrm{L}_X v$ is the infinitesimal version of the representation $G \times V \to V$. It is also an example of a representation of a Lie algebra. In general, if L is a Lie algebra and V is a vector space, then any bilinear map $L \times V \to V, (X, v) \mapsto Xv$, satisfying $X(Yv) - Y(Xv) = [X, Y]v$ is called a ***representation of the Lie algebra L on V***. In this context V is also called an ***L*-module**. Of course, a representation of L has an equivalent definition as a homomorphism of Lie algebras $L \to \mathrm{End}(V)$.

If L happens to be the Lie algebra $\mathrm{L}G$ of a simply connected Lie group G, then a homomorphism $\tau\colon \mathrm{L}G \to \mathrm{End}(V)$ uniquely determines a representation $\rho\colon G \to \mathrm{Aut}(V)$ such that replacing $\mathrm{L}\rho$ by τ in (9.2) results in a commutative diagram (see Tits [2], §4.2, or Chevalley [1], Ch. VI, §VI, Th. 2). This opens the way for dealing with representations algebraically. We will not pursue this point, but it may nonetheless serve to illuminate the following exposition.

We now study the infinitesimal versions of torus representations. This leads to infinitesimal weights and an infinitesimal characterization of weight spaces.

Let V be a complex T-module for a torus T. Recall from (8.2) that a homomorphism $\vartheta\colon T \to \mathrm{U}(1)$ is called a ***weight*** of V if the corresponding weight space

$$V(\vartheta) = \{v \in V \mid xv = \vartheta(x)\cdot v \text{ for all } x \in T\}$$

is nonzero. As explained above, we have the infinitesimal version of the T-module V

$$\mathrm{L}T \times V \to V, \qquad (X, v) \mapsto \mathrm{L}_X v$$

and the differential

$$\Theta = \mathrm{L}\vartheta\colon \mathrm{L}T \to \mathrm{LU}(1).$$

Recall from I, (2.16) that $\mathrm{LU}(1) = \{a \in \mathbb{C} \mid a + \bar{a} = 0\} = i\mathbb{R}$.

(9.3) Definition. An $\mathbb{R}$-linear form $\Theta\colon \mathrm{L}T \to \mathrm{LU}(1) = i\mathbb{R} \subset \mathbb{C}$ is called an ***infinitesimal weight*** of the T-module V if the corresponding ***weight space***

$$V(\Theta) = \{v \in V \mid \mathrm{L}_X v = \Theta(X)\cdot v \text{ for all } X \in \mathrm{L}T\}$$

is nonzero.

(9.4) Proposition. *The map $\vartheta \to \mathrm{L}\vartheta$ is a bijection between weights and infinitesimal weights of the T-module V. For every weight ϑ we have $V(\vartheta) = V(\mathrm{L}\vartheta)$.*

Proof. Let ϑ be a weight of V and let $v \in V(\vartheta)$. By differentiating $t \mapsto (\exp(tX))v = \vartheta(\exp(tX))\cdot v$ at $t = 0$ we obtain the equation

$$\mathrm{L}_X v = \mathrm{L}\vartheta(X)\cdot v.$$

Hence $v \in V(\mathrm{L}\vartheta)$. Therefore $V(\vartheta) \subset V(\mathrm{L}\vartheta)$ and $\mathrm{L}\vartheta$ is an infinitesimal weight.

Conversely, let $\Theta\colon LT \to LU(1)$ be an infinitesimal weight of V and let $0 \neq v \in V(\Theta)$ be given. Then for all $X \in LT$ we have $L_X v = \Theta(X) \cdot v$, so $e^{L_X} \cdot v = e^{\Theta(X)} \cdot v$. Combining this with (9.2), we see that $\exp(X) \cdot v = e^{\Theta(X)} \cdot v$; hence the map $LT \to U(1)$, $X \mapsto e^{\Theta(X)}$ factors through $\exp\colon LT \to T$ and yields a map $\vartheta_\Theta\colon T \to U(1)$. Furthermore, $\{0\} \neq V(\Theta) \subset V(\vartheta_\Theta)$, so ϑ_Θ is actually a weight of V. It is now easy to check that $\vartheta_{L\vartheta} = \vartheta$ and that $L(\vartheta_\Theta) = \Theta$, so $\Theta \mapsto \vartheta_\Theta$ is inverse to $\vartheta \to L\vartheta$. □

Since V is a complex vector space, every representation $LT \times V \to V$ of the Lie algebra LT on V extends to the complexified Lie algebra $LT_{\mathbb{C}} = \mathbb{C} \otimes_{\mathbb{R}} LT$. The extended action $LT_{\mathbb{C}} \times V \to V$ is also written $(X, v) \mapsto L_X v$.

(9.5) Definition. A $\mathbb{C}$-linear form $\Phi\colon LT_{\mathbb{C}} \to \mathbb{C}$ is called a ***complex infinitesimal weight*** of the T-module V if the corresponding ***weight space***

$$V(\Phi) = \{v \in V \mid L_H v = \Phi(H) \cdot v \text{ for all } H \in LT_{\mathbb{C}}\}$$

is nonzero.

Note that $\mathbb{C} \otimes_{\mathbb{R}} LU(1) \to \mathbb{C}$, $z \otimes u \mapsto zu$, is an isomorphism. An $\mathbb{R}$-linear form $\Theta\colon LT \to LU(1)$ therefore induces a $\mathbb{C}$-linear form

$$\Phi_\Theta\colon \mathbb{C} \otimes LT \to \mathbb{C} \otimes LU(1) \cong \mathbb{C}$$

mapping $X + iY$ to $\Theta(X) + i\Theta(Y)$ for $X, Y \in LT$.

(9.6) Proposition. *The assignment $\Theta \mapsto \Phi_\Theta$ is a bijection between infinitesimal and complex infinitesimal weights of the T-module V. Furthermore, $V(\Theta) = V(\Phi_\Theta)$.*

Proof. Given Θ and $v \in V(\Theta)$ we have

$$L_H v = L_X v + iL_Y v = \Theta(X) \cdot v + i\Theta(Y) \cdot v = \Phi_\Theta(H) \cdot v$$

for $H = X + iY \in LT_{\mathbb{C}}$. Hence Φ_Θ is a complex infinitesimal weight with $V(\Theta) \subset V(\Phi_\Theta)$. Now suppose we start with a complex infinitesimal weight Φ. Let $0 \neq v \in V(\Phi)$ be a corresponding weight-vector. Then *a fortiori* for any X in the real Lie algebra LT we have $L_X v = \Phi(X) \cdot v$. However the LT-module V decomposes into weight spaces (see (1.14), (9.4)):

$$V = \bigoplus_j V(\Theta_j), \qquad v = \sum_j v_j.$$

Therefore $L_X v = \sum_j \Theta_j(X) \cdot v_j$, and hence $\Phi(X) \cdot v_j = \Theta_j(X) \cdot v_j$. We renumber so that $v_1 \neq 0$. Then $\Phi(X) = \Theta_1(X)$ for all $X \in LT$. Therefore $v_j = 0$ for $j \neq 1$ and, since Φ is determined by $\Phi \mid LT$, we have $\Phi = \Phi_{\Theta_1}$. This shows that the assignment $\Theta \mapsto \Phi_\Theta$ is surjective and $V(\Phi_\Theta) \subset V(\Theta)$. Since $\Theta \mapsto \Phi_\Theta$ is obviously injective and $V(\Theta) \subset V(\Phi_\Theta)$, the proposition follows. □

For the purpose of general formulas it is convenient to use a slightly different parametrization of the circle U(1) and its Lie algebra. The isomorphism $\mathbb{R}/\mathbb{Z} \to U(1)$, $t + \mathbb{Z} \mapsto e^{2\pi i t}$, induces an isomorphism of exact sequences

$$\begin{array}{ccccccccc} 0 & \longrightarrow & \mathbb{Z} & \longrightarrow & \mathbb{R} & \longrightarrow & \mathbb{R}/\mathbb{Z} & \longrightarrow & 0 \\ & & \downarrow{\scriptstyle 2\pi i} & & \downarrow{\scriptstyle 2\pi i} & & \downarrow{\scriptstyle \cong} & & \\ 0 & \longrightarrow & 2\pi i\mathbb{Z} & \longrightarrow & i\mathbb{R} & \xrightarrow[\exp]{} & U(1) & \longrightarrow & 1. \end{array}$$

Weights $T \to U(1)$ then correspond to homomorphisms $T \to \mathbb{R}/\mathbb{Z}$ and infinitesimal weights $LT \to i\mathbb{R}$ correspond to $\mathbb{R}$-linear forms $\alpha: LT \to \mathbb{R}$.

(9.7) Notation. An $\mathbb{R}$-linear form α: $LT \to \mathbb{R}$ is a ***real*** (infinitesimal) ***weight*** of the T-module V if $2\pi i\alpha$ is an infinitesimal weight of V. The weight space of α is then

$$V(\alpha) = \{v \in V \mid L_X v = 2\pi i\alpha(X) \cdot v \text{ for all } X \in LT\}$$

The nonzero vectors in $V(\alpha)$ are called ***weight vectors*** of α.

So all told we have weights and three different types of infinitesimal weights. We close by mentioning that for a ***real*** T-module W we define the infinitesimal (or real) weights to be those of its complexification $W_{\mathbb{C}} = \mathbb{C} \otimes_{\mathbb{R}} W$.

(9.8) Exercises

1. Show that if Θ is an infinitesimal weight of a real T-module W (in one of its three forms) then $-\Theta$ is also a weight.

2. Consider $S^1 \subset \mathbb{C}^* = \mathbb{C}\backslash\{0\}$ as multiplicative groups. Then there is a canonical isomorphism $LS^1_{\mathbb{C}} \cong L\mathbb{C}^*$. Let V be a complex S^1-module. Show it has an extension to a $\mathbb{C}^*$-module $\mathbb{C}^* \times V \to V$. Show that the corresponding infinitesimal action $L\mathbb{C}^* \times V \to V$ is the complexification of $LS^1 \times V \to V$. Are there representations of $\mathbb{C}^*$ which do not come from representations of S^1 in this way?

3. Let α be a nonconstant weight of the real T-module W with $W(\alpha)$ the corresponding weight space. Show that $W(\alpha)$ decomposes as $W(\alpha) = W_1 \oplus W_2$ with $t \in T$ acting via
$$\begin{pmatrix} w_1 \\ w_2 \end{pmatrix} \mapsto \begin{pmatrix} \alpha_{11}(t) & \alpha_{12}(t) \\ \alpha_{21}(t) & \alpha_{22}(t) \end{pmatrix}\begin{pmatrix} w_1 \\ w_2 \end{pmatrix} = \alpha(t)\begin{pmatrix} w_1 \\ w_2 \end{pmatrix},$$
where $\alpha: T \to SO(2)$. Describe this in terms of
$$L\alpha: LT \to LSO(2) = \left\{\begin{pmatrix} 0 & a \\ -a & 0 \end{pmatrix} \middle| a \in \mathbb{R}\right\}.$$

4. Let $B: V \times V \to \mathbb{R}$ be a G-invariant inner product on the real G-module V. Show that
$$B(L_X u, v) + B(u, L_X v) = 0$$
for all $X \in LG$ and $u, v \in V$.

10. The Lie Algebra sl(2, ℂ)

The theory of representations of the groups SU(2) and SO(3) provides a simple and instructive example of the general theory. Furthermore, these groups play a fundamental role in the structure theory of compact Lie groups (see Chapter V). We therefore devote this section to studying the infinitesimal forms of their representations and their applications to spherical functions and the associated special functions (Legendre polynomials).

The Lie algebra su(2) of SU(2) consists of the skew-Hermitian (2×2)-matrices with trace zero. An $\mathbb{R}$-basis is given by

$$iH = \begin{pmatrix} i & 0 \\ 0 & -i \end{pmatrix}, \quad Y = \begin{pmatrix} 0 & i \\ i & 0 \end{pmatrix}, \quad \text{and} \quad Z = \begin{pmatrix} 0 & 1 \\ -1 & 0 \end{pmatrix}.$$

These elements appeared in I, (1.9), (6.20) as a real basis for the pure quaternions, but here we consider them just as (2×2)-matrices. The relations between these elements in their new guise are

$$Y(iH) = -(iH)Y = Z, \qquad (iH)Z = -Z(iH) = Y, \qquad ZY = -YZ = iH$$

and hence their Lie products are

$$[iH, Y] = -2Z, \qquad [iH, Z] = 2Y, \qquad [Y, Z] = -2iH.$$

Now, the Lie algebra sl(2, ℂ) of SL(2, ℂ) consists of the complex (2×2)-matrices with trace zero. The complexification

$$\mathbb{C} \otimes_{\mathbb{R}} \mathrm{su}(2) \to \mathrm{sl}(2, \mathbb{C}), \qquad z \otimes U \mapsto zU$$

is an isomorphism of complex Lie algebras. Note that the action of ℂ on a matrix U is given by ordinary componentwise multiplication, which is not at all the same as the action induced by considering ℂ as a subalgebra of ℍ. A complex basis of sl(2, ℂ) is given by

$$H = \begin{pmatrix} 1 & 0 \\ 0 & -1 \end{pmatrix}, \qquad X^+ = \tfrac{1}{2}(Z - iY) = \begin{pmatrix} 0 & 1 \\ 0 & 0 \end{pmatrix},$$

and

$$X^- = \tfrac{1}{2}(Z + iY) = \begin{pmatrix} 0 & 0 \\ -1 & 0 \end{pmatrix}.$$

These elements satisfy

(10.1) $\quad [H, X^+] = 2X^+, \qquad [H, X^-] = -2X^-, \qquad [X^+, X^-] = -H.$

The corresponding one-parameter groups are given by

$$\exp(tH) = \begin{pmatrix} \exp(t) & 0 \\ 0 & \exp(-t) \end{pmatrix}, \qquad \exp(tX^+) = \begin{pmatrix} 1 & t \\ 0 & 1 \end{pmatrix},$$

$$\exp(tX^-) = \begin{pmatrix} 1 & 0 \\ -t & 1 \end{pmatrix}.$$

Next, we consider representations of the Lie algebra sl(2, ℂ) on complex vector spaces E as defined in §9:

$$\mathrm{sl}(2, \mathbb{C}) \times E \to E, \qquad (A, x) \mapsto Ax.$$

Our first goal is to describe the simple sl(2, ℂ)-modules. They correspond to irreducible SU(2)-modules.

Slightly abusing the language from §9, we will call an element $x \in E$ satisfying $Hx = \lambda x$ an ***element of weight*** λ. This is, of course, an element in some appropriate weight space.

(10.2) Lemma. *If x is an element of weight λ, then X^+x is an element of weight $\lambda + 2$ and X^-x is an element of weight $\lambda - 2$.*

Proof. $HX^+x = [H, X^+]x + X^+Hx = 2X^+x + X^+\lambda x = (\lambda + 2)X^+x$ and similarly for X^-x. □

(10.3) Definition. A nonzero element x of the sl(2, ℂ)-module E is called ***primitive*** if it is an eigenvector of $y \mapsto Hy$, and if $X^+x = 0$.

(10.4) Proposition. *Let x be a primitive element of the* sl(2, ℂ)*-module E of weight λ. For $j \geq 0$ set $x_j = ((-1)^j/j!)(X^-)^j x$ and set $x_{-1} = 0$. Then $Hx_j = (\lambda - 2j)x_j$, $X^-x_j = -(j + 1)x_{j+1}$, and $X^+x_j = (\lambda - j + 1)x_{j-1}$.*

Proof. The first formula follows from (10.2) and the second from the definition of x_j. The third is proved by induction on j starting with $j = 0$ where it is true because $x_{-1} = 0$ and $X^+x = 0$. For $j > 0$ we have

$$\begin{aligned} jX^+x_j &= -X^+X^-x_{j-1} = -[X^+, X^-]x_{j-1} - X^-X^+x_{j-1} \\ &= Hx_{j-1} - X^-(\lambda - j + 2)x_{j-2} \quad \text{(induction)} \\ &= (\lambda - 2j + 2 + (j - 1)(\lambda - j + 2))x_{j-1} = j(\lambda - j + 1)x_{j-1}. \end{aligned}$$ □

Noting that $x = x_0$, we see that the sl(2, ℂ)-submodule of E generated by x is the vector space spanned by the x_j. Moreover, if $x_j \neq 0$ for $j > 0$, then $x_{j-1} \neq 0$. This means that $\{j \geq 0 \mid x_j \neq 0\}$ is an interval of $\mathbb{N}_0$. The corresponding x_j are linearly independent since they are eigenvectors of $y \mapsto Hy$ associated to distinct eigenvalues.

(10.5) Proposition. *Suppose the* sl(2, ℂ)*-submodule V of E generated by the primitive element x is finite-dimensional. Then*

(i) *The weight λ associated to x is the nonnegative integer* $\dim V - 1$.
(ii) *The elements $x_0, x_1, \ldots, x_\lambda$ form a basis of V and $x_j = 0$ for $j > \lambda$.*
(iii) *We have $Hx_j = (\lambda - 2j)x_j$. Hence the eigenvalues of $y \mapsto Hy$ are λ, $\lambda - 2, \ldots, -\lambda$ and the corresponding eigenspaces are one-dimensional.*
(iv) *If $y \in V$ is primitive, then $y = ax$ for some $a \in \mathbb{C}$.*

Proof. Let m be the largest integer such that $x_m \neq 0$. Then $0 = X^+ x_{m+1} = (\lambda - m)x_m$, so $m = \lambda$. As remarked above, the nonzero x_j form a basis for V, so (i) and (ii) are proved. The fourth statement follows from (ii) and the fact that $X^+ x_j = c \cdot x_{j-1}$, $c \neq 0$ by (10.4). □

Note that $x = x_0$ generates the kernel of $X^+ \mid V$.

(10.6) Lemma. *Let $E \neq 0$ be a finite-dimensional* sl(2, ℂ)*-module. Then E contains a primitive element.*

Proof. We know that there is some nonzero weight vector $v \in E$. By (10.2) the iterates $v_n = (X^+)^n v$ are either zero or linearly independent weight vectors. It follows that v_n is primitive for some $n < \dim E$. □

If a finite-dimensional sl(2, ℂ)-module $E \neq 0$ is simple in the sense that it has no proper nonzero submodules V, then E must be generated by a primitive element. Thus (10.5) describes the structure of E. Specific models of the modules appearing in (10.5) arise from the representations V_m of SL(2, ℂ) on the vector space of homogeneous polynomials of degree m in variables z_1 and z_2 (§5). In fact, if we let $x_j \in V_m$ be the basis element

$$x_j(z_1, z_2) = \binom{m}{j} z_1^{m-j} z_2^j,$$

then we find:

(10.7) Proposition. *$Hx_j = (m - 2j)x_j$, $X^+ x_j = (m - j + 1)x_{j-1}$ and $X^- x_j = -(j + 1)x_{j+1}$, with $x_{-1} = x_{m+1} = 0$. Thus the elements $x_0, \ldots, x_m$ constitute the canonical basis derived from the primitive element x_0.*

Proof. We use the one-parameter groups described after (10.1) and the definition of the Lie derivative in §9. We also recall that $g \in$ SL(2, ℂ) acts on a polynomial $P \in V_m$ via

$$(gP)(z_1, z_2) = P((z_1, z_2) \cdot g),$$

see II, §5. Thus $\exp(tH) \cdot x_j = \exp((m - 2j)t) \cdot x_j$ and hence

$$Hx_j = \lim_{t \to 0} \frac{1}{t} (\exp(tH) \cdot x_j - x_j) = (m - 2j)x_j.$$

Verifying the other two equations is equally straightforward, for example, $X^+ x_j$ is calculated by computing that

$$(\exp(tX^+) \cdot x_j)(z_1, z_2) = \binom{m}{j} z_1^{m-j}(z_1 t + z_2)^j,$$

differentiating with respect to t, and evaluating at $t = 0$. □

Our next task is to relate spherical functions to classical special functions, as promised in §5. The irreducible representations of SO(3) were realized by a natural action of SO(3) on the complex vector space $\mathfrak{H}_l$ of harmonic polynomials $\mathbb{R}^3 \to \mathbb{C}$ of degree l. The Lie derivative will make this into a module over $\mathrm{so}(3)_{\mathbb{C}} \cong \mathrm{sl}(2, \mathbb{C})$. We may then apply (10.5) to find a basis of $\mathfrak{H}_l$ and try to explicitly determine the functions of this basis. This program will occupy the rest of this section.

It turns out that the functions we are seeking are best expressed in spherical coordinates. One *a priori* reason for this is the homogeneity of the polynomials involved. A second reason is that a weight vector for the zero weight corresponds to a function which is invariant under rotations about a suitable axis.

The Lie algebra so(3) of SO(3) consists of the skew-symmetric real (3×3)-matrices. A basis is given by the infinitesimal rotations about the coordinate axes

$$Z(1) = \begin{pmatrix} 0 & 0 & 0 \\ 0 & 0 & -1 \\ 0 & 1 & 0 \end{pmatrix}, \quad Z(2) = \begin{pmatrix} 0 & 0 & 1 \\ 0 & 0 & 0 \\ -1 & 0 & 0 \end{pmatrix}, \quad Z(3) = \begin{pmatrix} 0 & -1 & 0 \\ 1 & 0 & 0 \\ 0 & 0 & 0 \end{pmatrix}.$$

The following Lie brackets may be easily computed (cf. I, (1.16), Ex. 13; (2.22), Ex. 8)); this is our old acquaintance the basic pure quaternions in another guise:

(10.8)
$$[Z(2), Z(3)] = Z(1), \qquad [Z(3), Z(1)] = Z(2), \quad \text{and}$$
$$[Z(1), Z(2)] = Z(3).$$

The group SO(3) acts on smooth functions $f\colon \mathbb{R}^3 \to \mathbb{C}$ by $(Af)(x) = f(xA)$, $A \in SO(3)$, $x \in \mathbb{R}^3$. The Lie derivative is

$$R_X f(x) = \frac{d}{dt}\bigg|_{t=0} f(x \cdot \exp(tX)), \qquad X \in \mathrm{so}(3).$$

We will frequently denote $R_X f$ by Xf to simplify notation. We have

$$XYf - YXf = [X, Y]f.$$

The operator R_X is a linear differential operator on smooth functions $\mathbb{R}^3 \to \mathbb{C}$. In terms of coordinates (x_1, x_2, x_3) on $\mathbb{R}^3$, this operator is given at the point $x \in \mathbb{R}^3$ by

(10.9)
$$a_1 \frac{\partial}{\partial x_1} + a_2 \frac{\partial}{\partial x_2} + a_3 \frac{\partial}{\partial x_3} \quad \text{with} \quad (a_1, a_2, a_3) = x \cdot X.$$

To see this, note that R_X is the directional derivative with respect to the velocity vector of the curve $t \mapsto x \cdot \exp(tX)$ at $t = 0$ and this velocity vector is $x \cdot X$.

(10.10) Remark. Recall that there is an isomorphism of Lie algebras $\varphi\colon \mathrm{so}(3) \to \mathbb{R}^3$ where the Lie bracket in $\mathbb{R}^3$ is given by the vector product $(x, y) \mapsto x \times y$ (see I, (2.22), Ex. 8). The image of $Z(j)$ under φ is the jth standard basis vector of $\mathbb{R}^3$, i.e., $\varphi Z(1) = (1, 0, 0)$, etc. Moreover, one may easily verify on the basis vectors that $x \cdot g = x \times \varphi(g)$ for $x \in \mathbb{R}^3$ and $g \in \mathrm{so}(3)$. Thus if $\varphi(X) = u = (u_1, u_2, u_3)$ we may formally write R_X at the point (x_1, x_2, x_3) as

$$\mathrm{R}_X = \langle \nabla, x \times u \rangle = \det(\nabla, x, u) \quad \text{with} \quad \nabla = \left(\frac{\partial}{\partial x_1}, \frac{\partial}{\partial x_2}, \frac{\partial}{\partial x_3}\right).$$

In other words,

(10.11)
$$\mathrm{R}_X = \begin{vmatrix} \dfrac{\partial}{\partial x_1} & \dfrac{\partial}{\partial x_2} & \dfrac{\partial}{\partial x_3} \\ x_1 & x_2 & x_3 \\ u_1 & u_2 & u_3 \end{vmatrix}.$$

In particular,

(10.12)
$$\mathrm{R}_{Z(1)} = x_3 \frac{\partial}{\partial x_2} - x_2 \frac{\partial}{\partial x_3},$$
$$\mathrm{R}_{Z(2)} = x_1 \frac{\partial}{\partial x_3} - x_3 \frac{\partial}{\partial x_1},$$
$$\mathrm{R}_{Z(3)} = x_2 \frac{\partial}{\partial x_1} - x_1 \frac{\partial}{\partial x_2}.$$

We may describe an isomorphism $\mathrm{so}(3)_{\mathbb{C}} \cong \mathrm{sl}(2, \mathbb{C})$ by observing that the vectors

(10.13) $\quad H = 2iZ(3), \qquad X^+ = Z(1) + iZ(2), \qquad X^- = Z(1) - iZ(2)$

satisfy the standard sl(2, ℂ)-relations (10.1), i.e.,

$$[H, X^+] = 2X^+, \qquad [H, X^-] = -2X^-, \qquad [X^+, X^-] = -H.$$

We now wish to apply the differential operators corresponding to H, X^+, and X^- to the functions in the space $\mathfrak{H}_l$ of harmonic polynomials of degree l. From §5 we know that $\mathfrak{H}_l$ is an irreducible SO(3)-module, and so we want to find a basis of the form considered in (10.5). The basis vectors will have to be eigenvectors for the operator H. Since $\dim \mathfrak{H}_l = 2l + 1$, a primitive element will have weight $2l$. Making a "lucky guess," we consider the polynomial $Y(x_1, x_2, x_3) = (x_1 + ix_2)^l$. It is clearly harmonic of degree l, and applying $H = 2i(x_2(\partial/\partial x_1) - x_1(\partial/\partial x_2))$ yields $HY = 2lY$. Similarly, $X^+(Y) = 0$ so Y is primitive. From (10.5) we conclude that $\mathfrak{H}_l$ has a basis consisting of the functions

(10.14) $\qquad Y_{l-j} = (iX^-)^j Y, \qquad 0 \le j \le 2l,$

which is, up to constants, the canonical basis. Our intention is to identify these functions with classical spherical functions, and this requires the use of spherical coordinates.

The spherical coordinates

$$x_1 = r \sin \vartheta \cos \varphi,$$
$$x_2 = r \sin \vartheta \sin \varphi, \qquad 0 \le r, 0 \le \varphi \le 2\pi, 0 \le \vartheta \le \pi,$$
$$x_3 = r \cos \vartheta,$$

obey the transformation rules (see Korn and Korn [1], table 6.5-1, p. 179 (gradient))

$$\frac{\partial}{\partial x_1} = \sin \vartheta \cos \varphi \frac{\partial}{\partial r} + \frac{\cos \vartheta \cos \varphi}{r} \frac{\partial}{\partial \vartheta} - \frac{\sin \varphi}{r \sin \vartheta} \frac{\partial}{\partial \varphi},$$

$$\frac{\partial}{\partial x_2} = \sin \vartheta \sin \varphi \frac{\partial}{\partial r} + \frac{\cos \vartheta \sin \varphi}{r} \frac{\partial}{\partial \vartheta} + \frac{\cos \varphi}{r \sin \vartheta} \frac{\partial}{\partial \varphi},$$

$$\frac{\partial}{\partial x_3} = \cos \vartheta \frac{\partial}{\partial r} - \frac{\sin \vartheta}{r} \frac{\partial}{\partial \vartheta}.$$

Substituting this into (10.12) and (10.13) yields

$$H = -2i \frac{\partial}{\partial \varphi},$$

(10.15)
$$X^+ = e^{i\varphi}\left(-i \frac{\partial}{\partial \vartheta} + \cot \vartheta \frac{\partial}{\partial \varphi}\right),$$

$$X^- = e^{-i\varphi}\left(i \frac{\partial}{\partial \vartheta} + \cot \vartheta \frac{\partial}{\partial \varphi}\right).$$

The polynomial Y is transformed into $Y = r^l e^{il\varphi} \sin^l \vartheta$. Since the operators H, X^+, and X^- do not involve r, we may fix $r = 1$ and simply compute with $Y = e^{il\varphi} \sin^l \vartheta$ in what follows.

Since $H = -2i(\partial/\partial\varphi)$, the relation $HY_m = 2mY_m$, $-l \le m \le l$, shows that Y_m has the form

$$Y_m(\vartheta, \varphi) = e^{im\varphi} f_m(\vartheta).$$

From $iX^-(Y_m) = Y_{m-1}$ and from the expression for X^- in (10.15), we obtain the differential equation

$$-\frac{df_m}{d\vartheta} - m \cot \vartheta\, f_m = f_{m-1}.$$

This equation may be solved by substituting $s = \cos \vartheta$. Denoting $f_m(\vartheta)$ by $p_m(s)$, this substitution gives

$$\frac{df_m}{d\vartheta}(\vartheta) = -\frac{dp_m}{ds}(s) \cdot \sin \vartheta$$

and the differential equation becomes

$$(1 - s^2)^{1/2}\left(\frac{dp_m}{ds} - \frac{ms}{1 - s^2} p_m\right) = p_{m-1}.$$

Notice that this substitution is justified because $0 \le \vartheta \le \pi$ and $\cos \vartheta$ is monotonic in this interval.

Setting $u_m(s) = (1 - s^2)^{m/2} \cdot p_m(s)$, we compute that

$$\frac{du_m}{ds} = u_{m-1}.$$

Next, we observe that $u_l(s) = (1 - s^2)^{l/2} p_l(s)$ and that $p_l(s) = f_l(\vartheta) = \sin^l \vartheta = (1 - \cos^2 \vartheta)^{l/2} = (1 - s^2)^{l/2}$. Thus $u_l(s) = (1 - s^2)^l$, so

$$u_m(s) = \frac{d^{l-m}}{ds^{l-m}} (1 - s^2)^l,$$

$$p_m(s) = (1 - s^2)^{-m/2} \frac{d^{l-m}}{ds^{l-m}} (1 - s^2)^l.$$

In this context one uses a conventional normalization by defining

$$P_l^m(s) = \frac{(1 - s^2)^{-m/2}}{l!\,2^l} \frac{d^{l-m}}{ds^{l-m}} (s^2 - 1)^l.$$

This yields the basis

(10.16) $$\tilde{Y}_m(\varphi, \vartheta) = e^{im\varphi} P_l^m(\cos \vartheta), \qquad -l \le m \le l,$$

for $\mathfrak{H}_l$. The function

(10.17) $$P_l(t) = \frac{1}{2^l l!} \frac{d^l}{dt^l} (t^2 - 1)^l$$

is called the lth ***Legendre polynomial***, and the functions $(1 - t^2)^{m/2}(d^m/dt^m)P_l(t)$ its ***associated Legendre functions***. The eigenfunctions $\tilde{Y}_m$ for $m = 0$ have weight 0 and are hence annihilated by $H = -2i(\partial/\partial\varphi)$. They are independent of φ. The function $\tilde{Y}_0(\varphi, \vartheta) = P_l(\cos \vartheta)$ is invariant under rotations about the x_3-axis (zonal spherical function). For spherical functions on S^n, see Dieudonné [3], 38.

(10.18) Exercises

1. Show that the Lie algebras su(2) and sl(2, ℝ) are not isomorphic but that $\mathrm{su}(2)_{\mathbb{C}} \cong \mathrm{sl}(2, \mathbb{C}) \cong \mathrm{sl}(2, \mathbb{R})_{\mathbb{C}}$ (these are the real forms of sl(2, ℂ)).
2. We made $\mathfrak{H}_l$ into an so(3)-module using the Lie derivative. But $\mathfrak{H}_l$ is also an so(3)-module via the infinitesimal representations. How are these two structures related?

3. Show that the isomorphism $\mathrm{so}(3)_{\mathbb{C}} \cong \mathrm{so}(3, \mathbb{C}) \cong \mathrm{sl}(2, \mathbb{C})$ used in this section has the form

$$\begin{pmatrix} 0 & 2ia & i(b+c) \\ -2ia & 0 & c-b \\ -i(b+c) & b-c & 0 \end{pmatrix} \leftrightarrow \begin{pmatrix} a & b \\ c & -a \end{pmatrix}$$

4. Define a differential operator C on the smooth functions $\mathbb{R}^3 \to \mathbb{C}$ by

$$C = -Z(1)^2 - Z(2)^2 - Z(3)^2.$$

This is the ***Casimir operator***.

(i) Show that C commutes with each $Z(i)$, $i = 1, 2, 3$.

(ii) Show that

$$C = -x^2\Delta + L(L+1),$$

where

$$\Delta = \frac{\partial^2}{\partial x_1^2} + \frac{\partial^2}{\partial x_2^2} + \frac{\partial^2}{\partial x_3^2} \quad \text{is the Laplace operator,}$$

$$L = x_1\frac{\partial}{\partial x_1} + x_2\frac{\partial}{\partial x_2} + x_3\frac{\partial}{\partial x_3} \quad \text{is the Euler operator,}$$

and x^2 is multiplication by $x_1^2 + x_2^2 + x_3^2$.

(iii) Show that $Cf = l(l+1)f$ for $f \in \mathfrak{H}_l$.

(iv) Use (i) to show that C must be a scalar multiple of the identity on each irreducible module of functions $\mathbb{R}^3 \to \mathbb{C}$ (Schur's lemma).

(v) Show that $4C = H^2 - 2(X^+X^- + X^-X^+) = H^2 + 2H - 4X^-X^+$ where H, X^+, and X^- are as in (10.13).

(vi) Show that in spherical coordinates $-C$ has the form

$$\frac{1}{\sin^2\vartheta}\frac{\partial^2}{\partial\varphi^2} + \frac{1}{\sin\vartheta}\frac{\partial}{\partial\vartheta}\left(\sin\vartheta\frac{\partial}{\partial\vartheta}\right).$$

(vii) Show that $H^2 - 2(X^+X^- + X^-X^+)$ acts as multiplication by $m(m+2)$ on the space V_m (see (10.7)). How is this related to (iii)?

For the Casimir element in general, see Bourbaki [1], Ch. I, §3.7, and Dieudonné [3], p. 277.

5. Let V be a finite-dimensional $\mathrm{sl}(2, \mathbb{C})$-module and let $V_j \subset E$ be an eigenspace of $H \in \mathrm{sl}(2, \mathbb{C})$ with corresponding eigenvalue j. Show that X^- is injective on $\bigoplus_{j \geq 1} V_j$.

CHAPTER III

Representative Functions

The individual entries of a matrix representation of a Lie group G are continuous functions on G. They generate the ring of representative functions. The celebrated theorem of Peter and Weyl asserts that the representative functions are dense in the space of all continuous functions. This central result is proved in §3 with the help of some functional analysis which is reviewed in §2. We devote §1 to definitions and to showing that any (left or right) G-translation invariant finite-dimensional subspace of the ring of continuous functions on G actually consists of representative functions.

Applications of the theorem of Peter and Weyl are given in §4, and §5 deals with some of the formal generalizations to the decomposition of infinite-dimensional representations. In §6 we introduce induced representations. This is an important concept in representation theory, but, as its applications are mainly to finite groups and infinite-dimensional representations of locally compact groups, it does not play much of a role in this book.

The ring of representative functions determines the structure of G. This is shown in §7 and is part of the so-called Tannaka–Kreĭn duality theory. The ring of representative functions may also be used to define the complexification of a compact Lie group, as is indicated in §8. This opens the way for looking at Lie groups as algebraic groups.

1. Algebras of Representative Functions

Let G be a compact Lie group and $G \to \mathrm{GL}(n, \mathbb{C})$, $g \mapsto (a_{ij}(g))$, a matrix representation. Viewed as functions on G, the a_{ij} are continuous, and one should anticipate that such functions play a special role among all continuous functions on G. For example, the orthogonality relations (II, §4) express some of the distinguished properties enjoyed by the a_{ij}. In fact, the investigation of functions arising as entries in matrix representations is an analytic side of representation theory, and the purpose of this section is to collect some results concerning the linear algebra of these functions. Before proceeding, we point out that we have already encountered analytic aspects of our theory in II, §5 (spherical functions).

Matrix multiplication

$$a_{ij}(gh) = \sum_k a_{ik}(g)a_{kj}(h)$$

shows that the translated functions $g \mapsto a_{ij}(gh)$ are linear combinations of the functions a_{ik}, $1 \le k \le n$, and similarly for the functions $g \mapsto a_{ij}(hg)$. Thus the vector space generated by the a_{ij} is invariant under left and right translation. In other words, if it contains f it also contains every function of the form $x \mapsto f(gx)$ and $x \mapsto f(xg)$, $g \in G$. In this manner we are led to undertake a more systematic study of the ring of continuous functions on G and its finite-dimensional G-invariant subspaces.

Let $C^0(G, K)$ denote the ring of continuous functions $G \to K$ into the field K of real or complex numbers. Left and right translation in G induce actions of G on $C^0(G, K)$ as follows:

$$\begin{aligned} L\colon G \times C^0(G, K) &\to C^0(G, K), \\ L(g, f)(x) &= f(g^{-1}x). \\ R\colon G \times C^0(G, K) &\to C^0(G, K), \\ R(g, f)(x) &= f(xg). \end{aligned}$$

Note that using g^{-1} instead of g in the definition of L leads to a left action on $C^0(G, K)$—the reader should check that both L and R are left actions.

(1.1) Definition. Let G act on $C^0(G, K)$ via R. A function $f \in C^0(G, K)$ is called a (K-valued) ***representative function*** for G if f generates a finite-dimensional G-subspace of $C^0(G, K)$, i.e., if the smallest G-subspace containing f is finite-dimensional.

We have already seen that the vector space generated by the entries of a matrix representation consists of representative functions. We will soon see that representative functions always stem from matrix representations, so (1.1) gives an axiomatic characterization of functions arising in this fashion.

The next step is to understand how representations lead to representative functions on a more abstract level. Let V be a finite-dimensional continuous representation of G over K and let $V^* = \operatorname{Hom}(V, K)$ be the dual representation on the space of K-linear maps $V \to K$. Recall that G acts on V^* via

$(g \cdot f)(v) = f(g^{-1}v)$. Given $v \in V$ and $f \in V^*$ we define $d_{f,v} \in C^0(G, K)$ by $d_{f,v}(g) = f(gv)$. Thus we obtain a linear map

$$s_V \colon V^* \otimes_K V \to C^0(G, K), \qquad f \otimes v \mapsto d_{f,v}.$$

It is easy to verify that $L(g, d_{f,v}) = d_{g\cdot f,v}$ and $R(g, d_{f,v}) = d_{f,gv}$, so the action L corresponds to the action on V^* and the action R corresponds to that on V. The space

$$S(V) = \text{image of } s_V$$

is a finite-dimensional G-subspace of $(C^0(G, K), R)$ and of $(C^0(G, K), L)$—this notation indicates a vector space and a specific action of G. Thus $S(V)$ consists of representative functions. If $e_1, \ldots, e_n$ is a basis of V, and $e_1^*, \ldots, e_n^*$ is the dual basis of V^*, then $d_{e_i^*, e_j} = a_{ij}$ where $ge_j = \sum_i a_{ij}e_i$. Hence $S(V)$ is simply the space generated by the a_{ij} coming from the matrix representation corresponding to V.

(1.2) Proposition. *If f is a representative function, then f generates a finite-dimensional G-subspace of $(C^0(G, K), L)$. The representative functions form a K-subalgebra $\mathscr{T}(G, K)$ of $C^0(G, K)$ closed under complex conjugation.*

PROOF. Let f be a representative function, so f generates a finite-dimensional G-subspace V of $(C^0(G, K), R)$. Let $e_1, \ldots, e_n$ be a basis of V and $e_1^*, \ldots, e_n^*$ be the dual basis of V^*. If $R(g, f) = \sum_j a_j(g)e_j$, we have

$$s_V(e_j^*, f)(g) = a_j(g).$$

Therefore

$$f(g) = R(g, f)(1) = \sum_j a_j(g)e_j(1) = \sum_j s_V(e_j^*, f)(g)e_j(1).$$

It follows that $f = \sum e_j(1)s_V(e_j^*, f) \in S(V)$, which is known to be finite-dimensional and invariant under the L-action of G. This establishes the first assertion.

For the second assertion, we need only consider representative functions coming from matrix representations, since we have seen above that $\mathscr{T}(G, K)$ is generated by these as a K-vector space. But if $g \mapsto (a_{ij}(g))$ and $g \mapsto (b_{kl}(g))$ are two matrix representations, considering their direct sum and tensor product will show that $a_{ij} + b_{kl}$ and $a_{ij}b_{kl}$ are representative functions. Considering a dual representation shows that $\bar{a}_{ij}$ is a representative function, completing the proof. □

(1.3) Example. The algebra $\mathscr{T}(S^1, \mathbb{C})$ consists of the trigonometric polynomials on S^1. Namely, if V is a complex representation of S^1, then we can find a basis $e_1, \ldots, e_n$ of V such that $z \in S^1$ acts by sending e_i to $z^{n(i)}e_i$, $n(i) \in \mathbb{Z}$ (decomposition into irreducibles, see II, (8.1)). Thus $\mathscr{T}(S^1, \mathbb{C})$ is generated by the functions $z \mapsto z^n$, $n \in \mathbb{Z}$.

We will consider two topologies on $C^0(G, K)$—that defined by the supremum norm and that defined by the inner product $\langle f_1, f_2\rangle = \int_G f_1 \bar{f}_2$. The representations $\mathscr{T}(G, K)$ for both actions R and L are, in general, infinite-dimensional, but they do admit the pleasant decomposition into isotypical submodules which we encountered in II, §1. Moreover, they are completely describable in terms of irreducible G-modules.

(1.4) Proposition. *Let B be a G-submodule of $(\mathscr{T}(G, K), R)$. Then the following hold*:

(i) *B is the orthogonal direct sum of the submodules $B \cap S(U)$, where U ranges over* $\mathrm{Irr}(G, K)$.
(ii) *The submodule $B \cap S(U)$ is the U-isotypical part of B.*
(iii) *B is closed in $\mathscr{T}(G, K)$ with respect to both the inner product and supremum norm topologies.*

PROOF. The orthogonality relations of II, §4 and II, (6.10), Ex. 3 imply that the spaces $S(U)$, $U \in \mathrm{Irr}(G, K)$, are pairwise orthogonal. And from the proof of (1.2) we see that the $S(U)$ generate $\mathscr{T}(G, K)$. Thus $\mathscr{T}(G, K)$ is the direct sum of the $S(U)$. Since $S(U)$ is the image of $U^* \otimes U$, we see that $S(U)$ is U-isotypical in the sense of II, §2. Applying II, (2.2) to the semisimple module $\mathscr{T}(G, K)$ yields both (i) and (ii).

Now let f be an element of $\mathscr{T}(G, K)$ lying in the closure of B in the inner product topology. Let π_U be orthogonal projection from $\mathscr{T}(G, K)$ onto $S(U)$. Since

$$\langle \pi_U(f - b), \pi_U(f - b)\rangle \le \langle f - b, f - b\rangle$$

for $b \in B$ we see that $\pi_U(f)$ lies in the closure of $\pi_U(B)$ in $S(U)$. But $S(U)$ is finite-dimensional, so $\pi_U(f) \in \pi_U(B)$. It follows that f is contained in the sum of the spaces $\pi_U(B)$ where U ranges over $\mathrm{Irr}(G, K)$, and hence $f \in B$. Furthermore, $\langle f, f\rangle^{1/2} \le |f|$ shows that sets closed in the inner product norm topology are also closed in the supremum norm topology, and this establishes all of (iii). □

Needless to say, analogous results hold for submodules of $(\mathscr{T}(G, K), L)$, but we now want to study submodules of $\mathscr{T}(G, K)$ simultaneously invariant under both actions R and L. In other words, we may consider $\mathscr{T}(G, K)$ to be a $(G \times G)$-module with the action $(g, h)f = L(g, R(h, f))$, which is well defined since R and L commute. The simultaneously invariant submodules which interest us are precisely the $(G \times G)$-submodules.

(1.5) Proposition. *The maps s_U: $U^* \otimes U \to C^0(G, K)$ induce an isomorphism of $(G \times G)$-modules*

$$(s_U): \bigoplus_{U \in \mathrm{Irr}(G, K)} U^* \otimes_{D(U)} U \to \mathscr{T}(G, K),$$

where $D(U)$ is the endomorphism algebra of U over K. Also, any $(G \times G)$-submodule B of $\mathscr{T}(G, K)$ is the direct sum of the $B \cap S(U)$, and if $B \cap S(U) \neq \{0\}$, then $B \cap S(U) = S(U)$.

(1.6) Remark. It is no accident that the isomorphism of this proposition resembles the canonical isomorphisms in II, (1.14) and II, (6.9). Suitably rewritten, they turn out to be the same: For each G-module E there is a canonical isomorphism of vector spaces

$$\mathrm{Hom}(E, K) \cong \mathrm{Hom}_G(E, C^0(G, K))$$

mapping $\varphi \in \mathrm{Hom}(E, K)$ to the map $\tilde{\varphi}\colon E \to C^0(G, K)$, $\tilde{\varphi}(x)(g) = \varphi(gx)$. The inverse maps ψ to $\tilde{\psi}\colon x \mapsto \psi(x)(1)$. This isomorphism makes the following diagram commute, where d_E is evaluation:

$$\begin{array}{ccc} \mathrm{Hom}(E, K) \otimes E & \cong & \mathrm{Hom}_G(E, C^0(G, K)) \otimes E \\ & {\scriptstyle s_E}\searrow \quad \swarrow{\scriptstyle d_E} & \\ & C^0(G, K) & \end{array}$$

This is a special case of Frobenius reciprocity for induced representations, see (6.2) with $H = \{1\}$.

Proof of Proposition (1.5). Let $\varphi\colon U \to U$ be an endomorphism. Then $s_U(f, \varphi v)$ maps g to $f(g\varphi v)$ and $s_U(f\varphi, v)$ maps g to $f\varphi(gv)$. Since φ is a G-map, $\varphi(gv) = g\varphi(v)$ and these two functions are the same. Thus s_U gives a $(G \times G)$-map $s_U\colon U^* \otimes_{D(U)} U \to \mathscr{T}(G, K)$. Using the commutative diagram from (1.6), the map s_U is transferred to the map

$$d_U\colon \mathrm{Hom}_G(U, \mathscr{T}(G, K)) \otimes_{D(U)} U \to \mathscr{T}(G, K).$$

We may now apply II, (1.14) and II, (6.9) to show that (s_U) is an isomorphism. (The reader may note that in the results referred to we only considered finite-dimensional vector spaces, and $\mathscr{T}(G, K)$ could well be infinite-dimensional. However, the argument used to prove II, (1.14) remains valid for infinite direct sums of irreducible G-modules, i.e., for V semisimple in the sense of II, §2).) This proves the first statement.

The second statement is equivalent to the assertion that $U^* \otimes_{D(U)} U$ is an irreducible $(G \times G)$-module. In the case $K = \mathbb{C}$, this follows from II, (4.14), and for the real case, see II, (6.10), Ex. 9. □

We close with some general remarks concerning infinite-dimensional representations.

Let G be a compact Lie group and consider a K-vector space V with G-action $\rho\colon G \times V \to V$ such that all the left translations $v \mapsto \rho(g, v)$ are K-linear. We do not assume that ρ is continuous. The finite-dimensional continuous G-submodules of V generate a subspace V_s of V which is G-invariant. Each element of V_s is contained in a finite-dimensional continuous G-submodule. A representation V is called ***locally finite*** if $V = V_s$.

Given V, the G-module V_s is the unique maximal locally finite subrepresentation of V. Furthermore, V_s has an algebraic decomposition into isotypical summands as in the finite-dimensional case II, (1.14), II, (6.9). Let E be a finite-dimensional continuous G-module, and consider the vector space $\mathrm{Hom}_G(E, V)$ and the representation $\mathrm{Hom}_G(E, V) \otimes E$. Then the evaluation map

$$c_E\colon \mathrm{Hom}_G(E, V) \otimes E \to V, \qquad \varphi \otimes x \mapsto \varphi(x)$$

is G-equivariant and linear. Moreover, its image $\mathrm{c}(E)$ is contained in V_s.

(1.7) Proposition.

(i) *The maps c_E induce an isomorphism*

$$c = (c_E)\colon \bigoplus_{E \in \mathrm{Irr}(G, K)} \mathrm{Hom}_G(E, V) \otimes_{D(E)} E \to V_s$$

(ii) *If $W \subset V_s$ is a G-subspace, then $W = \bigoplus_{E \in \mathrm{Irr}(G,K)} (\mathrm{c}(E) \cap W)$.*

(iii) *If $E \in \mathrm{Irr}(G, K)$, then $W \subset \mathrm{c}(E)$ if and only if the irreducible submodules of W are isomorphic to E.*

PROOF. See Exercise 4. □

The subspace $\mathrm{c}(E) \subset V_s$ for $E \in \mathrm{Irr}(G, K)$ is again called the E-isotypical part of V_s.

(1.8) Exercises

1. Let $V \subset C^0(G, K)$ be a finite-dimensional subspace which is G-invariant with respect to the action R. Show that the induced action of G on V is continuous.

2. Let U be a continuous finite-dimensional G-module and W a G-invariant subspace. Show that W and U/W are continuous.

3. Show that if G is finite, then $V = V_s$ for any representation V of G, but that if G is not finite, then there exists a representation V such that $V \neq \{0\}$ and $V_s = \{0\}$.

4. Give a proof of (1.7). *Hint*: If $x \in W \subset V_s$ (i.e., W is finite-dimensional), then $x \in \mathrm{c}(W)$ so c is surjective. Show $\mathrm{Hom}_G(E, V) = \mathrm{Hom}_G(E, V_s)$. Each element in the domain of c is contained in a subspace $\bigoplus \mathrm{Hom}_G(E, U) \otimes E$ with $U \subseteq V_s$ finite-dimensional. Now apply II, (1.14) and II, (6.9).

5. Prove or give a counterexample: Let f be a representative function with $f(e) = 1$. Then there is a continuous homomorphism $G \to \mathrm{GL}(n, K)$, $g \mapsto (a_{ij}(g))$, such that $a_{11} = f$.

6. Let $f \in \mathcal{T}(G, \mathbb{C})$. Show that the real and imaginary part of f are contained in $\mathcal{T}(G, \mathbb{R})$. Conclude that there is an isomorphism of algebras $\mathcal{T}(G, \mathbb{R}) \otimes_{\mathbb{R}} \mathbb{C} \cong \mathcal{T}(G, \mathbb{C})$.

7. Show that representative functions are smooth (C^∞).
8. Show that convolution of functions on G makes $\mathscr{T}(G, K)$ into a K-algebra. Use II, (4.16) and a real analogue.
9. Composing the canonical isomorphism $\mathrm{Hom}(V, V) \cong V^* \otimes V$ with the map s_V yields a homomorphism $t_V\colon \mathrm{Hom}(V, V) \to \mathscr{T}(G, \mathbb{C})$. Show that $t_V(f)$ maps g to $\mathrm{Tr}(l_g \circ f)$, and that composition of homomorphisms in $\mathrm{Hom}(V, V)$ is transformed into convolution by t_V. Conclude that the algebra $\mathscr{T}(G, \mathbb{C})$ with convolution is isomorphic to the direct sum of endomorphism algebras $\bigoplus_{V \in \mathrm{Irr}(G, \mathbb{C})} \mathrm{Hom}(V, V)$.
10. Let G act on itself by left translation. This gives rise to the notion of Lie derivative $\mathrm{L}_X f$, $X \in \mathrm{L}(G)$, $f \in C^\infty(G, \mathbb{C})$. Show that L_X maps $\mathscr{T}(G, \mathbb{C})$ into itself.
11. Let $\mathscr{T}(G, \mathbb{C})$ be the convolution algebra of Exercise 9. Show that its center is spanned as a vector space by the characters χ_V, $V \in \mathrm{Irr}(G, \mathbb{C})$.
12. Let G be finite and abelian and let $\mathbb{C}[G] = C^0(G, \mathbb{C})$ be the group ring of G; see II, §1. Show that, as a representation (with the action L or R), $\mathbb{C}[G]$ is isomorphic to the direct sum of the distinct irreducible representations of G (each with multiplicity one). Give an explicit decomposition of $\mathbb{C}[\mathbb{Z}/n]$ into irreducible $\mathbb{Z}/n$-modules.

2. Some Analysis on Compact Groups

This section is preparatory to the next, in which we will prove that representative functions are dense in the space of all continuous functions. We need a few notions from analysis.

Let $H = C^0(G, \mathbb{C})$ be the vector space of continuous complex functions on the compact group G. We equip H with the supremum norm

$$|u| = \sup\{|u(g)| \mid g \in G\}.$$

We also have the inner product and the corresponding norm

$$\langle u, v\rangle = \int_G u\bar{v}, \qquad \|u\| = \langle u, u\rangle^{1/2}.$$

Completion of H for this norm yields the Hilbert space $\hat{H} = L^2(G)$ of square integrable functions on G. From the Cauchy-Schwarz inequality

$$|\langle u, v\rangle| \le \|u\| \cdot \|v\|$$

using $v = 1$, and then using $\int u\bar{u} \le |u|^2$, we get

$$\int_G |u| \le \|u\| \le |u|.$$

By the second inequality, the identity map (and the inclusion)

$$(H, |\cdot|) \to (H, \|\cdot\|) \subset (\hat{H}, \|\cdot\|)$$

is continuous with respect to the topologies induced by the respective norms. It is a fact from functional analysis that a linear map (also called linear operator) $K: (E, |\cdot|) \to (F, \|\cdot\|)$ between two normed vector spaces is continuous if and only if there is a constant A such that $\|Kv\| \le A \cdot |v|$ for all $v \in E$. If $k: G \times G \to \mathbb{C}$ is a continuous function and $f \in \hat{H}$, the integral

$$Kf(g) = \int_G k(g, h) f(h)\, dh$$

defines a continuous function Kf on G. Moreover,

$$|Kf(g)| \le \int |k(g, h)|\, |f(h)|\, dh \le A \int |f(h)|\, dh \le A \|f\|,$$

where $A = \sup\{|k(g, h)| \mid g, h \in G\}$. Hence

$$K: (\hat{H}, \|\cdot\|) \to (H, |\cdot|), \qquad f \mapsto Kf$$

is continuous.

A subset L of a normed vector space F is called ***precompact*** if every sequence (f_n) in L contains a subsequence which converges in F. A linear map $K: E \to F$ between normed vector spaces is called ***compact*** if it maps every bounded subset $B \subset E$ into a precompact subset $K(B)$ of F.

The next theorem characterizes the precompact subsets of $(C^0(G, \mathbb{C}), |\cdot|)$ for a compact group (or just space) G.

(2.1) Theorem (Ascoli). *A subset L of the space of continuous functions $G \to \mathbb{C}$ with the supremum norm is precompact if and only if L is equicontinuous and bounded.*

For a proof of this theorem see, for example, Dieudonné [1] or Lang [1]. Recall that L is ***equicontinuous at*** x_0 if for each $\varepsilon > 0$ there is a neighborhood U of x_0 such that $|f(x) - f(x_0)| < \varepsilon$ for all $x \in U$ and $f \in L$. Then L is called ***equicontinuous*** if it is equicontinuous at every point of G.

The following proposition is an application of the Ascoli theorem to the map K defined above.

(2.2) Proposition. *The operator $K: (\hat{H}, \|\cdot\|) \to (H, |\cdot|)$ is compact and hence $K: (\hat{H}, \|\cdot\|) \to (\hat{H}, \|\cdot\|)$ is compact.*

Proof. Let $B \subset \hat{H}$ be $\|\cdot\|$-bounded by a constant $C > 0$. Given $\varepsilon > 0$, choose a neighborhood V of $e \in G$ such that given any $h \in G$ one has $|k(x, h) - k(x_0, h)| < \varepsilon C^{-1}$ as long as $x x_0^{-1} \in V$. Then

$$|Kf(x) - Kf(x_0)| = \left| \int (k(x, h) - k(x_0, h)) f(h)\, dh \right|$$

$$\le \varepsilon C^{-1} \int |f(h)|\, dh \le \varepsilon C^{-1} \|f\| \le \varepsilon.$$

This shows that $K(B)$ is equicontinuous, and since it is also bounded, $K(B)$ is precompact and K is compact. □

Now suppose that $k: G \times G \to \mathbb{C}$ is symmetric in the sense that $k(x, y) = \bar{k}(y, x)$. Then the operator K turns out to be ***symmetric***, i.e., we have

(2.3) Proposition. *If $k: G \times G \to \mathbb{C}$ is symmetric as explained above, then $\langle Kf_1, f_2\rangle = \langle f_1, Kf_2\rangle$ for all $f_1, f_2 \in L^2(G)$.*

PROOF. Use Fubini's theorem to switch the order of integration. □

Also recall that the ***norm of an operator*** is defined by

$$\|K\| = \sup\{\|Kf\| \mid \|f\| = 1\}.$$

Thus $\|Kf\| \leq \|K\|\cdot\|f\|$ for all f, and $\|K\|$ is the smallest constant A such that $\|Kf\| \leq A\|f\|$ for all f.

(2.4) Proposition. *Suppose K is a symmetric operator. Then*

$$\|K\| = \sup\{|\langle Kf, f\rangle| \mid \|f\| = 1\}.$$

PROOF. By the Cauchy–Schwarz inequality, for $\|f\| = 1$

$$|\langle Kf, f\rangle| \leq \|Kf\|\cdot\|f\| \leq \|K\|\cdot\|f\| = \|K\|,$$

so that the supremum (call it M) exists and is at most $\|K\|$.

In order to show that $\|K\| \leq M$, i.e., that $\|Kx\| \leq M$ for $\|x\| = 1$, we may assume that $Kx \neq 0$ and let $y = Kx/\|Kx\|$. Then $\langle Kx, y\rangle = \|Kx\| = \langle x, Ky\rangle$ and

$$\langle K(x + y), x + y\rangle = \langle Kx, x\rangle + 2\|Kx\| + \langle Ky, y\rangle$$
$$\langle K(x - y), x - y\rangle = \langle Kx, x\rangle - 2\|Kx\| + \langle Ky, y\rangle.$$

Subtracting these two equations we get

$$\begin{aligned} 4\|Kx\| &= \langle K(x + y), x + y\rangle - \langle K(x - y), x - y\rangle \\ &\leq M\cdot\|x + y\|^2 + M\cdot\|x - y\|^2 = 4M. \end{aligned}$$ □

(2.5) Proposition. *Let K be a symmetric compact operator. Then $\|K\|$ or $-\|K\|$ is an eigenvalue of K.*

PROOF. By (2.4) we can find a sequence $x_n \in H$ such that $\|x_n\| = 1$ and $\lim |\langle Kx_n, x_n\rangle| = \|K\|$. Passing to a subsequence if necessary, we may assume $\langle Kx_n, x_n\rangle$ converges to an α which is either $\|K\|$ or $-\|K\|$. Then we have

$$\begin{aligned} 0 \leq \|Kx_n - \alpha x_n\|^2 &= \langle Kx_n - \alpha x_n, Kx_n - \alpha x_n\rangle \\ &= \|Kx_n\|^2 - 2\alpha\langle Kx_n, x_n\rangle + \alpha^2\|x_n\|^2 \\ &\leq \alpha^2 - 2\alpha\langle Kx_n, x_n\rangle + \alpha^2. \end{aligned}$$

The right-hand side converges to zero, and since K is compact we may assume (after passing to a subsequence again) that (Kx_n) converges to some element y. The inequality above shows that (αx_n) also converges to y. If $\alpha = 0$, then $\|K\| = 0$ so $K = 0$, and the proposition is trivial. If $\alpha \neq 0$, then (x_n) converges to x, where $x = \alpha^{-1}y \neq 0$. But then $Kx = \alpha x$, showing that α is an eigenvalue of K. □

We are now ready to prove the spectral theorem for compact symmetric operators $K: \hat{H} \to \hat{H}$. Note that every eigenvalue λ of such an operator is real; this follows from $\lambda\langle x, x\rangle = \langle \lambda x, x\rangle = \langle Kx, x\rangle = \langle x, Kx\rangle = \langle x, \lambda x\rangle = \bar{\lambda}\langle x, x\rangle$, where $Kx = \lambda x$ and $x \neq 0$. Let $H_\lambda = \{x \mid Kx = \lambda x\}$ be the eigenspace of λ. Then for $\lambda \neq \mu$ the spaces H_λ and H_μ are orthogonal. In fact, for $x \in H_\lambda$ and $y \in H_\mu$ we have $\lambda\langle x, y\rangle = \langle Kx, y\rangle = \langle x, Ky\rangle = \mu\langle x, y\rangle$, which implies orthogonality.

(2.6) Theorem. *Let $K: (\hat{H}, \|\cdot\|) \to (\hat{H}, \|\cdot\|)$ be a compact symmetric operator. Then for each $\varepsilon > 0$ the subspace*

$$\bigoplus_{|\lambda| \geq \varepsilon} H_\lambda$$

is finite-dimensional and $\bigoplus_\lambda H_\lambda$ is dense in $\hat{H}$.

Proof. Were the space in question not finite-dimensional, it would contain a sequence $(x_n, n \in \mathbb{N})$ of orthonormal eigenvectors. But then the equations $Kx_n = \lambda_n x_n$, together with the inequalities $\|\lambda_n x_n - \lambda_m x_m\|^2 = \lambda_n^2 + \lambda_m^2 \geq 2\varepsilon^2$ for $n \neq m$, would contradict the compactness of K.

Now let E be the closure of $\bigoplus_\lambda H_\lambda$ in $\hat{H}$. Then $K(E) \subset E$. Let F be the orthogonal complement of E in H. For $e \in E$ and $f \in F$ we have $0 = \langle Ke, f\rangle = \langle e, Kf\rangle$, so Kf is orthogonal to E. Consequently, K induces a linear map $K': F \to F$ which is still symmetric and compact and as such has an eigenvalue by (2.5). If $F \neq \{0\}$, this contradicts the construction of E. □

In closing we recall the Stone–Weierstrass approximation theorem. Let X be a compact space and let $C^0(X, K)$ be the space of continuous functions $X \to K$ with the supremum norm. (Here K is $\mathbb{R}$ or $\mathbb{C}$.) Pointwise multiplication of functions makes this into an algebra. A set $B \subset C^0(X, K)$ is said to ***separate points*** if, given any two distinct points $x_1, x_2 \in X$, there is an $f \in B$ such that $f(x_1) \neq f(x_2)$. The real and complex Stone–Weierstrass approximation theorem goes as follows:

(2.7) Theorem.

(i) *Let $B \subset C^0(X, \mathbb{R})$ be a subalgebra which contains all real constants and separates points. Then B is dense in $C^0(X, \mathbb{R})$.*
(ii) *Let $B \subset C^0(X, \mathbb{C})$ be a subalgebra which contains all complex constants, separates points, and is closed under complex conjugation. Then B is dense in $C^0(X, \mathbb{C})$.*

PROOF. See, for example, Lang [1], III.1 or Dieudonné [1], 7.3. □

(2.8) Exercises

1. Let $E_0 \xrightarrow{f_0} E_1 \xrightarrow{f_1} E_2 \xrightarrow{f_2} E_3$ be continuous linear operators. Show that $f_2 \circ f_1 \circ f_0$ is compact if f_1 is compact. Show that an operator with finite-dimensional image is compact.

2. Show that a subset L in a normed vector space is precompact if and only if its closure is compact.

3. Recall the geometry behind (2.4) and (2.5): If $A \in \mathrm{GL}(n, \mathbb{R})$, then

$$C = \{y \mid y = Ax, \|x\| = 1\}$$

is an $(n-1)$-dimensional ellipsoid defined by $C = \{y \in \mathbb{R}^n \mid {}^t y {}^t BBy = 1\}$, where $B = A^{-1}$. The matrix ${}^t AA$ is symmetric and positive-definite, so ${}^t AA = P^2$ for some symmetric positive-definite P. Hence $C = \{y \mid Px = y, \|x\| = 1\}$ and

$$\|A\| = \max\{\|y\| \mid y \in C\} = \|P\|.$$

If A is symmetric, then for a suitable $U \in \mathrm{O}(n)$ the matrix ${}^t UAU = D$ is diagonal with entries $(\lambda_1, \ldots, \lambda_n)$ on the diagonal. Thus

$$\|A\|^2 = \max\{\langle Ax, Ax\rangle \mid \langle x, x\rangle = 1\} = \max\left\{\sum_i \lambda_i^2 z_i^2 \,\Big|\, \sum_i z_i^2 = 1\right\} = \max\{\lambda_i^2\}.$$

Moreover, in this case

$$\{\langle Ax, x\rangle \mid \|x\| = 1\} = \{\langle Dx, x\rangle \mid \|x\| = 1\} = \{\sum \lambda_i x_i^2 \mid \|x\| = 1\},$$

and consequently

$$\max\{|\lambda_i x_i^2| \mid \|x\| = 1\} = \max\{|\lambda_i|\}.$$

Show that the tangent space at y to $\{y \in \mathbb{R}^n \mid {}^t yMy = 1\}$ for a symmetric M is perpendicular to y if and only if y is an eigenvector of M.

3. The Theorem of Peter and Weyl

In §1 we introduced the ring $\mathscr{T}(G) = \mathscr{T}(G, \mathbb{C})$ of complex-valued representative functions. This is a subring of the ring $C^0(G) = C^0(G, \mathbb{C})$ of continuous complex-valued functions on the compact group G. As a vector space, $C^0(G)$ is complete with respect to the supremum norm

$$|f| = \sup\{|f(g)| \mid g \in G\}$$

and hence is a Banach space with this norm. The actions of G on this space given by left and right translation are continuous.

On the other hand, we may complete $C^0(G)$ with respect to the inner product metric

$$\langle u, v\rangle = \int_G u\bar{v}.$$

This yields the Hilbert space $L^2(G)$ of square-integrable functions on G. The group G acts on $L^2(G)$ by left and right translation. Invariance of the integral shows that these actions are unitary.

The goal of this section is to prove the following basic theorem due to Peter and Weyl [1]:

(3.1) Theorem.

(i) *The representative functions are dense in both $C^0(G)$ and $L^2(G)$.*
(ii) *The irreducible characters generate a dense subspace of the space of continuous class functions.*

To gain some insight into the meaning of this result, we examine the special case of the classical groups before giving a proof. In this case, the result essentially follows from the Stone–Weierstrass approximation theorem (2.7).

Let G be a subgroup of $\mathrm{GL}(n, \mathbb{C})$. Then the standard representation of $\mathrm{GL}(n, \mathbb{C})$ on $\mathbb{C}^n$ restricts to a faithful representation of G. Hence the representative functions $\mathcal{T}(G, \mathbb{C})$ separate points. Thus by theorem (2.7)(ii), we see that $\mathcal{T}(G, \mathbb{C})$ is dense in $C^0(G, \mathbb{C})$. On the other hand, we will later deduce from (3.1) that every compact Lie group admits a faithful representation on some $\mathbb{C}^n$. Thus the existence of a faithful representation and the theorem of Peter and Weyl are equivalent via the Stone–Weierstrass approximation theorem.

Proof of (3.1). We begin by showing that (ii) follows from (i). Suppose we are given a φ which is a class function, so $\varphi(gxg^{-1}) = \varphi(x)$ for all $g, x \in G$. Given $\varepsilon > 0$, by (i) we may find a representative function f such that $|\varphi - f| < \varepsilon$. Then the function $\psi: x \mapsto \int_G f(gxg^{-1})\, dg$ is a class function satisfying $|\varphi - \psi| < \varepsilon$. Thus it suffices to show that ψ is a linear combination of irreducible characters.

But f is a representative function, and hence

$$f(g) = \sum_i f_i(ge_i)$$

for some $e_i \in E_i, f_i \in \mathrm{Hom}(E_i, \mathbb{C})$, and E_i irreducible ((1.5)(i)). Consequently

$$\int f(gxg^{-1})\, dg = \sum_i f_i\left(\int gxg^{-1} e_i\, dg\right).$$

Letting τ be the representation $\tau: G \to \mathrm{Aut}(E_i)$, we have seen in II, (4.4) that

$$\int \tau(gxg^{-1})\, dg = (\dim E_i)^{-1} \chi_i(x) \cdot \mathrm{id}(E_i),$$

where χ_i is the character of E_i. Thus $f_i(\int gxg^{-1} e_i\, dg) = (\dim E_i)^{-1} f_i(e_i) \chi_i(x)$, and we have shown that (i) implies (ii).

We must now prove (i). So let $f: G \to \mathbb{C}$ be continuous and let $\varepsilon > 0$ be given. Then by uniform continuity of f there is a neighborhood U of e in

G such that $U = U^{-1}$ and such that $|f(x) - f(y)| < \varepsilon$ whenever $x^{-1}y \in U$. Let $\delta\colon G \to [0, \infty[$ be a continuous "bump function" with support contained in U such that $\delta(x) = \delta(x^{-1})$ and $\int_G \delta = 1$.

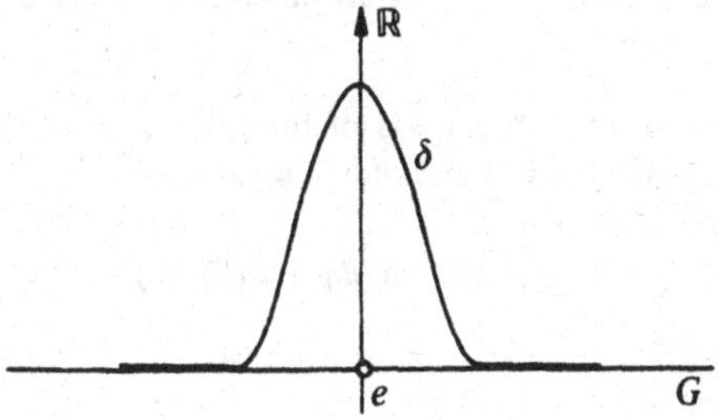

Figure 17

This δ should be thought of as an approximation to the Dirac δ "function."

We now consider the linear operator

$$K\colon L^2(G) \to C^0(G), \qquad f \mapsto Kf$$

defined by

$$\begin{aligned} Kf(x) &= \int \delta(g)f(xg)\,dg = \int \delta(x^{-1}g)f(g)\,dg \\ &= \int f(g)\delta(g^{-1}x)\,dg = f * \delta(x). \end{aligned}$$

Then we have

$$|Kf - f| = \sup \left| \int \delta(g)(f(xg) - f(x))\,dg \right| \le \int \varepsilon\delta(g)\,dg = \varepsilon$$

because $|f(xg) - f(x)| < \varepsilon$ whenever $\delta(g) \ne 0$. Thus for the proof of (i) it suffices to approximate Kf by representative functions.

Observe that K is of the type considered in §2, namely symmetric and compact. Furthermore, K is equivariant since (using left translation)

$$\begin{aligned} K(hf)(x) &= \int f(h^{-1}g)\delta(g^{-1}x)\,dg \\ &= \int f(g)\delta(g^{-1}h^{-1}x)\,dg = (hKf)(x). \end{aligned}$$

It follows that the eigenspaces of K are G-invariant.

Let λ_n for $n \ge 1$ and $\lambda_0 = 0$ be the eigenvalues and H_n be the corresponding eigenspaces. Applying theorem (2.6), we have $\bigoplus_{n \ge 0} H_n$ dense in $\hat{H} = L^2(G)$ and hence $\bigoplus_{n \ge 0} KH_n$ dense in $K\hat{H}$. But $KH_0 = 0$ and $KH_n = H_n$ is finite-dimensional for $n \ge 1$, so $H_n \subset \mathcal{T}(G, \mathbb{C})$ for $n \ge 1$. Since Kf is contained in the closure of $\bigoplus_{n \ge 1} H_n$, this completes the proof. □

(3.2) Exercises

1. Show that $\mathcal{T}(G, \mathbb{R})$ is dense in $C^0(G, \mathbb{R})$.

2. Let $f: G \to \mathbb{R}$ be a class function such that $f(g) = f(g^{-1})$ for all $g \in G$. Show that f can be uniformly approximated by $\mathbb{R}$-linear combinations of characters of real representations.

3. Show that the function $\psi \in C^0(G, \mathbb{C})$ is of the form $\psi = \dim(V)^{-1} \cdot \chi_V$ for some character χ_V of an irreducible V if and only if ψ satisfies

$$\int_G \psi(ghg^{-1}k)\, dg = \psi(h)\psi(k)$$

for all $h, k \in G$.

4. Let $(V^\alpha | \alpha \in A)$ be a set of irreducible unitary representations of G. Let (r_{ij}^α) be the matrix form of V^α with respect to some orthonormal basis. Show that if the set of finite linear combinations of the functions r_{ij}^α is dense in $C^0(G, \mathbb{C})$ or $L^2(G)$, then every irreducible unitary representation is unitarily equivalent to one of the V^α.

5. Let $f \in L^2(S^1)$. Show that the Fourier series of f converges to f in the L^2 topology. Show that every continuous function $f: S^1 \to \mathbb{C}$ may be uniformly approximated by trigonometric polynomials.

4. Applications of the Theorem of Peter and Weyl

We remind the reader that our group G is supposed to be compact throughout. A representation V of G is called ***faithful*** if the associated homomorphism $G \to \operatorname{Aut}(V)$ is injective. Thus a faithful representation is a realization of G as a subgroup of $\operatorname{Aut}(V)$.

(4.1) Theorem. *Every compact Lie group G admits a faithful representation.*

Before giving the proof, we mention a property of compact Lie groups—the ***descending chain property***—which will prove useful in other places as well. This property simply states that every sequence $K_1 \supset K_2 \supset K_3 \supset \cdots$ of closed subgroups K_i of G is eventually constant. To prove this, note that the K_i are closed manifolds, and if $K_{i+1} \neq K_i$, either $\dim K_{i+1} < \dim K_i$ or K_{i+1} has fewer components than K_i.

PROOF OF THEOREM 4.1. Let $g \in G \backslash \{1\}$. Then there exists a continuous function $f: G \to \mathbb{R}$ such that $f(g) \neq f(1)$, and, by the theorem of Peter and Weyl, we are able to find a representative function u with $u(g) \neq u(1)$. This implies the existence of a representation whose kernel is a proper closed subgroup K_1 of G, assuming $G \neq \{1\}$. If $K_1 \neq \{1\}$, we may choose $g \in K_1 \backslash \{1\}$ and find a representation of G whose kernel is K_2 with $K_2 \cap K_1$ properly contained in K_1. Continuing in this fashion and applying the descending chain property yields a finite number of representations whose kernels have trivial intersection. The direct sum of these representations is faithful. □

(4.2) Proposition. *Given any proper closed subgroup H of G, there is a non-trivial irreducible representation of G whose restriction to H contains the trivial representation as a summand.*

Proof. Let f be a representative function belonging to a representation containing no trivial summand. Then, by II, (4.8), $\int_G f = 0$. Thus, assuming (4.2) to be false, we see that $\int_H f = \int_G f$ for every $f \in \mathscr{T}(G, \mathbb{C})$. By the theorem of Peter and Weyl, this means $\int_H F = \int_G F$ for all $F \in C^0(G, \mathbb{C})$. But we know there are functions $F \in C^0(G, \mathbb{C})$ such that $F|H = 0$ and $\int_G F > 0$, a contradiction. □

(4.3) Proposition. *Let $G \to \mathrm{GL}(n, \mathbb{C})$, $g \mapsto (r_{ij}(g))$, be a faithful representation. Then the functions r_{ij} and $\bar{r}_{ij}$ generate $\mathscr{T}(G, \mathbb{C})$ as a $\mathbb{C}$-algebra.*

Proof. The $\mathbb{C}$-algebra A generated by the r_{ij} and $\bar{r}_{ij}$ satisfies the hypotheses of the Stone–Weierstrass theorem (2.7) and is therefore dense in $\mathscr{T}(G, \mathbb{C})$. Since A is a G-submodule of $\mathscr{T}(G, \mathbb{C})$, the assertion follows from (1.4). □

Note that the combination of (4.1) and (4.3) shows that $\mathscr{T}(G, \mathbb{C})$ is a finitely generated $\mathbb{C}$-algebra.

Next, we let V be a faithful representation of G and $\bar{V}$ its conjugate representation. Define

$$V(k, l) = (V \otimes \cdots \otimes V) \otimes (\bar{V} \otimes \cdots \otimes \bar{V})$$

with k factors of V and l factors of $\bar{V}$. Then we have

(4.4) Theorem. *Every irreducible representation of G is contained in some $V(k, l)$.*

Proof. Think of V as a matrix representation $g \mapsto (r_{ij}(g))$. Then the matrix entries of $V(k, l)$ are the various monomials of degree k in the r_{ij} and l in the $\bar{r}_{ij}$. By (4.3), these monomials generate $\mathscr{T}(G, \mathbb{C})$ as k and l run through the natural numbers. Now, suppose the irreducible representation U were not contained in any $V(k, l)$. Then $\int_G uv = 0$ for every representative function u of U and v of $V(k, l)$ by orthogonality II, (4.8). Yet such v are dense in $C^0(G, \mathbb{C})$, so $u = 0$, a contradiction. □

(4.5) Theorem. *Let $H \subset G$ be a closed subgroup. Then each irreducible representation U of H is contained in the restriction to H of an irreducible representation of G.*

Proof. Let V be a faithful representation of G and hence of H. By (4.4), U is contained in some $V(k, l)$ and hence in one of the irreducible G-summands of $V(k, l)$. □

(4.6) Theorem. *Every closed subgroup H of G appears as the isotropy group of an element of some G-module.*

PROOF. Using the descending chain property it suffices to show that, given $g \in G \backslash H$, there exists a G-module V and an element $v \in V$ such that $v \in V^H$ and $gv \neq v$.

By the theorem of Peter and Weyl there is a representative function f such that $f(H) \subset [0, 2]$ and $f(Hg^{-1}) \subset [3, 5]$. In fact, the function f is an approximation to a $u \in C^0(G, \mathbb{C})$ with the property that $u|H = 1$ and $u|Hg^{-1} = 4$. Let $F(g) = \int_H f(hg)\, dh$. Then $F(H) \subset [0, 2]$ and $F(Hg^{-1}) \subset [3, 5]$. Hence $F(1) \leq 2$ and $(g \cdot F)(1) = F(g^{-1}) \geq 3$. Consequently $g \cdot F \neq F$. The function F is itself a representative function, since it is contained in the G-subspace of $\mathscr{T}(G, \mathbb{C})$ generated by f. Thus we simply take V to be the G-submodule of $\mathscr{T}(G, \mathbb{C})$ generated by F and take $v = F$. □

(4.7) Exercises

1. Show that every compact Lie group is isomorphic to a closed subgroup of $O(n)$ or $U(n)$ for some n.

2. Show that compact Lie groups are real analytic manifolds (use I, (3.11), the fact that exp is analytic, and (4.1)). Show that representative functions are analytic.

3. Prove that an infinite compact Lie group has a countably infinite number of non-equivalent irreducible representations.

4. Suppose that all irreducible representations of G are one-dimensional. Show that G is abelian.

5. Let $H \subset G$ be a closed subgroup. Show that there exists a neighborhood U of 1 such that each closed sugroup K satisfying $H \subset K \subset HU$ is equal to H.

6. Let G be a compact group satisfying the descending chain condition for closed subgroups. Show that G is a Lie group.

7. Let $H \subset G$ be a closed subgroup and let W be an H-module. Show that there is a G-module V which contains $G \times_H W$ as a differentiable G-submanifold
$$(G \times_H W = G \times W/\sim, (g, w) \sim (gh, h^{-1}w) \text{ for } h \in H).$$

8. Prove or disprove: Let V be a G-module and $H \subset G$ a closed subgroup. Then there exists an isotropy group $K = \{g \in G \mid gv = v\}$ for some $v \in V$ with the property that $V^H = V^K$. In other words, is there a point $v \in V^H$ such that $gv = v$ if and only if g operates as the identity on V^H?

5. Generalizations of the Theorem of Peter and Weyl

We will only consider compact groups G and will discuss two types of generalizations of the theorem of Peter and Weyl. First, we give more general conditions under which the finite-dimensional G-submodules of a continuous G-module A generate a dense subspace of A. Second, we show that any unitary representation may be decomposed into a Hilbert sum of isotypical pieces. The section following this one presents some applications.

It seems appropriate at this stage to take a look at some of the analytic methods of representation theory already mentioned in II, §1. In particular, we want to treat representations as modules over convolution algebras. Since we will need to integrate vector-valued continuous functions, we start by collecting the necessary tools.

Let A be an Hausdorff locally convex complete vector space over $\mathbb{C}$. Recall that a locally convex space has a topology definable through a family of seminorms $p: A \to \mathbb{R}$. Let $C^0(G, A)$ be the vector space of continuous functions $G \to A$ endowed with the compact-open topology (see (5.11), Ex. 1 and 2). Then invariant integration enjoys the following properties:

(5.1) Invariant Integration. *There exists a continuous linear map, called* ***integration***,

$$\int : C^0(G, A) \to A, \qquad f \mapsto \int_G f = \int_G f(g)\, dg$$

which is both left and right invariant and which satisfies

(i) $\int a\, dg = a$ *for all* $a \in A$,
(ii) $\int f(g)\, dg = \int f(g^{-1})\, dg$,
(iii) $p(\int_G f) \le \int_G pf$ *for each continuous seminorm* $p: A \to \mathbb{R}$ *and any* $f \in C^0(G, A)$, *and*
(iv) *if* $L: A_1 \to A_2$ *is a continuous linear map between locally convex complete Hausdorff spaces, then* $L \int f = \int Lf$ *for all* $f \in C^0(G, A_1)$.

PROOF. The case where A is a Hilbert space has been treated in the remarks after theorem I, (5.13). In general, one may approximate continuous functions by A-valued step functions to construct the integral—see Lang [1], for example. □

Next, let $C^0(G) = C^0(G, \mathbb{C})$ be the Banach space of continuous functions $f: G \to \mathbb{C}$ with the supremum norm $|f|$. There is a convolution product II, (1.3) on this space

$$(f_1, f_2) \mapsto f_1 * f_2, \qquad (f_1 * f_2)(g) = \int_G f_1(gh^{-1}) f_2(h)\, dh.$$

The next lemma states some of the properties of this product.

(5.2) Lemma. *The convolution product makes* $C^0(G)$ *into a Banach algebra, i.e.,* $(f_1, f_2) \mapsto f_1 * f_2$ *is bilinear and associative and satisfies* $|f_1 * f_2| \le |f_1| \cdot |f_2|$.

Proof. The proof is simply a matter of verification and is left to the reader in (5.11), Ex. 3. Note that $C^0(G)$ does not, in general, have a unit. □

Now let A be a locally convex complete complex vector space as above, together with a continuous action $G \times A \to A$ such that each left translation is linear—in other words, a ***continuous G-module***. Our aim is to make A into a module over the convolution algebra $C^0(G)$. Define

$$C^0(G) \times A \to A, \qquad (f, a) \mapsto f * a \quad \text{by} \quad f * a = \int_G f(g)ga\,dg.$$

Note that $g \mapsto f(g)ga$ is a continuous map from G to A, so $f * a$ exists by (5.1). Some algebraic properties of this operation are stated in the next lemma.

(5.3) Lemma. *The map $(f, a) \mapsto f * a$ is bilinear and satisfies*

$$(f_1 * f_2) * a = f_1 * (f_2 * a).$$

Hence it makes A into a module over $C^0(G)$.

Proof. Again, verification is left to the reader. □

A special case arises when A is a complex Hilbert space with inner product $\langle -, - \rangle$, and the left translations of G are unitary. In this situation we have:

(5.4) Lemma $\langle f * a, b\rangle = \langle a, f^{\sim} * b\rangle$ *for $a, b \in A$, where $f^{\sim}(g) = \overline{f(g^{-1})}$.*

Proof. This time we write out the details. Using (5.1)(iv) and (ii),

$$\left\langle \int f(g)ga\,dg, b \right\rangle = \int \langle f(g)ga, b\rangle\,dg$$

$$= \int \langle a, \overline{f(g)}g^{-1}b\rangle\,dg = \int \langle a, \overline{f(g^{-1})}gb\rangle\,dg. \qquad \square$$

The $C^0(G)$-module structure of A is connected with the action of G on A as follows:

(5.5) Lemma. *For $u \in C^0(G)$, $a \in A$, and $g \in G$ we have $g(u * a) = (g \cdot u) * a$, where $(g \cdot u)(x) = u(g^{-1}x)$.*

Proof. $g(u * a) = g \int u(x)xa\,dx = \int u(x)gxa\,dx = \int u(g^{-1}x)xa\,dx = (g \cdot u) * a.$ □

(5.6) Corollary. *Let u be a representative function. Then $u * a$ is contained in a finite-dimensional G-subspace of A.*

Proof. See (5.11), Ex. 4. □

The next theorem is the first main result of this section.

(5.7) Theorem. *Let A be a continuous G-module as defined above, and let A_s be the subspace generated by the finite-dimensional G-subspaces of A. Then A_s is dense in A. (Also see (5.11), Ex. 5.)*

PROOF. First, note that the result is plausible because of (5.6), which says that the operation $C^0(G) \times A \to A$ induces a map $\mathscr{T}(G, \mathbb{C}) \times A \to A_s$. The actual strategy for the proof involves first choosing a function $v \in C^0(G)$ which acts as an approximate identity—i.e., approximates the Dirac measure—and then further approximating v by a representative function u.

Let $V = \bigcap_{i=1}^n V(p_i, a_i)$ be a basic neighborhood of 0 in A, where $p_1, \ldots, p_n$ are continuous seminorms and $V(p_i, a_i) = \{x \in A \mid p_i(x) < a_i\}$. Let $a \in A$ be given. By continuity of the G-action, there is a neighborhood U of 1 in G such that $ga - a \in \frac{1}{2}V$ for all $g \in U$. Let $v \in C^0(G, \mathbb{R})$ be a nonnegative function with support in U and $\int v(g)\, dg = 1$. Then $v * a - a = \int v(g)(ga - a)\, dg$ by (5.1)(i), and $v * a - a \in \frac{1}{2}V$ by (5.1)(iii).

Now integration $\int : C^0(G, A) \to A$ is continuous, so there is a neighborhood W of the function $g \mapsto v(g)ga$ which integration maps into $V + a$. Scalar multiplication $\mathbb{C} \times A \to A$ induces another continuous map

$$\lambda: C^0(G, \mathbb{C}) \times C^0(G, A) \to C^0(G, A)$$

with $\lambda(w, f)(g) = w(g)f(g)$. Thus if $r_a(g) = ga$ as usual, then $\lambda(v, r_a)$ is the function $g \mapsto v(g)ga$, and hence $\lambda(v, r_a) \in W$. From the continuity of λ, and from the theorem of Peter and Weyl, it follows that we may find a $u \in \mathscr{T}(G, \mathbb{C})$ such that $\lambda(u, r_a) \in W$. But this implies that $u * a \in V + a$, so $u * a - a \in V$. By (5.6), $u * a \in A_s$, so we have succeeded in approximating a by an element of A_s. □

(5.8) Corollary. *Let A be a Hausdorff locally convex complete G-module (e.g., a Banach or Hilbert space). If A is irreducible, then A is finite-dimensional.* □

We now consider continuous unitary representations of G on an Hilbert space H. We decompose H as a $C^0(G)$-module by using idempotent elements of the convolution algebra $C^0(G)$ which come from the orthogonality relations. Everything we need was verified in II, (4.16): Let χ be an irreducible character and set $e_\chi = \dim \chi \cdot \bar{\chi}$. Then by II, (4.16)

(5.9)
$$e_\chi * e_\chi = e_\chi, \qquad e_\chi * e_\psi = 0 \quad \text{if } \chi \neq \psi.$$

Thus the e_χ form a system of orthogonal idempotent elements of $C^0(G)$, and to this system there corresponds a decomposition of the $C^0(G)$-module H.

Define

$$P_\chi: H \to H \quad \text{by} \quad v \mapsto e_\chi * v.$$

(5.10) Theorem.

(i) *P_χ is an orthogonal projection operator onto a subspace H_χ.*
(ii) *H_χ and H_ψ are orthogonal if $\chi \neq \psi$.*
(iii) *H is the Hilbert sum of the H_χ for $\chi \in \mathrm{Irr}(G, \mathbb{C})$.*
(iv) *H_χ is the smallest closed subspace of H containing all irreducible subspaces with character χ.*

PROOF. P_χ is a projection operator by (5.9), while (5.4) and the equation $e_\chi = e_\chi^\sim$ imply that $\langle P_\chi v, w\rangle = \langle v, P_\chi w\rangle$ and hence that P_χ is an orthogonal projection. Moreover,

$$\langle P_\chi v, P_\psi w\rangle = \langle v, P_\chi P_\psi w\rangle,$$

and $P_\chi P_\psi = 0$ for $\chi \neq \psi$ by (5.9). Thus H_χ and H_ψ are orthogonal, and we have verified (i) and (ii).

To prove (iii), note that the convolutions $\varphi * v$ with $\varphi \in \mathscr{T}(G, \mathbb{C})$ and $v \in H$ are dense in H due to (5.7). Also, if φ is a representative function for the irreducible representation V with character χ, then $\varphi * e_\chi = e_\chi * \varphi = \varphi$ by II, (4.16). As a consequence, the $\varphi * v$ are in H_χ for each such φ. Thus $\bigoplus_\chi H_\chi, \chi \in \mathrm{Irr}(G, \mathbb{C})$, is dense in H.

As for (iv), if W is an irreducible G-submodule of H with character $\psi \neq \chi$, then $e_\chi * w = 0$ for $w \in W$ by II, (4.16). Hence H_χ can only contain irreducible submodules with character χ, and the result follows from II, §2. □

(5.11) Exercises

1. Let X and Y be topological spaces, and let Y^X be the set of continuous maps from X to Y. The compact-open topology on Y^X is determined by a subbasis consisting of the sets

$$W(K, U) = \{f \mid f(K) \subset U\}, \qquad K \subset X \text{ compact}, \quad U \subset Y \text{ open}.$$

   (i) Given $f: X \times Y \to Z$, define the adjoint map $\bar{f}: X \to Z^Y$ by $\bar{f}(x)(y) = f(x, y)$. Show that $\bar{f}$ is continuous if f is continuous.
   (ii) As in (i), $f \mapsto \bar{f}$ defines a map $\alpha: Z^{X \times Y} \to (Z^Y)^X$. Show that α is continuous if X is Hausdorff, that α is surjective if Y is locally compact, and that α is an embedding if X and Y are both Hausdorff. Thus, if X and Y are Hausdorff and Y is locally compact, then α is a homeomorphism.
   (iii) Show that the evaluation map $Y^X \times X \to Y$, $(f, x) \mapsto f(x)$, is continuous if X is locally compact.
   (iv) Show that $(Y \times Z)^X$ and $Y^X \times Z^X$ are canonically isomorphic if X is Hausdorff.
   (v) Show that a continuous map $f: X \to Y$ induces continuous maps $X^Z \to Y^Z$, $g \mapsto f \circ g$, and $Z^Y \to Z^X$, $g \mapsto g \circ f$.

   For details see Hu [1], III.9.

2. Show that $C^0(G, A)$ is a topological vector space.

3. Demonstrate the associativity of the convolution product in lemma (5.2). Use Fubini's theorem and invariance of the integral.

4. Convince yourself that (5.6) is true.

5. Prove (5.7) for continuous real A-modules.

6. Let A be a continuous G-module and $B \subset A$ a closed subspace such that $\varphi * w \in B$ whenever $\varphi \in C^0(G)$ and $w \in B$. Show that B is a G-subspace of A. Use Dirac approximations.

7. Show that $G \times C^0(G, A) \to C^0(G, A)$, $(g, f) \mapsto g \cdot f$, $(g \cdot f)(x) = f(g^{-1}x)$, is continuous.

8. Let M be a smooth compact G-manifold, and let $C^\infty(M, \mathbb{R}^k)$ be the space of smooth maps $M \to \mathbb{R}^k$ with the C^∞ topology. Thus two functions are "close" if their first n derivatives are "close" for a "large" n. This space is a Fréchet space, and it is well known (and not too difficult to prove) that the set of embeddings $M \to \mathbb{R}^k$ forms a dense open subspace of $C^\infty(M, \mathbb{R}^k)$ if k is large enough ($k \geq 2 \dim M + 1$), see, e.g., Narasimhan [1], Th. 2.15.8, p. 146. Using all this, prove that there is an equivariant embedding of M into a finite-dimensional G-module.

6. Induced Representations

In this section we describe an important class of infinite-dimensional representations which may be studied using the results in §5. These representations arise by considering representations of a closed subgroup H of the compact Lie group G. Each representation of H induces a representation (the induced representation) of G as follows.

We start with a given continuous representation of H on some topological K-vector space E. Let

$$i_H^G E = iE$$

be the vector space of all continuous functions $f\colon G \to E$ satisfying

$$f(gh) = h^{-1}f(g), \qquad h \in H, \quad g \in G.$$

We endow iE with the compact-open topology and define a G-action on iE by

$$(g \cdot f)(u) = f(g^{-1}u).$$

The reader may verify that this action is continuous, cf. (5.11), Ex. 2 and 7. This G-module (representation) $i_H^G E$ is called the ***induced G-module (representation)***.

There is another more conceptual definition of iE. Let $G \times_H E$ be the quotient space of $G \times E$ under the equivalence relation $(g, v) \sim (gh, h^{-1}v)$ for $h \in H$. The map $(g, v) \mapsto gH$ induces a projection $p\colon G \times_H E \to G/H$ which is a continuous G-map. In fact, p is a G-vector bundle with fiber $p^{-1}\{x\}$ isomorphic to E for every $x \in G/H$. And given a function $f\colon G \to E$ with $f(gh) = h^{-1}f(g)$ as above, we obtain a section $s_f\colon G/H \to G \times_H E$ of p (i.e., $ps_f = \mathrm{id}$) defined by $s_f(gH) = (g, f(g))$.

(6.1) Lemma. *The map $f \mapsto s_f$ is an isomorphism between $i_H^G E$ and the space of continuous sections of p.*

PROOF. It is clear that s_f is continuous, so we need to construct an inverse to $f \mapsto s_f$. Since the diagram

$$\begin{array}{ccc} G \times E & \longrightarrow & G \times_H E \\ {\scriptstyle \mathrm{pr}_1}\downarrow & & \downarrow{\scriptstyle p} \\ G & \longrightarrow & G/H \end{array}$$

is a pullback diagram, any section s of p induces a section S of pr_1. We thus define $f_s = \mathrm{pr}_2 \circ S$ and check that this is in iE. The map $s \mapsto f_s$ is the desired inverse; see (6.5), Ex. 1. □

The G-action on iE corresponds to the action on sections defined by $(g \cdot s)(x) = gs(g^{-1}x)$.

The main property of this construction is the following result, known as ***Frobenius reciprocity***. It says that the functor $E \mapsto i_H^G E$ is right adjoint to the functor res_H^G which assigns to every representation of G its restriction to H.

(6.2) Proposition. *Let E be an H-module and V a G-module over K. Then there is a canonical isomorphism*

$$\mathrm{Hom}_G(V, i_H^G E) \cong \mathrm{Hom}_H(\mathrm{res}_H^G V, E).$$

PROOF. Of course, Hom_G denotes continuous linear G-maps. To prove this result, we will simply construct inverse isomorphisms between the two vector spaces in question.

Let $F: V \to iE$ be given. The evaluation map $\eta: iE \to E$, $f \mapsto f(1)$, is a continuous H-map. The first of our isomorphisms is the map $F \mapsto \eta \circ F$.

Now let $f: V \to E$ be given. We may then define a map F from V into the space of continuous functions $G \to E$ by $F(v)(g) = f(g^{-1}v)$ for $g \in G$ and $v \in V$. We have $F(v)(gh) = f(h^{-1}g^{-1}v) = h^{-1}f(g^{-1}v) = h^{-1}F(v)(g)$. Thus F is a G-map $V \to iE$ if f is an H-map. If f is continuous, then so is the map $V \times G \to E$ defined by $(v, g) \mapsto f(g^{-1}v)$. Hence the adjoint F is also continuous, and the map $f \mapsto F$ is our second isomorphism.

All that remains is for the reader to check that these two isomorphisms are inverses. □

(6.3) Example. Let E be the trivial one-dimensional representation of H over K. Then $i_H^G E \cong C^0(G/H, K)$, the space of continuous functions $G/H \to K$ with G-action given by left translation. In this case Frobenius reciprocity says that the multiplicity of $V \in \mathrm{Irr}(G, \mathbb{C})$ in $C^0(G/H, \mathbb{C})$ is equal to $\dim_{\mathbb{C}} V^H$.

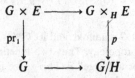

In particular, V appears in $C^0(G/H, \mathbb{C})$ if and only if $\operatorname{res}_H^G V$ contains the trivial representation. The case $G/H = SO(n+1)/SO(n) = S^n$ leads to spherical functions, see II, §5 for $n = 2$.

If E is finite-dimensional, then $i_H^G E$ is infinite-dimensional unless H has finite index in G. Also, any $V \in \operatorname{Irr}(G, K)$ appears with finite multiplicity, although this multiplicity is, in general, hard to compute. We may also choose a norm on E, in which case the compact-open topology on iE is the same as the sup-norm topology. Thus iE becomes a Banach space with this norm, and the generalized theorem of Peter and Weyl is applicable. This shows that the locally finite part iE_s of iE is dense in iE.

The algebraic description of the space of representative functions $\mathcal{T}(G, \mathbb{C})$ given in §1 may be generalized to iE_s as follows. For $V \in \operatorname{Irr}(G, V)$, we have the map

$$d_V\colon \operatorname{Hom}_H(V, E) \otimes V \to iE$$

sending $\varphi \otimes v$ to $g \mapsto \varphi(g^{-1}v)$. It is easy to check that $d_V(\varphi \otimes v)$ is actually in iE and that d_V is G-equivariant. Frobenius reciprocity tells us that $\dim \operatorname{Hom}_H(V, E)$ is the multiplicity of V in iE, and the map d_V exhibits the V-isotypical part of iE_s. This leads to an isomorphism

(6.4) $$\bigoplus_{V \in \operatorname{Irr}(G,\mathbb{C})} \operatorname{Hom}_H(V, E) \otimes V \to iE_s.$$

A similar isomorphism exists for real representations.

We mention in closing that induced representations may also be understood in the framework of Hilbert spaces, see, e.g., Robert [1], Ch. I.8 or Pukansky [1].

The fundamental role that induced representations play for finite groups can be seen by looking at, for example, Curtis and Reiner [1] or Serre [4]; see, for example, the Brauer induction theorem. Generalities about induced representations for locally compact groups in Warner [1], Ch. 5. Induced representations can also be studied in the context of representation rings, see Bott [2] and Segal [1].

(6.5) Exercises

1. Fill in the details in the proof of (6.1): The diagram is a pullback, what is an induced section, $f_s \in iE$, and $s \mapsto s_f$, $s \mapsto f_s$ are inverses.

2. Fill in the details in the proof of (6.2).

3. Show that (6.4) is an isomorphism.

4. What is the analogue of (6.4) for real representations?

5. Let E be a finite-dimensional H-module. Let $i_\infty E \subset i_H^G E$ be the G-subspace of C^∞ sections of $G \times_H E \to G/H$. Show that $iE_s \subset i_\infty E$.

7. Tannaka–Krein Duality

Duality theory is primarily concerned with reconstructing the compact Lie group G from the algebra $\mathscr{T}(G, \mathbb{R})$ of representative functions (or from the category of representations). The strategy for reconstructing G goes as follows: Let $G_{\mathbb{R}}$ be the set of all $\mathbb{R}$-algebra homomorphisms $\mathscr{T}(G, \mathbb{R}) \to \mathbb{R}$. Each $g \in G$ determines an evaluation homomorphism $e_g\colon \mathscr{T}(G, \mathbb{R}) \to \mathbb{R}$, $f \mapsto f(g)$. We thus obtain a map

$$i\colon G \to G_{\mathbb{R}}, \qquad g \mapsto e_g$$

which we will show to be bijective. In fact, $G_{\mathbb{R}}$ may be naturally considered to be a topological group, and i turns out to be a topological isomorphism.

It is an elementary result of point set topology that the algebra homomorphisms $C(X, \mathbb{R}) \to \mathbb{R}$ for a compact space X are precisely the evaluations at points of X (see, e.g., Lang [1], p. 52). However, in our case it is not *a priori* clear that a homomorphism $\mathscr{T}(G, \mathbb{R}) \to \mathbb{R}$ extends to $C(G, \mathbb{R})$.

Now, the $\mathscr{T}(G, K)$ have more structure than just that of a K-algebra ($K = \mathbb{R}, \mathbb{C}$). This additional structure may be used to make the set G_K of K-algebra homomorphisms $\mathscr{T}(G, K) \to K$ into a group. For this we need an elementary lemma whose proof is relegated to (7.16), Ex. 1.

(7.1) Lemma. *The K-algebra homomorphism*

$$t\colon \mathscr{T}(G, K) \otimes_K \mathscr{T}(H, K) \to \mathscr{T}(G \times H, K)$$

sending $u \otimes v$ to $(g, h) \mapsto u(g)v(h)$ is an isomorphism.

Our intention is to use $\mathscr{T}(G, K)$ as a kind of model for the group G. We are thus forced to translate the group axioms into statements about this algebra. For instance, group multiplication $G \times G \to G$, $(g, h) \mapsto gh$, induces a homomorphism $\mathscr{T}(G, K) \to \mathscr{T}(G \times G, K)$. The reader should check that the map $(g, h) \mapsto f(gh)$ is really contained in $\mathscr{T}(G \times G, K)$ if $f \in \mathscr{T}(G, K)$. Using this and (7.1), we obtain a K-algebra homomorphism

(7.2) $$d\colon \mathscr{T}(G, K) \to \mathscr{T}(G \times G, K) \cong \mathscr{T}(G, K) \otimes_K \mathscr{T}(G, K)$$

called ***comultiplication***.

Similarly, the map $g \mapsto g^{-1}$ induces a K-algebra homomorphism

(7.3) $$c\colon \mathscr{T}(G, K) \to \mathscr{T}(G, K), \qquad \mathrm{c}(f)(g) \mapsto f(g^{-1}),$$

called ***coinverse***, and evaluation at the unit element $1 \in G$ gives rise to the algebra homomorphism

(7.4) $$\varepsilon\colon \mathscr{T}(G, K) \to K$$

called ***counit***. The prefix "co" here and in what follows refers to the fact that these operations and their governing axioms are obtained by taking the analogous defining diagrams for groups and reversing all the arrows. Thus the homomorphisms (7.2)–(7.4) satisfy properties (7.5)–(7.7) below which are simply the translation of the standard axioms for a group, see (7.16), Ex. 2. The associativity of group multiplication yields the ***coassociativity*** of d:

(7.5) $$(d \otimes \mathrm{id}) \circ d = (\mathrm{id} \otimes d) \circ d.$$

The counit satisfies:

(7.6) $$(\varepsilon \otimes \mathrm{id}) \circ d = \mathrm{id} = (\mathrm{id} \otimes \varepsilon) \circ d$$

and coinverse satisfies

(7.7) $$m \circ (c \otimes \mathrm{id}) \circ d = \eta \circ \varepsilon$$

In (7.7), the m and η refer to the multiplication and unit in $\mathscr{T}$ as a K-algebra which we view as K-algebra homomorphisms

(7.8) $$m\colon \mathscr{T} \otimes_K \mathscr{T} \to \mathscr{T}, \qquad f \otimes h \mapsto f \cdot h$$

and

(7.9) $$\eta\colon K \to \mathscr{T}, \qquad 1 \mapsto 1.$$

The algebra $\mathscr{T}(G, K)$, together with comultiplication d and the counit ε, is an example of what is called a ***Hopf algebra***. The coinverse c lends additional structure to this Hopf algebra.

With the aid of (7.2) we may multiply two homomorphisms $s, t \in G_K$ as follows:

(7.10) $$s \cdot t\colon \mathscr{T} \xrightarrow[d]{} \mathscr{T} \otimes_K \mathscr{T} \xrightarrow[s \otimes t]{} K \otimes K \cong K.$$

This map is again a K-algebra homomorphism.

(7.11) Proposition. *The composition law* (7.10) *makes* G_K *into a group.*

PROOF. Associativity follows from (7.5), and by (7.6) the counit ε serves as the unit element of G_K. Finally, the element sc is inverse to s:

$$sc \cdot s = m(sc \otimes s)d = m(s \otimes s)(c \otimes \mathrm{id})d = sm(c \otimes \mathrm{id})d = s\eta\varepsilon = \varepsilon. \quad \square$$

(7.12) Proposition. *The map* $i\colon G \to G_K, g \mapsto e_g$, *is an injective homomorphism.*

PROOF. If $f \in \mathscr{T}(G, K)$ and $d(f) = \sum_j f_j' \otimes f_j''$, then

$$f(gh) = \sum_j f_j'(g) f_j''(h), \quad \text{and}$$

$$(s \cdot t)(f) = \sum_j s(f_j') t(f_j'')$$

by the definitions of d and $s \cdot t$. Hence

$$(i(g) \cdot i(h))(f) = \sum_j f_j'(g) f_j''(h) = f(gh) = i(gh)(f),$$

showing i to be a homomorphism. If g is in the kernel of i, then $f(g) = f(1)$ for all $f \in \mathcal{T}(G, K)$. Since the f separate points (Peter and Weyl), we conclude that $g = 1$. □

This is a good juncture at which to convince oneself of the analogy between the map $i: G \to G_{\mathbb{R}}$ and the more familiar embedding of a vector space in its double dual. This lends credence to the hope that i will be an isomorphism.

Next, we define a topology for G_K. We take the weakest topology for which all the evaluation maps

$$\lambda_f: G_K \to K, \qquad s \mapsto s(f)$$

are continuous. This topology is characterized by the following property: A map $\varphi: X \to G_K$ from a topological space X into G_K is continuous if and only if all compositions $\lambda_f \varphi$ are continuous.

Here are two immediate consequences of this definition.

(7.13) Proposition.

(i) *G_K is a topological group.*
(ii) *$i: G \to G_K$ is continuous.*

PROOF. (i) We have to show that the composition of multiplication $G_K \times G_K \to G_K$ with λ_f is continuous. If $d(f) = \sum_j f_j' \otimes f_j''$, then this map reduces to $(s, t) \mapsto \sum_j s(f_j')t(f_j'')$ and is therefore continuous. The inverse in G_K may be treated similarly.

(ii) $\lambda_f i$ is just the map $g \mapsto f(g)$, which certainly is continuous. □

The group $G_{\mathbb{R}}$ is a compact Lie group. This is shown by identifying it with a compact subgroup of some $O(n)$ (see I, (3.12)). The proof uses the following construction which will also be used in the next section where we describe $G_{\mathbb{C}}$ as an algebraic group.

Let $r: G \to GL(n, K)$, $g \mapsto (r_{ij}(g))$ be a matrix representation. This gives us a map

$$r_K: G_K \to GL(n, K), \qquad s \mapsto (s(r_{ij})).$$

(7.14) Proposition.

(i) *r_K is a continuous homomorphism making the diagram*

$$\begin{array}{ccc} G & \xrightarrow{\ i\ } & G_K \\ & {\scriptstyle r}\searrow \quad \swarrow{\scriptstyle r_K} & \\ & GL(n, K) & \end{array}$$

commutative.

(ii) *If the r_{ij} generate $\mathscr{T}(G, K)$ as a K-algebra (e.g., if $r: G \to \mathrm{GL}(n, \mathbb{R})$ is faithful by* (4.3)) *then r_K is injective.*
(iii) *If r maps G into $\mathrm{O}(n)$, then $r_{\mathbb{R}} G_{\mathbb{R}} \subset \mathrm{O}(n)$ as a closed subgroup.*

PROOF. Suppose $s, t \in G_K$ are given. Since

$$r_{ij}(gh) = \sum_k r_{ik}(g) r_{kj}(h),$$

we get

$$r_K(s \cdot t) = ((s \cdot t)(r_{ij})) = \left(\sum_k s(r_{ik}) t(r_{kj})\right) = r_K(s) r_K(t).$$

Thus r_K is a homomorphism. The map r_K is continuous if and only if all the maps $s \mapsto s(r_{ij})$, $i, j = 1, \ldots, n$, are continuous, but these are continuous by the definition of the topology on G_K. This shows (i).

If the r_{ij} generate $\mathscr{T}(G, K)$, a homomorphism $s: \mathscr{T}(G, K) \to K$ is determined by its values $s(r_{ij})$, so (ii) is clear. For (iii) we note that if $r(G) \subset \mathrm{O}(n)$, then $(r_{ij})^t(r_{ij}) = E$. It follows that $(s(r_{ij}))^t(s(r_{ij})) = E$, so $r_{\mathbb{R}}(s) = (s(r_{ij})) \in \mathrm{O}(n)$. It remains to show that $r_{\mathbb{R}} G_{\mathbb{R}}$ is closed, which we do by showing that $G_{\mathbb{R}}$ is, in fact, compact.

For this purpose we consider any faithful representation $r: G \to \mathrm{O}(n)$. Since the r_{ij} generate $\mathscr{T}(G, K)$ by (4.3), a typical element $f \in \mathscr{T}(G, K)$ is a polynomial $f = P(r_{ij})$ in n^2 variables corresponding to the n^2 entries of a matrix in $\mathrm{O}(n)$. Since $\mathrm{O}(n)$ is compact, the image of $\mathrm{O}(n)$ in $\mathbb{R}$ under P is contained in some compact interval I_f. We thus obtain a continuous map

$$\Lambda = (\lambda_f): G_{\mathbb{R}} \to \prod_f I_f, \qquad f \in \mathscr{T}(G, \mathbb{R}),$$

of $G_{\mathbb{R}}$ into a compact space. An element $(t_f) \in \prod_f I_f$ is in the image of Λ if and only if it satisfies the conditions of an algebra homomorphism, i.e., $t_1 = 1$, $t_{fg} = t_f \cdot t_g$, and $t_{rf} = r \cdot t_f$ for $1, f, g \in \mathscr{T}(G, \mathbb{R})$ and $r \in \mathbb{R}$. Thus the image of $G_{\mathbb{R}}$ is closed and hence compact. Since Λ is clearly an embedding, $G_{\mathbb{R}}$ is compact. □

(7.15) Theorem. *The duality map $i: G \to G_{\mathbb{R}}$ of* (7.12) *is an isomorphism of Lie groups.*

PROOF. The major step is showing that i induces an isomorphism of the algebras of representative functions.

Let $f \in \mathscr{T}(G, \mathbb{R})$. Then $\lambda_f: G_{\mathbb{R}} \to \mathbb{R}$ is a representative function of $G_{\mathbb{R}}$, in fact, if f is a matrix coefficient of some representation r, then λ_f is the corresponding coefficient of $r_{\mathbb{R}}$. The map

$$\lambda: \mathscr{T}(G, \mathbb{R}) \to \mathscr{T}(G_{\mathbb{R}}, \mathbb{R}), \qquad f \mapsto \lambda_f,$$

is a homomorphism of algebras, as the reader may verify. If $f \in \mathscr{T}(G, \mathbb{R})$ and $t \in G_{\mathbb{R}}$, then $t \cdot \lambda_f$ maps s to

$$\lambda_f(s \cdot t) = (s \cdot t)(f) = \sum_j s(f_j')t(f_j'') = \sum_j \lambda_{f_j'}(s)\lambda_{f_j''}(t),$$

showing that the image of λ is $G_{\mathbb{R}}$-invariant. The functions λ_f separate points in $G_{\mathbb{R}}$, so by (2.7) the image of λ is dense in the supremum norm topology. By (1.4)(iii), the image of λ must be all of $\mathscr{T}(G_{\mathbb{R}}, \mathbb{R})$. Since the composition of λ with the map

$$i^*: \mathscr{T}(G_{\mathbb{R}}, \mathbb{R}) \to \mathscr{T}(G, \mathbb{R}), \qquad s \mapsto si,$$

is the identity, both λ and i^* must be isomorphisms.

Since $C(G_{\mathbb{R}}, \mathbb{R})$ and $C(G, \mathbb{R})$ are the completions of $\mathscr{T}(G_{\mathbb{R}}, \mathbb{R})$ and $\mathscr{T}(G, \mathbb{R})$, we may conclude that $i^*: C(G_{\mathbb{R}}, \mathbb{R}) \to C(G, \mathbb{R})$ is also an isomorphism. But this implies that $i: G \to G_{\mathbb{R}}$ is surjective, which together with (7.12) completes the proof. □

Theorem (7.15) is half of the duality theory. For the other half, one starts with a suitable Hopf algebra $\mathscr{T}$ and considers the group G of all algebra homomorphisms $\mathscr{T} \to \mathbb{R}$. One must show that the canonical map $\lambda: \mathscr{T} \to \mathscr{T}(G, \mathbb{R})$ is an isomorphism (see Hochschild [1], II.3, Theorem 3.5). Another presentation of duality is in Robert [1], I.9, and a more extensive treatment appears in Hewitt and Ross [2], §30. The original sources are Tannaka [1] and Kreĭn [1]. Our presentation is based on Hochschild [1].

(7.16) Exercises

1. Prove lemma (7.1) as follows: First show that $C(G, K) \otimes_K C(H, K) \to C(G \times H, K)$ is injective. To show surjectivity, let $f \in \mathscr{T}(G \times H, K)$ be given and let $S \subset \mathscr{T}(H, K)$ be generated by the functions $h \mapsto f(g, h)$. This space is finite-dimensional, so it has a basis $e_1, \ldots, e_n$ such that there are elements $h_1, \ldots, h_n \in H$ with $e_i(h_j) = \delta_{ij}$. Write $f(g, h) = \sum_i u_i(g)e_i(h)$ and show that $u_i \in \mathscr{T}(G, K)$.

2. Write the group axioms in terms of commutative diagrams and express (7.5)–(7.7) as diagrams. Use this to verify (7.5)–(7.7). Study the notion of an Hopf algebra and its relevance to the material in this section (Hochschild [2], Ch. I).

3. Show that $\mathscr{T}(S^1, \mathbb{C}) \cong \mathbb{C}[x, x^{-1}]$, where x corresponds to the inclusion $S^1 \subset \mathbb{C}$. Show that comultiplication is the map $\mathbb{C}[x, x^{-1}] \to \mathbb{C}[y, y^{-1}] \otimes \mathbb{C}[z, z^{-1}]$ determined by $x \mapsto y \otimes z$. Show that $(S^1)_{\mathbb{C}} \cong \mathbb{C}^*$, the multiplicative group $\mathbb{C}\backslash\{0\}$.

4. Show that $\mathscr{T}(S^1, \mathbb{R}) \cong \mathbb{R}[u, v]/(u^2 + v^2 - 1)$ using the standard representation $S^1 \subset O(2)$. Give an explicit isomorphism

$$\mathbb{C}[x, x^{-1}] \cong \mathbb{R}[u, v]/(u^2 + v^2 - 1) \otimes_{\mathbb{R}} \mathbb{C}$$

(cf. (1.8), Ex. 6). Describe the comultiplication in $\mathscr{T}(S^1, \mathbb{R})$.

5. Let $G'_{\mathbb{R}} \subset G_{\mathbb{C}}$ be the subgroup of $\mathbb{C}$-algebra homomorphisms $s: \mathscr{T}(G, \mathbb{C}) \to \mathbb{C}$ satisfying $s(\bar{f}) = \overline{s(f)}$.
   Show:
   (i) The restriction of $s \in G'_{\mathbb{R}}$ to $\mathscr{T}(G, \mathbb{R})$ lies in $G_{\mathbb{R}}$. This gives a homomorphism $\rho: G'_{\mathbb{R}} \to G_{\mathbb{R}}$.
   (ii) ρ is an isomorphism. Use (1.8), Ex. 6.
   (iii) $G'_{\mathbb{R}} \subset G_{\mathbb{C}}$ is closed.
   (iv) The image of G under the homomorphism $i: G \to G_{\mathbb{C}}$ is contained in $G'_{\mathbb{R}}$ and $\rho i: G \to G_{\mathbb{R}}$ is the isomorphism of (7.15). It follows that the image of i is all of $G'_{\mathbb{R}}$.

6. Show that $\mathscr{T}(\mathrm{U}(n), \mathbb{C}) \cong \mathbb{C}[X_{11}, X_{12}, \ldots, X_{nn}, t]/I$, where I is the ideal generated by $t \cdot \det(X_{ij}) - 1$. Show that the comultiplication is given by $d(X_{ij}) = \sum_k X_{ik} \otimes X_{kj}$. *Hint*: The reader may find it convenient to apply a dimension argument using the material in the next section.

7. Let Γ be the group of those algebra automorphisms of $\mathscr{T}(G, \mathbb{C})$ which commute with the action of G on $\mathscr{T}(G, \mathbb{C})$ via left translation. Given $\sigma \in \Gamma$, define $w_\sigma \in G_{\mathbb{C}}$ by $w_\sigma(f) = (\sigma f)(e)$. Show that the map $\Gamma \to G_{\mathbb{C}}, \sigma \mapsto w_\sigma$, is an isomorphism.

8. Let $H \subset G$ be a closed subgroup. Let $\mathscr{T}(G/H, K) \subset \mathscr{T}(G, K)$ be the subring of those functions which are compositions $G \to G/H \to K$. Show that G/H may be identified with the algebra homomorphisms $\mathscr{T}(G/H, \mathbb{C}) \to \mathbb{R}$ (see Iwahori and Sugiura [1]).

9. Let $i: H \to G$ be a homomorphism of compact Lie groups inducing an isomorphism $i^*: \mathscr{T}(G, \mathbb{R}) \to \mathscr{T}(H, \mathbb{R})$. Show that i is an isomorphism. More precisely, if i^* is surjective than i is injective (since $\mathscr{T}(H, \mathbb{R})$ separates points) and if i^* is injective then i is surjective. *Hint*: If i is not surjective, use (4.2) or the proof of (4.6) to construct a nonzero $f \in \mathscr{T}(G, \mathbb{R})$ with $i^* f = 0$.

8. The Complexification of Compact Lie Groups

In the previous section we associated a commutative diagram

(8.1)

$$
\begin{array}{ccc}
G & \xrightarrow{\ i\ } & G_{\mathbb{C}} \\
& {\scriptstyle r}\searrow & \downarrow{\scriptstyle r_{\mathbb{C}}} \\
& & \mathrm{GL}(n, \mathbb{C})
\end{array}
$$

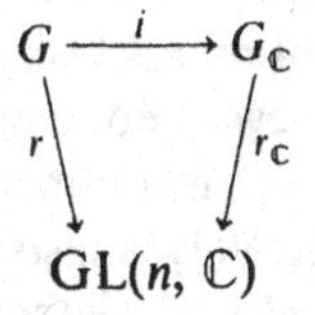

to each representation $r: G \to \mathrm{GL}(n, \mathbb{C})$. The group $G_{\mathbb{C}}$ consists of all $\mathbb{C}$-algebra homomorphisms $\mathscr{T}(G, \mathbb{C}) \to \mathbb{C}$ and $i(g)$ is the evaluation map at g. In this section we will show that $G_{\mathbb{C}}$ is a complex analytic Lie group with Lie algebra $L(G_{\mathbb{C}}) \cong \mathbb{C} \otimes_{\mathbb{R}} LG$ and that $r_{\mathbb{C}}$ is a holomorphic representation. In the process we essentially describe $G_{\mathbb{C}}$ as an algebraic group.

Let r be a representation such that the entries r_{kj} of r generate $\mathcal{T}(G, \mathbb{C})$ as a $\mathbb{C}$-algebra, for example, a faithful real representation, by (4.3). Then

$$\mathcal{T}(G, \mathbb{C}) \cong \mathbb{C}[X_{kj}]/I,$$

where I is the kernel of the map $\mathbb{C}[X_{kj}] \to \mathcal{T}(G, \mathbb{C})$, $X_{kj} \mapsto r_{kj}$.

At this point we recall some elementary notions from algebraic geometry. Suppose $J \subset \mathbb{C}[X_1, \ldots, X_n]$ is an ideal in a polynomial ring. Then the affine variety $\mathrm{V}(J)$ of J is defined to be the set of common zeros of the polynomials in J,

$$\mathrm{V}(J) = \{z \in \mathbb{C}^n \mid p(z) = 0 \text{ for all } p \in J\}.$$

Evaluating polynomials at $z \in \mathrm{V}(J)$ defines a $\mathbb{C}$-algebra homomorphism $\mathbb{C}[X_1, \ldots, X_n]/J \to \mathbb{C}$. On the other hand, it is easy to see that any such homomorphism is evaluation at the point of $\mathrm{V}(J)$ whose coordinates are the images of the $X_\nu + J$. If A is any finitely generated $\mathbb{C}$-algebra, we may express A as a quotient of the form $\mathbb{C}[X_1, \ldots, X_n]/J$, and the set of all $\mathbb{C}$-algebra homomorphisms $A \to \mathbb{C}$ provides an intrinsic description of the associated affine variety $\mathrm{V}(J)$.

We may apply these remarks to $\mathcal{T}(G, \mathbb{C}) \cong \mathbb{C}[X_{kj}]/I$. This gives us a bijection

$$\sigma\colon \mathrm{V}(I) \to G_{\mathbb{C}}$$

which sends z to the map $\sigma_z\colon p + I \mapsto p(z)$. The composition of σ with $r_{\mathbb{C}}$ is given by

$$z \mapsto \sigma_z \mapsto (\sigma_z(r_{kj})) = (X_{kj}(z)) = z,$$

so $r_{\mathbb{C}} \circ \sigma = \mathrm{id}$. This proves

(8.2) Proposition. *If $r\colon G \to \mathrm{GL}(n, \mathbb{C})$ is a representation such that the r_{kj} generate $\mathcal{T}(G, \mathbb{C})$, then $r_{\mathbb{C}}$ maps $G_{\mathbb{C}}$ bijectively onto $\mathrm{V}(I) \subset \mathrm{GL}(n, \mathbb{C}) \subset \mathbb{C}^{n \cdot n}$ with inverse σ. In particular, $r_{\mathbb{C}}$ is injective.* □

The set $\mathrm{V}(I)$ is a closed subgroup of $\mathrm{GL}(n, \mathbb{C})$ and is therefore a Lie group. In (7.13) we defined a topology on $G_{\mathbb{C}}$, and this coincides with the subspace topology induced by $r_{\mathbb{C}}$. Indeed, if the λ_f of §7 are continuous for $f = r_{kj}$, then all the λ_f are continuous because the r_{kj} generate $\mathcal{T}(G, \mathbb{C})$.

It is a general fact that, for every nonempty variety $\mathrm{V}(I)$, there is an open set $U \subset \mathbb{C}^{n \cdot n}$ such that $\mathrm{V}(I) \cap U$ is a nonempty analytic submanifold of U, see Mumford [1], Ch. I. Since $\mathrm{V}(I)$ is also a group, we conclude by homogeneity that all of $\mathrm{V}(I)$ is an analytic submanifold. Furthermore, group multiplication and inversion in $\mathrm{GL}(n, \mathbb{C})$ are globally defined rational maps. Their restrictions to $\mathrm{V}(I)$ are therefore certainly holomorphic.

We also point out that $G_{\mathbb{C}}$ possesses an involution defined by sending $s \in G_{\mathbb{C}}$ to the map

$$\bar{s}\colon \mathcal{T}(G, \mathbb{C}) \to \mathbb{C}, \qquad f \mapsto \overline{s(\bar{f})}.$$

It is easily checked that $\tilde{s} \in G_{\mathbb{C}}$, and $s \mapsto \tilde{s}$ is an automorphism. The fixed points form a set $G'_{\mathbb{R}}$ which we saw ((7.17), Ex. 5) to be canonically isomorphic to $G \cong G_{\mathbb{R}}$. In terms of algebraic geometry, $G'_{\mathbb{R}}$ consists of the real points of $G_{\mathbb{C}}$, cf. (8.8), Ex. 10.

Our next aim is to clarify the topological nature of $G_{\mathbb{C}}$. For this purpose we assume that $r: G \to \mathrm{GL}(n, \mathbb{C})$ is a *unitary* representation such that the r_{kj} generate $\mathscr{T}(G, \mathbb{C})$. If $A \in \mathrm{GL}(n, \mathbb{C})$, we define

$$\tilde{A} = {}^t\bar{A}^{-1}.$$

Then the map $s \mapsto \tilde{s}$ corresponds to the map $A \mapsto \tilde{A}$ via the embedding $r_{\mathbb{C}}$. In fact, applying s componentwise to the identity $(\bar{r}_{kj})^t(r_{kj}) = E$ yields $(s\bar{r}_{kj}) = {}^t(sr_{kj})^{-1}$, from which the identity $r_{\mathbb{C}}(\tilde{s}) = r_{\mathbb{C}}(s)^{\sim}$ follows. Since $A = \tilde{A}$ precisely if $A \subset \mathrm{U}(n)$, we find that $r_{\mathbb{R}} G_{\mathbb{R}} = \mathrm{U}(n) \cap r_{\mathbb{C}} G_{\mathbb{C}}$.

Let $\mathrm{P}(n)$ denote the set of positive-definite Hermitian $(n \times n)$-matrices. Then there is a homeomorphism

$$\mathrm{U}(n) \times \mathrm{P}(n) \to \mathrm{GL}(n, \mathbb{C}), \qquad (H, P) \mapsto HP,$$

see I, (1.16), Ex. 9. Let $\tilde{G} = r_{\mathbb{C}} G_{\mathbb{C}}$.

(8.3) Proposition.

(i) *If we express an $A \in \tilde{G}$ as $A = HP$ with $H \in \mathrm{U}(n)$ and $P \in \mathrm{P}(n)$, then both H and P are in $\tilde{G}$. Furthermore, the map*

$$(\tilde{G} \cap \mathrm{U}(n)) \times (\tilde{G} \cap \mathrm{P}(n)) \to \tilde{G}, \qquad (H, P) \mapsto HP$$

is a homeomorphism.

(ii) *$\tilde{G} \cap \mathrm{P}(n)$ is homeomorphic to a Euclidean space of dimension* $\dim G = \dim(\tilde{G} \cap \mathrm{U}(n))$.

(iii) *$\tilde{G} \cap \mathrm{U}(n)$ is a maximal compact subgroup of $\tilde{G}$.*

PROOF. (i) Let $A = HP$. Then there is a $U \in \mathrm{U}(n)$ such that $UPU^{-1} = D$, where D is a nonsingular diagonal matrix with nonnegative real entries. Thus $D = \exp(Z)$, where Z is a diagonal real matrix with diagonal $(a_1, \ldots, a_n)$.

Now $\tilde{G} \subset \mathrm{GL}(n, \mathbb{C})$ is an algebraic variety and the map $X_{kj} \mapsto (UXU^{-1})_{kj}$ defines an algebraic isomorphism $\mathbb{C}^{n \cdot n} \to \mathbb{C}^{n \cdot n}$. The image $U\tilde{G}U^{-1}$ of $\tilde{G}$ under this map is therefore an algebraic variety, say $V(J)$ for some ideal $J \subset \mathbb{C}[X_{kj}]$. If $A = HP \in \tilde{G}$, then ${}^t\bar{A} = \tilde{A}^{-1} \in \tilde{G}$, so ${}^t\bar{A}A = P^2 \in \tilde{G}$. Hence $D^2 \in U\tilde{G}U^{-1}$. If $Q(X_{kj}) \in J$, the substitution $X_{kj} \mapsto 0$ if $k \neq j$ and $X_{jj} \mapsto X_j$ yields a new polynomial $q(X_1, \ldots, X_n)$. Since $D^{2k} \in U\tilde{G}U^{-1}$ for $k \in \mathbb{Z}$, we have

$$q(\exp(2ka_1), \ldots, \exp(2ka_n)) = 0$$

for all $Q \in J$. It follows ((8.8), Ex. 8) that $q(\exp(ta_1), \ldots, \exp(ta_n)) = 0$ for all $Q \in J$. Setting $t = 1$, we conclude that $D \in V(J) = U\tilde{G}U^{-1}$.

This also shows that $P \in \tilde{G}$ and $H = AP^{-1} \in \tilde{G}$. Thus the map in (i) is a continuous bijection and, being the restriction of the homeomorphism $U(n) \times P(n) \to GL(n, \mathbb{C})$, it is also a homeomorphism.

(ii) Let L be the Lie algebra of $\tilde{G} \cap U(n)$. We show that the map $X \mapsto \exp(iX)$ establishes a homeomorphism $L \to \tilde{G} \cap P(n)$.

If $X \in L$, then $\exp(tX) \in \tilde{G} \cap U(n)$ for all $t \in \mathbb{R}$. Writing the variety $\tilde{G}$ as $V(I)$, let $p(X_{kj}) \in I$. Substituting the entries of $\exp(tX)$ for the X_{kj} yields an entire analytic function in t which vanishes for $t \in \mathbb{R}$ and hence for $t \in \mathbb{C}$. Thus $\exp(itX) \in \tilde{G}$. Since $L \subset LU(n)$ consists of skew-Hermitian matrices, the matrix itX is Hermitian, thus $\exp(itX)$ lies in $P(n)$. Hence $X \mapsto \exp(iX)$ defines a map $L \to \tilde{G} \cap P(n)$ which we must show to be a homeomorphism.

The proof of (i) shows that every $P \in \tilde{G} \cap P(n)$ has the form $P = \exp(Y)$, where Y is Hermitian and $\exp(tY) \in \tilde{G}$ for all $t \in \mathbb{C}$. Consequently the map in question is surjective. But then it must be a homeomorphism, since it is the restriction of the homeomorphism

$$SH(n) \to P(n), \qquad X \mapsto \exp(iX)$$

from the set $SH(n)$ of skew-Hermitian $(n \times n)$-matrices to $P(n)$.

(iii) $\tilde{G} \cap U(n)$ is closed in $U(n)$ and therefore compact. Were K a larger compact subgroup of $\tilde{G}$, it would contain an element of $\tilde{G} \cap P(n)$ other than E by (i). This is clearly impossible by (ii).

□

From the proof of (8.3) we see that the Lie algebra $L(\tilde{G})$ is the direct sum of its subspaces $L = L(\tilde{G} \cap U(n))$ and iL. We know that $\tilde{G} \cap U(n)$ is the image of $G_{\mathbb{R}}$ under $r_{\mathbb{R}}$, and hence isomorphic to G. Thus $L \cong L(G)$ and we obtain a linear isomorphism

$$\mathbb{C} \otimes_{\mathbb{R}} L(G) \cong \mathbb{C} \otimes_{\mathbb{R}} L \cong L \oplus iL = L(\tilde{G}) \cong L(G_{\mathbb{C}}).$$

(8.4) Proposition. *$\mathbb{C} \otimes_{\mathbb{R}} L(G) \cong L(G_{\mathbb{C}})$ as Lie algebras.*

Proof. The isomorphism of vector spaces just described preserves Lie brackets. □

(8.5) Definition. The homomorphism $i: G \to G_{\mathbb{C}}$, as well as the Lie group $G_{\mathbb{C}}$ itself, is called the ***complexification*** of G.

The complexification enjoys the following universal property:

(8.6) Proposition. *Given a representation $r: G \to GL(m, \mathbb{C})$ there is a unique holomorphic representation $r_{\mathbb{C}}: G_{\mathbb{C}} \to GL(m, \mathbb{C})$ with $r_{\mathbb{C}} \circ i = r$.*

Proof. We first prove existence. We have already constructed $r_{\mathbb{C}}$, namely $r_{\mathbb{C}}(s) = (s(r_{kj}))$ if $r = (r_{kj})$. Suppose

$$\mathscr{T}(G, \mathbb{C}) = \mathbb{C}[a_1, \ldots, a_d] \cong \mathbb{C}[X_1, \ldots, X_d]/I.$$

Identifying $G_{\mathbb{C}}$ with $V(I) \subset \mathbb{C}^d$ as in (8.2), each r_{kj} becomes an algebraic function $V(I) \to \mathbb{C}$. Thus $r_{\mathbb{C}}$ is an algebraic map $\mathbb{C}^d \to \mathbb{C}^{m \cdot m}$ and is, in particular, holomorphic.

We now deal with uniqueness. We use (8.3) and identify $G_{\mathbb{C}} \cong \tilde{G} = (\tilde{G} \cap \mathrm{U}(n)) \times (\tilde{G} \cap \mathrm{P}(n))$. Suppose we are given the values of a holomorphic representation on $\tilde{G} \cap \mathrm{U}(n)$. If $P \in \tilde{G} \cap \mathrm{P}(n)$ has the form $\exp(iX)$, we may then determine the values of this representation on $\exp(tX)$ for all $t \in \mathbb{R}$. Since $\exp\colon \mathrm{L}(\tilde{G}) \to \tilde{G}$ is holomorphic, this gives us the values for all $t \in \mathbb{C}$. □

We close by describing the complexifications of the classical groups in terms of other classical groups.

(8.7) Proposition. *The following inclusions of matrix groups are complexifications:*

$$\mathrm{U}(n) \subset \mathrm{GL}(n, \mathbb{C}), \qquad \mathrm{SU}(n) \subset \mathrm{SL}(n, \mathbb{C}),$$

$$\mathrm{O}(n) \subset \mathrm{O}(n, \mathbb{C}), \qquad \mathrm{SO}(n) \subset \mathrm{SO}(n, \mathbb{C}),$$

$$\text{and} \quad \mathrm{Sp}(n) \subset \mathrm{Sp}(n, \mathbb{C}).$$

Note: The group $\mathrm{O}(n, \mathbb{C})$ is defined as $\{A \in \mathrm{GL}(n, \mathbb{C}) \mid {}^tAA = E\}$, and $\mathrm{SO}(n, \mathbb{C}) = \mathrm{O}(n, \mathbb{C}) \cap \mathrm{SL}(n, \mathbb{C})$. The remaining groups are discussed in I, (1.6)–(1.12).

Proof. Let $r\colon \mathrm{U}(n) \to \mathrm{GL}(n, \mathbb{C})$ be the canonical embedding. Then $r_{\mathbb{C}}$ is injective ((8.8), Ex. 1) and must be bijective by reasons of dimension.

If we start with $\mathrm{SU}(n) \subset \mathrm{GL}(n, \mathbb{C})$, then the image of the injection $r_{\mathbb{C}}$ is contained in $\mathrm{SL}(n, \mathbb{C})$, which is the variety defined by $\det(X_{kj}) = 1$. Again, comparing dimensions shows that $r_{\mathbb{C}}\colon \mathrm{SU}(n)_{\mathbb{C}} \to \mathrm{SL}(n, \mathbb{C})$ is an isomorphism.

Starting with $r\colon \mathrm{O}(n) \subset \mathrm{GL}(n, \mathbb{C})$, we see (as with $\mathrm{SU}(n)$) that $r_{\mathbb{C}}\colon \mathrm{O}(n)_{\mathbb{C}} \to \mathrm{GL}(n, \mathbb{C})$ is an injection onto a subvariety. The functions r_{kj} satisfy ${}^t(r_{kj})(r_{kj}) = E$. Hence the image of $r_{\mathbb{C}}$ lies in $\mathrm{O}(n, \mathbb{C})$. Examining components and dimensions shows that $r_{\mathbb{C}}$ maps $\mathrm{O}(n)_{\mathbb{C}}$ isomorphically onto $\mathrm{O}(n, \mathbb{C})$. The same goes for $\mathrm{SO}(n)_{\mathbb{C}} \cong \mathrm{SO}(n, \mathbb{C})$.

Finally, we turn to the group $\mathrm{Sp}(n)$ and its standard representation $r\colon \mathrm{Sp}(n) \subset \mathrm{GL}(2n, \mathbb{C})$ as the set of unitary matrices A satisfying ${}^tAJA = J$, see I, (1.12). Taking determinants in this equation, we get $(\det A)^2 = 1$. Since $\mathrm{Sp}(n)$ is connected, $\det A = 1$. Using Cramer's rule, which says that $A \cdot \mathrm{adj}(A) = \det(A) \cdot E$, we find that the r_{kj} generate $\mathscr{T}(\mathrm{Sp}(n), \mathbb{C})$. Consequently $r_{\mathbb{C}}\colon \mathrm{Sp}(n)_{\mathbb{C}} \to \mathrm{GL}(2n, \mathbb{C})$ is an embedding onto a subvariety which is contained in $\mathrm{Sp}(n, \mathbb{C})$. Checking dimensions, $\mathrm{Sp}(n)_{\mathbb{C}} \cong \mathrm{Sp}(n, \mathbb{C})$. □

The methods and results of this section point out the value of looking at $G_{\mathbb{C}}$ as an ***algebraic group***. See Humphreys [2] for an elementary exposition.

The presentation in this section is based on Chevalley [1], Ch. VI. For a further discussion of complexifications, see Hochschild [1], XVII.

(8.8) Exercises

1. Let $r: G \to \mathrm{GL}(n, \mathbb{C})$ be injective. Show that $r_{\mathbb{C}}$ is injective. *Hint*: $r_{\mathbb{C}} \oplus \bar{r}_{\mathbb{C}}$ is injective (8.2).

2. Given $r: G \to \mathrm{GL}(n, \mathbb{C})$ and $A \in \mathrm{GL}(n, \mathbb{C})$, show that $(ArA^{-1})_{\mathbb{C}} = Ar_{\mathbb{C}}A^{-1}$.

3. Prove or disprove: If $r: G \to \mathrm{GL}(n, \mathbb{C})$ is injective, then the image of $r_{\mathbb{C}}$ is an affine subvariety of $\mathbb{C}^{n,n}$.

4. Show that the assignment $G \mapsto G_{\mathbb{C}}$ is functorial in the following sense: A continuous homomorphism $\varphi: H \to G$ induces an algebra homomorphism $\varphi^*: \mathscr{T}(G, \mathbb{C}) \to \mathscr{T}(H, \mathbb{C})$ and the induced map $\varphi_{\mathbb{C}}: H_{\mathbb{C}} \to G_{\mathbb{C}}$ is an (algebraic and hence holomorphic) group homomorphism. The functorial relations $\mathrm{id}_{\mathbb{C}} = \mathrm{id}$ and $(\varphi\psi)_{\mathbb{C}} = \varphi_{\mathbb{C}}\psi_{\mathbb{C}}$ hold.

5. Show that $(G \times H)_{\mathbb{C}} \cong G_{\mathbb{C}} \times H_{\mathbb{C}}$.

6. Let $\varphi: G' \to G$ be a finite cover with kernel K. The action of G' on $\mathscr{T}(G', \mathbb{C})$ by right translation restricts to an action of K on $\mathscr{T}(G', \mathbb{C})$. Let $\mathscr{T}(G', \mathbb{C})^K$ be the ring of functions fixed by this action. Show that φ^* induces an algebra isomorphism $\mathscr{T}(G, \mathbb{C}) \cong \mathscr{T}(G', \mathbb{C})^K$. Show that $G'_{\mathbb{C}} \to G_{\mathbb{C}}$ is also a finite cover with kernel K.

7. Let $H \subset \mathrm{GL}(n, \mathbb{C})$ be a subgroup such that ${}^t\bar{A}^{-1} \in H$ for all $A \in H$ and which is also an affine variety. Show that H is the complexification of $H \cap \mathrm{U}(n)$ (use (8.3) and its proof). Use this result to show that $\mathrm{Sp}(n, \mathbb{C})$ is the complexification of $\mathrm{Sp}(n)$.

8. Let $a_j, b_j \in \mathbb{R}$ and define $g(t) = \sum_j a_j \exp(tb_j)$. Show that if $g(t) = 0$ for all $t \in \mathbb{Z}$, then $g(t) = 0$ for all $t \in \mathbb{R}$.

9. Representations of G correspond to holomorphic representations of $G_{\mathbb{C}}$. Show that this correspondence preserves direct sums and tensor products. Show that holomorphic representations of $G_{\mathbb{C}}$ are semisimple in the sense of II, §2.

10. Let $r: G \to \mathrm{GL}(n, \mathbb{R}) \subset \mathrm{GL}(n, \mathbb{C})$ be faithful. Show that $r_{\mathbb{C}}(G'_{\mathbb{R}}) = r_{\mathbb{C}}(G_{\mathbb{C}}) \cap \mathbb{R}^{n,n}$.

CHAPTER IV

The Maximal Torus of a Compact Lie Group

In this chapter we show that every connected compact Lie group G contains a maximal torus T. This maximal torus is unique up to conjugation, and its conjugates cover G. If N is the normalizer of T, then the Weyl group $W = N/T$ is finite and operates effectively on T. Thus there is a one-to-one correspondence between functions on G which are invariant under conjugation and functions on T which are invariant under the action of W. In particular, the characters of G are the W-invariant characters of T. For this reason understanding the operation of the Weyl group on the maximal torus is important to representation theory. In the third section we compute the maximal tori and Weyl groups of the classical Lie groups, and in the last section we give a generalization which handles the case of nonconnected groups.

1. Maximal Tori

In this section G will be a compact connected Lie group.

(1.1) Definition. A subgroup $T \subset G$ is a ***maximal torus*** if T is a torus and there is no other torus T' with $T \subsetneqq T' \subset G$. By a torus we mean a Lie group isomorphic to $\mathbb{R}^k/\mathbb{Z}^k$ for some k.

Since tori are compact and connected, if $T \subsetneqq T'$, then $\dim T < \dim T'$. This shows that maximal tori exist. In fact, a maximal torus is the same thing as a maximal connected abelian subgroup, since the closure of such a subgroup is also connected and abelian and hence is a torus (I, (3.7)).

Recall that an automorphism $\varphi: T^k \to T^k$ of a torus induces a commutative diagram of homomorphisms

$$\begin{array}{ccccccccc}
0 & \longrightarrow & \mathbb{Z}^k & \xrightarrow{\subset} & \mathbb{R}^k & \xrightarrow{\exp} & T^k & \longrightarrow & 0 \\
 & & \downarrow{\scriptstyle L\varphi} & & \downarrow{\scriptstyle L\varphi} & & \downarrow{\scriptstyle \varphi} & & \\
0 & \longrightarrow & \mathbb{Z}^k & \xrightarrow{\subset} & \mathbb{R}^k & \xrightarrow{\exp} & T^k & \longrightarrow & 0,
\end{array}$$

where we identify T^k with $\mathbb{R}^k/\mathbb{Z}^k$ and LT with $\mathbb{R}^k$. Therefore the automorphism φ corresponds to an invertible matrix $L\varphi$ with integral entries: we have an isomorphism

(1.2) $$L: \operatorname{Aut}(T^k) \cong \operatorname{Aut}(\mathbb{Z}^k) = GL(k, \mathbb{Z}), \qquad \varphi \mapsto L\varphi \mid \mathbb{Z}^k.$$

(1.3) Definition. Let T be a maximal torus in G and $N = N(T)$ the normalizer of T in G, so

$$N = \{g \in G \mid gTg^{-1} = T\}.$$

Then the group $W = N/T$ is called the ***Weyl group*** of G.

Note that according to (1.3) W depends upon the choice of a maximal torus T. But we will see that all maximal tori are conjugate, so different choices of T yield isomorphic Weyl groups. However, when we talk about *the* maximal torus in G and *the* Weyl group of G, we have implicitly chosen a particular T which is to remain fixed throughout the discussion.

The normalizer N of T operates on T by conjugation

$$N \times T \to T, \qquad (n, t) \mapsto ntn^{-1},$$

and since the operation of T on T is trivial, we obtain an induced operation of the Weyl group

(1.4) $$W \times T \to T, \qquad (nT, t) \mapsto ntn^{-1}.$$

(1.5) Theorem. *The Weyl group is finite.*

Proof. Let N_0 be the connected component of the identity in N. We will show that $N_0 = T$. Since N is compact, it will follow that $W = N/N_0$ is compact and discrete, and hence finite.

We view the action of N on T as a continuous map

$$N \to \operatorname{Aut}(T) \overset{L}{\cong} GL(k, \mathbb{Z}), \qquad n \mapsto \operatorname{Ad}(n) \mid LT.$$

Since $\mathrm{GL}(k, \mathbb{Z}) \subset \mathrm{Aut}\, LT$ is discrete and N_0 is connected, the image of N_0 is the identity, so N_0 acts trivially on T. Thus if $\alpha: \mathbb{R} \to N_0$ is a one-parameter group, the product $\alpha(\mathbb{R}) \cdot T$ is a connected abelian group containing T. This means $\alpha(\mathbb{R}) \cdot T = T$, and hence $\alpha(\mathbb{R}) \subset T$. But the groups $\alpha(\mathbb{R})$ cover an open neighborhood of the identity in N_0 and hence generate N_0. It follows that $N_0 = T$. □

The goal of this section is to prove the following theorem:

(1.6) Main Theorem on Tori. *Any two maximal tori in a compact connected Lie group G are conjugate, and every element of G is contained in a maximal torus.*

The proof of this theorem, as well as that of its generalization in (4.3), uses an argument based upon the mapping degree. Sard's theorem also plays an essential role in the proof in §4. Here we get by without it, but we need to use orientations on G and G/T. This involves the invariant volume forms. Theorem (1.6) will be easily proved from the following:

(1.7) Main Lemma. *Let G be a compact connected Lie group and T a maximal torus in G. Then the map*

$$q: G/T \times T \to G, \qquad (g, t) \mapsto gtg^{-1},$$

has mapping degree $\deg(q) = |W|$, *where* $|W|$ *is the order of the Weyl group associated to T. In particular, q is surjective.*

We first show how (1.6) follows from (1.7).

PROOF (of the main theorem on tori from the main lemma). Let T and T' be maximal tori and t' a generator of T', see Kronecker's theorem I, (4.13). By lemma (1.7), there is a $g \in G$ with $t' \in gTg^{-1}$. Hence $T' \subset gTg^{-1}$, and since gTg^{-1} is a torus, $T' = gTg^{-1}$. Since q is surjective every element of G is contained in some conjugate of T. □

In order to compute the mapping degree of $q: G/T \times T \to G$, we need to investigate how the canonical volume form dg is transformed by q. To this end we choose a metric on the Lie algebra LG which is invariant under the action of the adjoint representation I, (2.10)

$$\mathrm{Ad}: G \to \mathrm{Aut}(LG), \qquad g \mapsto Lc(g), \qquad c(g): h \mapsto ghg^{-1}.$$

The adjoint representation is real and the corresponding metric is a real Euclidean metric.

The Lie algebra LG is the tangent space to G at its unit. It decomposes into the tangent space to T at e, namely LT, and its orthogonal complement which we will call L(G/T):

$$\mathrm{L}G = \mathrm{L}(G/T) \oplus \mathrm{L}T.$$

This decomposition is invariant under the operation $\mathrm{Ad}\,|\,T$. The torus T acts trivially on LT and, since T is maximal, T acts nontrivially on every nonzero vector in L(G/T). The induced action on the first summand (see I, (5.14)) is denoted by

$$\mathrm{Ad}_{G/T}\colon T \to \mathrm{Aut}\,\mathrm{L}(G/T).$$

Now recall that there are left-invariant volume forms $d(gT)$, dt, and dg on G/T, T, and G. These forms are canonical up to a choice of sign.

The projection $\pi\colon G \to G/T$ induces a map between $\mathrm{L}G = \mathrm{L}(G/T) \oplus \mathrm{L}T$ and the tangent space to G/T at the point eT. This maps L(G/T) isomorphically onto this tangent space, so we will identify these two spaces via this map.

Let $n = \dim G$ and $k = \dim T$. We obtain a left-invariant alternating differential form $\pi^* d(gT) \in \Omega^{n-k}(G)$ from the volume form $d(gT)$ on G/T, and the alternating form $dt_e \in \mathrm{Alt}^k(\mathrm{L}T)$ gives rise to the alternating form $\mathrm{pr}_2^*\, dt_e \in \mathrm{Alt}^k(\mathrm{L}G)$, where $\mathrm{pr}_2\colon \mathrm{L}G = \mathrm{L}(G/T) \oplus \mathrm{L}T \to \mathrm{L}T$ is projection onto the second summand. But $\mathrm{pr}_2^*\, dt_e$ determines a left-invariant differential form $d\tau \in \Omega^k G$ such that $d\tau\,|\,T = dt$, and $\pi^* d(gT) \wedge d\tau$ is a left-invariant volume form on G. We may choose our signs so that $\pi^* d(gT) \wedge d\tau = c \cdot dg$ with $c > 0$.

For the proof of the main lemma, we could be satisfied with this state of affairs, but later we will need the not overly surprising fact that $c = 1$. In order to compute this constant, we integrate a suitable continuous function $\psi\colon G \to \mathbb{R}$ against both volume forms. We consider a bundle chart $\varphi\colon \pi^{-1}U \to U \times T$ of the T-principal bundle $\pi\colon G \to G/T$ and let ψ be a nonzero nonnegative continuous real-valued function on G with support in $\pi^{-1}U$. Then we have

$$0 \neq \int_G \psi\, dg = \int_{G/T}\left(\int_T \psi(gt)\, dt\right) d(gT) = \int_U \left(\int_T \psi(gt)\, dt\right) d(gT).$$

Here we have used the version of Fubini's theorem proved in I, (5.16). We may now make a change of coordinates based upon the following diagram associated to the bundle chart φ:

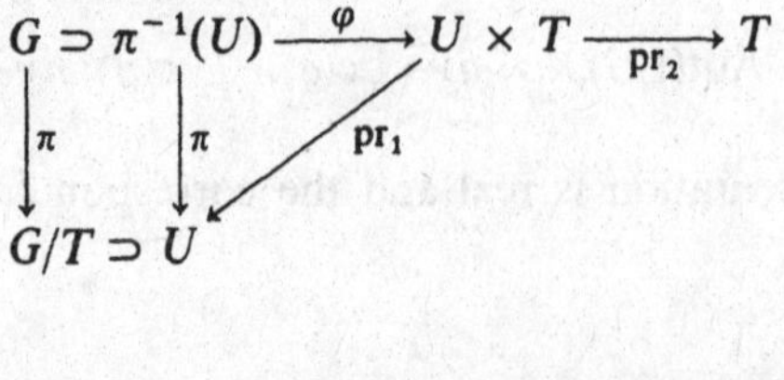

Note that, for a fixed $g \in \pi^{-1}U$, $\varphi(g) = (u, s)$ with $u = \pi(g)$ and some $s \in T$, and $\varphi(gt) = (u, ts)$, since φ is a right T-map.

To simplify matters, we may arrange that the decomposition $LG = L(G/T) \oplus LT$ above coincides with that induced by the bundle chart φ, so that $\mathrm{pr}_2^* \, dt = (\varphi^{-1})^* \, d\tau$. Then we have

$$\begin{aligned}
\int_U \left(\int_T \psi(gt)\, dt \right) d(gT) &= \int_U \left(\int_T \psi\varphi^{-1}(u, t)\, \mathrm{pr}_2^*\, dt \right) \mathrm{pr}_1^*\, d(gT) \\
&= \int_{U \times T} \psi\varphi^{-1}(u, t)\, \mathrm{pr}_1^*\, d(gT) \wedge \mathrm{pr}_2^*\, dt \\
&= \int_{U \times T} \psi\varphi^{-1} (\varphi^{-1})^*(\pi^* d(gT) \wedge d\tau) \\
&= c \cdot \int_{U \times T} \psi\varphi^{-1} \cdot (\varphi^{-1})^*\, dg \\
&= c \int \psi\, dg.
\end{aligned}$$

Thus we conclude $c = 1$. □

We now have the volume form

$$dg = \pi^* d(gT) \wedge d\tau, \qquad d\tau \,|\, T = dt,$$

on G and the volume form

$$\alpha = \mathrm{pr}_1^*\, d(gT) \wedge \mathrm{pr}_2^*\, dt$$

on $G/T \times T$. Identifying LG with $L(G/T) \oplus LT$ as we have done and evaluating at the unit element gives $\alpha_{(eT, e)} = dg_e$.

The ***determinant*** $\det(q)\colon G/T \times T \to \mathbb{R}$ of the conjugation map q is defined by the equation

$$q^*\, dg = \det(q) \cdot \alpha$$

(1.8) Proposition. *The determinant of the conjugation map $q\colon G/T \times T \to G$ is given by*

$$\det(q)(gT, t) = \det(\mathrm{Ad}_{G/T}(t^{-1}) - E_{G/T}),$$

where $E_{G/T}$ is the identity on $L(G/T)$.

Proof. The forms dg and $d(gT)$ are left-invariant under the action of G, and the form dt is left-invariant under the action of T. This allows us to compute our determinant at the point (gT, t) by first applying suitable left translations in $G/T \times T$ and G, and then computing the determinant at (T, e) of a map which sends (T, e) to e. Specifically, we consider the differential of the map $G/T \times T \to G$ which is induced by the following composition:

$$G \times T \xrightarrow{l_a} G \times T \xrightarrow{\tilde{q}} G \xrightarrow{l_b} G, \quad a = (g, t), \quad b = gt^{-1}g^{-1},$$
$$(x, y) \overset{l_a}{\mapsto} (gx, ty) \overset{\tilde{q}}{\mapsto} (gx)(ty)(gx)^{-1} \overset{l_b}{\mapsto} gt^{-1}xtyx^{-1}g^{-1}$$
$$= \mathrm{c}(g)(\mathrm{c}(t^{-1})(x) \cdot y \cdot x^{-1}).$$

Here $\mathrm{c}(g)\colon G \to G$, $x \mapsto gxg^{-1}$, is conjugation and $\tilde{q}$ is defined by $\tilde{q}\colon G \times T \to G$, $(g, t) \mapsto gtg^{-1}$. The determinant we want to compute is the determinant of the differential of this map at the point (e, e), restricted to the subspace $\mathrm{L}(G/T) \oplus \mathrm{L}(T)$ of $\mathrm{L}(G) \oplus \mathrm{L}(T) = \mathrm{L}(G \times T)$.

Now, $\mathrm{Lc}(g) = \mathrm{Ad}(g)$ has determinant 1, since $\mathrm{Ad}(g)$ is orthogonal and G is connected. Furthermore, the differential of a product is the sum of the differentials (I, (3.5)). Thus the determinant of q is the determinant of the linear endomorphism

$$(X, Y) \mapsto \mathrm{Ad}_{G/T}(t^{-1})X + Y - X$$

of $\mathrm{L}(G/T) \oplus \mathrm{L}(T)$. In matrix form this is

$$\left[\begin{array}{c|c} \mathrm{Ad}_{G/T}(t^{-1}) - E_{G/T} & 0 \\ \hline & E_T \end{array}\right],$$

from which the proposition follows. □

Proof of the Main Lemma (1.7). We may compute the mapping degree of q by counting the number of points (and keeping track of orientation) in the inverse image of any regular value of q; see I, (5.19). Thus (1.7) follows from the following:

(1.9) Lemma. *Let $t \in T$ be a generator (as in* I, (4.13)). *Then*

(i) $q^{-1}(t)$ *consists of* $|W|$ *points, and*
(ii) $\det(q) > 0$ *at each of these points.*

Proof. (i) Let N be the normalizer of T in G. Then $q(gT, s) = t$ if and only if $gsg^{-1} = t$, i.e., if and only if $s = g^{-1}tg \in T$. Of course, such an s exists if and only if $g^{-1}Tg \subset T$, which means $g \in N$. Thus we have

$$q^{-1}(t) = \{(gT, g^{-1}tg) | g \in N\},$$

and hence $q^{-1}(t)$ has $[N : T] = |W|$ elements.

(ii) The determinant we want is the determinant of the linear endomorphism $\mathrm{Ad}_{G/T}(t^{-1}) - E$ of $\mathrm{L}(G/T)$. We will show that this map has no real eigenvalues. It will follow that the eigenvalues come in complex conjugate pairs, and hence the determinant is positive.

So suppose $\mathrm{Ad}_{G/T}(t^{-1}) - E$ had a real eigenvalue. Then $\mathrm{Ad}_{G/T}(t^{-1})$ would also have a real eigenvalue, which would have to be ± 1 since $\mathrm{Ad}_{G/T}(t^{-1})$ is an orthogonal transformation. Observe that, in this case, $\mathrm{Ad}_{G/T}(t^{-2})$ has 1 as an eigenvalue. Since t^{-2} is also a generator of T, we see it suffices to show that $\mathrm{Ad}_{G/T}(t)$ cannot have 1 as an eigenvalue.

Now, T operates on G via conjugation, and linearly on $\mathrm{L}G$ via the adjoint representation. Since the exponential map $\exp\colon \mathrm{L}G \to G$ is natural (I, (3.2)) it is equivariant with respect to these T-actions. Thus if $\mathrm{Ad}_{G/T}(t)X = X$ for some $X \in \mathrm{L}(G/T)$, then X is fixed by the action of t and hence by all of T. But then the one-parameter group $H = \{\exp(sX) \mid s \in \mathbb{R}\}$ is left pointwise fixed under conjugation by elements of T. Consequently HT is abelian and connected, which implies that $H \subset T$ and $X \in \mathrm{L}T \cap \mathrm{L}(G/T) = 0$. □

We record a result which comes out of the proof above.

(1.10) Note. If t is a generating element of T, then $\mathrm{Ad}_{G/T}(t)$ has no real eigenvalues. Hence $\dim G/T$ is even.

The main theorem on tori is of fundamental significance to both the structure theory of Lie groups and representation theory. The following formula will prove to be an important tool in the representation theory:

(1.11) Weyl's Integral Formula. *Let G be a compact connected Lie group, T the maximal torus, and f a continuous function on G. Then*

$$|W| \cdot \int_G f(g)\, dg = \int_T \left[\det(E_{G/T} - \mathrm{Ad}_{G/T}(t^{-1})) \int_G f(gtg^{-1})\, dg\right] dt.$$

Proof. Let $f_t\colon G/T \to \mathbb{R}$ be given by $g \mapsto f(gtg^{-1})$, so in the last integral $f(gtg^{-1}) = f_t \circ \pi(g)$, where π is the canonical projection $G \to G/T$. Thus the last integral is $\int_{G/T} f_t\, d(gT)$ and, since $\dim(G/T)$ is even, we can use I, (5.19) as well as (1.7) and (1.8) to rewrite the right-hand side as

$$\int_{G/T}\left(\int_T f \circ q(gT, t) \cdot \det(q)(gT, t)\, dt\right) d(gT)$$

$$= \int_{G/T \times T} f \circ q(gT, t) \cdot q^* \, dg = \deg(q) \cdot \int_G f = |W| \int_G f(g)\, dg. \quad \square$$

Our proof of the conjugation theorem (1.6) is based on exposé 23 by Serre in Séminaire Sophus Lie [1], where also the Lie algebra proof based on work of E. Cartan, the proof of A. Weil [1] and other proofs are sketched; see also Hunt [1]. The use of maximal tori and the integration formula appear in the fundamental work of H. Weyl [1].

(1.12) Exercises

Where relevant, T denotes the maximal torus of the compact connected Lie group G.

1. Show that the subgroup of the diagonal matrices

$$\begin{bmatrix} \varepsilon_1 & & \\ & \ddots & \\ & & \varepsilon_n \end{bmatrix}, \qquad \varepsilon_\nu = \pm 1, \quad \varepsilon_1 \cdot \ldots \cdot \varepsilon_n = 1$$

is a maximal abelian subgroup of $\mathrm{SO}(n)$ but is not a torus.

2. Let $H \subset G$ be a closed subgroup containing the normalizer $\mathrm{N}(T)$ of T. Show that $\mathrm{N}(H) = H$.

3. A Lie algebra is called abelian if $[X, Y] = 0$ for all X and Y in the Lie algebra. Show that LT is a maximal abelian Lie subalgebra of LG.

4. Show that there is a linear decomposition $LG = LT \oplus \mathrm{L}(G/T)$ such that $[X, Y] \in \mathrm{L}(G/T)$ for every $X \in LT$ and $Y \in \mathrm{L}(G/T)$, and such that the linear map $\mathrm{ad}_{G/T}(H)\colon \mathrm{L}(G/T) \to \mathrm{L}(G/T)$, $Y \mapsto [H, Y]$ is an isomorphism for almost every $H \in LT$.

5. Let $f\colon G \to H$ be a surjective homomorphism of Lie groups and suppose H is abelian. Show that $f \mid T\colon T \to H$ is also surjective.

6. Show that every abelian normal subgroup of G lies in the center of G.

7. Let G be a (not necessarily compact) Lie group and $H \subset G$ be a one-parameter group which is not closed. Show that $\bar{H}$ is a torus. *Hint*: Use I, (3.6).

8. Let $S \subset G$ be a closed subgroup such that $G = \bigcup_{g \in G} gSg^{-1}$. Show that S contains a maximal torus.

9. Verify formula II, (5.2) for the invariant integral of a class function on SU(2) using the Weyl integral formula (1.11). Derive an analogous formula for SO(3).

2. Consequences of the Conjugation Theorem

We return to the main theorem on tori and collect a few consequences. Let G be a compact connected Lie group and T a fixed choice of a maximal torus in G with normalizer N and Weyl group W. This torus, together with the action of the Weyl group, is uniquely determined by G in the following sense: Suppose T' is another maximal torus with Weyl group W'. Then there is an inner automorphism φ of G such that $\varphi(T) = T'$. Hence $\varphi(N) = N'$, where N and N' are the normalizers of T and T'. Thus φ induces an isomorphism $\tilde{\varphi}\colon W = N/T \to N'/T' = W'$ and the following diagram commutes:

$$\begin{array}{ccc} W \times T & \longrightarrow & T \\ \Big\downarrow{\scriptstyle \tilde{\varphi} \times \varphi} & & \Big\downarrow{\scriptstyle \varphi} \\ W' \times T' & \longrightarrow & T' \end{array}$$

(2.1) Definition. The dimension of a maximal torus in G is called the ***rank*** of G and is denoted rank(G).

Henceforth any discussion of the maximal torus or operation of the Weyl group will refer to a particular torus chosen once and for all. A different choice simply leads to isomorphic structures and does not affect our results.

(2.2) Theorem. *The exponential map of a compact connected Lie group is surjective.*

PROOF. Every element of G is contained in a (maximal) torus $T \subset G$, and the exponential map of a torus is surjective. □

As has already been mentioned, this result is, in general, false for non-compact groups (I, (3.13), Ex. 1).

The centralizer $Z(H)$ of a subgroup $H \subset G$ is the subgroup

$$Z(H) = \{g \in G \mid gh = hg \text{ for all } h \in H\}.$$

(2.3) Theorem. *Let G be a compact connected Lie group and T its maximal torus.*

(i) *$Z(T) = T$, so T is a maximal abelian subgroup of G.*
(ii) *If $S \subset G$ is a connected abelian subgroup, then $Z(S)$ is the union of those (maximal) tori T containing S.*
(iii) *The center of G is the intersection of all maximal tori in G.*

PROOF. (i) follows from (ii) with $T = S$, so we skip right to the proof of (ii).

(ii) The closure $\bar{S}$ of S is compact, connected, and abelian. Thus it is a torus. Furthermore, if $g \in Z(S)$, then $g \in Z(\bar{S})$, so $Z(S) = Z(\bar{S})$. Thus we may assume that $S = \bar{S}$ is a torus. Now let $x \in Z(S)$, and let B be the closure of the subgroup generated by x and S. Then B is compact and abelian, so its connected component of the unit, B_0, is another torus. Since xB_0 generates B/B_0, we see that $B/B_0 \cong \mathbb{Z}/m$ for some cyclic group $\mathbb{Z}/m$. It follows that $B \cong B_0 \times \mathbb{Z}/m$ is the closure of a cyclic subgroup $\{g^n \mid n \in \mathbb{Z}\}$ for some $g \in B$ (see I, (4.14)). But g is contained in some maximal torus T in G, so $S \cup \{x\} \subset T$, which is what we need.

(iii) If x is in the center of G, then x is clearly in every maximal torus in G. Conversely, if x is in every maximal torus, then x must commute with every $g \in G$, since g is contained in some torus. □

The reader is reminded again that, while all maximal tori are maximal abelian subgroups, maximal abelian subgroups need not be tori ((1.12), Ex. 1).

The fact that $Z(T) = T$ may be restated as follows:

(2.4) Corollary. *The Weyl group acts effectively on the maximal torus, i.e., the homomorphism $W \to \mathrm{Aut}(T)$ defined by the action of W on T is injective.*

Thus we may interpret W as a group of automorphisms of T. The maximal torus decomposes into orbits under the action of W.

(2.5) Lemma. *Two elements of the maximal torus are conjugate in G if and only if they lie in the same orbit under the action of the Weyl group.*

Proof. Let $x, y \in T$, $g \in G$, and $gxg^{-1} = y$. Let $Z(x)$ and $Z(y)$ be the centralizers of x and y. Conjugation by g induces a map $c(g)\colon Z(x) \to Z(y)$, and since $T \subset Z(x)$ we have $c(g)T \subset Z(y)$. Thus both T and $c(g)T$ are maximal tori in the connected component of the unit $Z(y)_0$ of $Z(y)$. Hence there is an $h \in Z(y)_0$ with $T = c(h)(c(g)T) = c(hg)T$. But this means that $hg \in N$ and $c(hg)x = c(h)y = y$. Thus hgT is an element $w \in W$ with $wx = y$. □

We next define Con(G) to be the space of conjugacy classes of G. Thus Con(G) is the space of orbits under the action of G on the manifold G via conjugation $\tilde{q}\colon G \times G \to G$, $(g, x) \mapsto gxg^{-1}$. The topology on Con(G) is the quotient topology coming from the projection $G \to \mathrm{Con}(G)$.

(2.6) Proposition. *There is a canonical homeomorphism*

$$\kappa\colon T/W \xrightarrow{\approx} \mathrm{Con}(G), \qquad Wt \mapsto c(G)t$$

taking the orbit of t under the operation of W to the conjugacy class of t.

Proof. The map κ is well defined and continuous. It is surjective because the conjugates of T cover G (1.6) and injective because of (2.5). After convincing oneself that both spaces are Hausdorff (see (2.12), Ex. 8), one observes that they are compact, and so κ is a homeomorphism. □

We let $C^0(X) = C^0(X, \mathbb{C})$ denote the space of continuous complex-valued functions on a space X. Then conjugation induces an action of W on $C^0(T)$ and an action of G on $C^0(G)$:

$$W \times C^0(T) \to C^0(T), \qquad (w, f) \mapsto f \circ w^{-1}$$
$$G \times C^0(G) \to C^0(G), \qquad (g, f) \mapsto f \circ c(g^{-1}).$$

Furthermore,

$$C^0(T/W) = C^0(T)^W, \quad \text{and} \quad C^0(\mathrm{Con}(G)) = C^0(G)^G.$$

(2.7) Corollary. *There is a canonical isomorphism of normed complex vector spaces*

$$C^0(\mathrm{Con}(G)) \xrightarrow{\cong} C^0(T)^W, \qquad f \mapsto f \mid T$$

between the space of class functions on G and the space of continuous functions on the maximal torus invariant under the action of the Weyl group. □

The representation ring R(G) may be viewed as a ring of complex class functions on G, and we obtain:

(2.8) Corollary. *Restriction to the maximal torus induces an injective homomorphism*

$$\mathrm{R}(G) \to \mathrm{R}(T)^W, \qquad \chi \mapsto \chi | T.$$

In particular, R(G) *has no zero-divisors (i.e., it is an integral domain).* □

Later, in VI, (2.1) we will show that the map $\mathrm{R}(G) \to \mathrm{R}(T)^W$ is an isomorphism.

If we start with a W-invariant character χ on T, then (2.7) tells us that χ has a unique extension to a class function on G. Now the question is, does this class function have integer coefficients $\langle \chi, \chi_j \rangle \in \mathbb{Z}$ with respect to the basis $\{\chi_j\}$ of irreducible characters? In order to compute such an integral, we have Weyl's integral formula (1.11) at our disposal. However, to work out the right-hand side of (1.11), we have to analyze the function $\det(E - \mathrm{Ad}_{G/T}(t^{-1}))$. Thus we must study the restriction of the adjoint representation to the maximal torus. We will do this in the next chapter.

Of course, we also need to determine the W-invariant characters on T. To do this we need a precise geometric picture of the operation of the Weyl group on T and its Lie algebra. Therefore we continue with our study of the maximal torus. The next theorem describes the behavior of the maximal torus under homomorphisms.

(2.9) Theorem. *Let $f: G \to H$ be a surjective homomorphism of compact connected Lie groups. Then if $T \subset G$ is a maximal torus, so is $f(T) \subset H$. Furthermore,* $\ker(f) \subset T$ *if and only if* $\ker(f) \subset Z(G)$. *In this case f induces an isomorphism of the Weyl groups. In particular, this happens if* $\dim G = \dim H$.

Proof. Let S be a maximal torus in H with generating element s. Let $t \in f^{-1}\{s\}$. Then t is contained in some maximal torus T in G, and, since $f(T)$ is compact, abelian, and connected, $f(T)$ is a torus in H. But $S \subset f(T)$ because $s \in f(T)$, so we must have $f(T) = S$. To complete the proof of the first statement, simply note that any maximal torus in G is of the form $c(g)T$ for some $g \in T$, and $fc(g)T = c(f(g))S$.

Now suppose $\ker(f) \subset T$. Since $\ker(f) \subset G$ is a normal subgroup, it is fixed under conjugation, and hence $\ker(f)$ is contained in every maximal torus in G. Thus $\ker(f) \subset Z(G)$, see (2.4)(iii).

Next, we look at the Weyl groups. We always have $f(\mathrm{N}(T)) \subset \mathrm{N}(f(T))$. But if $\ker(f) \subset Z(G)$ and $f(T) = S$, then $\mathrm{N}(T) = f^{-1}\mathrm{N}(S)$. Indeed, if $f(g) \in \mathrm{N}(S)$ and $t \in T$, then $f(gtg^{-1}) = f(g)f(t)f(g)^{-1} \in S = f(T)$, so $f(gtg^{-1}) = f(t_1)$ for $t_1 \in T$. Now this means $gtg^{-1} = t_1 z$ for some $z \in Z(G) \subset T$, and hence $gtg^{-1} \in T$. Thus $g \in \mathrm{N}(T)$.

Let $W(T)$ and $W(S)$ be the Weyl groups associated to the tori $T \subset G$ and $S = f(T) \subset H$. If $\ker(f) \subset Z(G)$, the above says that f induces an isomorphism

$$W(T) = N(T)/T \to N(S)/S = W(S),$$

as claimed in the statement of the theorem. Finally, if $\dim G = \dim H$, then $\ker(f) = K$ is discrete. But a discrete normal subgroup—in fact, a totally disconnected normal subgroup—of a connected group is always in the center. This is seen by observing that the image of the map

$$f_k: G \to K, \qquad g \mapsto gkg^{-1},$$

lies in the connected component of k and is therefore constant for any $k \in K$. □

The equality $\dim G = \dim H$ means that $f: G \to H$ is a covering map with $K = \ker(f)$ the group of covering transformations. If $S \subset H$ is a maximal torus, then $T = f^{-1}S$ is a maximal torus in G, since some maximal torus T is mapped surjectively onto S and, since $\ker(f) \subset T$, the torus T contains the full preimage of S under the covering f. The projection $f|T: T \to S$ of maximal tori is also a covering map with the same group $K = \ker(f)$ of covering transformations. We met this situation when considering the covering $\mathrm{Spin}(n) \to SO(n)$ with kernel $\mathbb{Z}/2$.

(2.10) Definition. Let G be a compact connected Lie group. An element $g \in G$ is called ***general*** if g is a generating element for some maximal torus in G. Otherwise, g is called ***special***. The element g is called ***singular*** if g is contained in two distinct maximal tori and ***regular*** otherwise.

(2.11) Theorem. *Let G be a compact connected Lie group.*

(i) *Every general element of G is regular.*
(ii) *Almost every element of G is general.*
(iii) *An element $g \in G$ is singular if and only if*

$$\dim Z(g) > \operatorname{rank}(G) = \dim T.$$

PROOF. (i) is trivial. Statement (ii) means that the set of special elements has Lebesgue measure zero in G, and that, in turn, means that every chart of the manifold G maps the set of special elements in its domain to a set of measure zero in $\mathbb{R}^n$. Now, using induction on $\dim T$ and Fubini's theorem, it is easy to see that the set $U \subset T$ of special elements in the maximal torus T has measure zero. Therefore the same is true for the image $q(G/T \times U)$ under the conjugation map $q: G/T \times T \to G$. But this image is the set of special elements in G.

(iii) $Z(g)$ is the centralizer of g. Assuming that $g \in T \cap T_1$ with T and T_1 being distinct maximal tori, we have $T \cup T_1 \subset Z(g)$. Looking at the Lie algebras, $L(T) \neq L(T_1)$ and $L(T) + L(T_1) \subset LZ(g)$. Thus $\dim Z(g) \geq \dim(LT + LT_1) > \dim LT = \operatorname{rank}(G)$. In the other direction, assume that $\dim Z(g) > \operatorname{rank}(G)$. Since g is in some maximal torus, we may assume $g \in T$ and $\operatorname{rank}(G) = \dim T$. Now $LZ(g)$ contains a vector which does not lie in LT, so $Z(g)$ contains a one-parameter group H which does not lie in T. But $H \subset T_1$ for some maximal torus T_1 in $Z(g)$, and, since $T \subset Z(g)$, T_1 is actually a maximal torus in G. Furthermore, $g \in Z(T_1)$, from which it follows using (2.3)(i) that $g \in T_1$. Thus $g \in T_1 \cap T$ and $T \neq T_1$. □

This section is based on exposé 23 in Séminaire Sophus Lie [1]. See also exposé 24 for additional interesting results about subgroups of $N(T)$, monomial representations and Blichfeldt's induction theorem.

(2.12) Exercises

1. Show that a one- or two-dimensional compact connected Lie group is a torus.

2. Show that a compact connected Lie group is divisible, i.e., that given $g \in G$ and $n \in \mathbb{N}$ there is an $h \in G$ with $h^n = g$.

3. Show that the exponential map on $GL(n, \mathbb{C})$ is surjective. *Hint*: Jordan normal form.

4. Show that every compact Lie group contains a finitely generated dense subgroup. Give an upper bound for the number of generators needed.

5. If the compact Lie group G operates differentiably on M with fixed point p, then G also operates on the ring $\mathscr{E}_p$ of germs of differentiable functions at p via $g\varphi(x) = \varphi(g^{-1}x)$, where $\varphi \in \mathscr{E}_p$. Show that if $\varphi \in \mathscr{E}_p$ is G-invariant then φ has a G-invariant representative $\tilde{\varphi}: U \to \mathbb{R}$ with a G-invariant neighborhood U of p.
   Hint: Exercise 4.

6. Let G be a compact connected Lie group and $g \in G$. Show that $Z(g)_0$ is the union of the maximal tori in G containing g. Find a $g \in SO(3)$ such that $Z(g)$ is not connected.

7. Let G be a compact connected Lie group, $S \subset G$ connected and abelian, and $g \in Z(S)$. Show that there is a torus in G containing both g and S.

8. Let G be a compact Lie group and X a Hausdorff G-space. Show that X/G is Hausdorff.

9. Let $f: G \to H$ be a surjective homomorphism of compact connected Lie groups with $\ker(f) \subset Z(G)$. Show that f induces a diffeomorphism $G/T \to H/S$, where $S = f(T)$ is a maximal torus in H.

3. The Maximal Tori and Weyl Groups of the Classical Groups

The classical groups we consider are groups of linear maps, and the conjugation theorem says things about transformations of such maps which are familiar to the student in other forms.

We begin with the unitary group U(n) and the special unitary group SU(n), see I, (1.8). Let $\Delta(n) \subset \mathrm{U}(n)$ be the subgroup of diagonal matrices

$$D = \begin{bmatrix} z_1 & & \\ & \ddots & \\ & & z_n \end{bmatrix}, \qquad z_\nu \in S^1,$$

and let $\mathrm{S}\Delta(n) = \Delta(n) \cap \mathrm{SU}(n)$ be the subgroup of diagonal matrices as above with $z_1 \cdots z_n = 1$.

(3.1) Proposition. *The groups* $\Delta(n)$ *and* $\mathrm{S}\Delta(n)$ *are maximal tori in* U(n) *and* SU(n).

Proof. The inclusions $T \subset \mathrm{U}(n)$ and $T \subset \mathrm{SU}(n)$ of any tori are unitary representations of abelian groups. Thus by II, (1.13) they are conjugate to homomorphisms with image in $\Delta(n)$. Of course, we may achieve this through conjugation by an element in SU(n) and this sends SU(n) into itself. □

We will represent the diagonal matrix D above by the n-tuple $(\vartheta_1, \ldots, \vartheta_n) \in (\mathbb{R}/\mathbb{Z})^n$ with $z_\nu = \exp(2\pi i \vartheta_\nu)$. The element D generates the torus $\Delta(n)$ if and only if $1, \vartheta_1, \ldots, \vartheta_n$ are linearly independent over the rationals (I, (4.13)). In particular, the values of the ϑ_ν mod 1 are distinct for a generating element. Since eigenvalues stay fixed under conjugation—and hence also under the action of the Weyl group W of U(n)—W operates on a generating element $D = (\vartheta_1, \ldots, \vartheta_n)$ by permuting the ϑ_ν. This determines the operation of W on all of $\Delta(n)$, and it follows that W is contained in S(n), the group of permutations of $\{1, \ldots, n\}$. An element $\alpha \in \mathrm{S}(n)$ operates on $\Delta(n)$ by

$$\alpha^{-1}(\vartheta_1, \ldots, \vartheta_n) = (\vartheta_{\alpha(1)}, \ldots, \vartheta_{\alpha(n)}).$$

(3.2) Theorem. *The Weyl group of the unitary group* U(n) *is the full symmetric group* S(n).

Proof. Since

$$\begin{pmatrix} 0 & -1 \\ 1 & 0 \end{pmatrix} \begin{pmatrix} \zeta & 0 \\ 0 & \eta \end{pmatrix} \begin{pmatrix} 0 & 1 \\ -1 & 0 \end{pmatrix} = \begin{pmatrix} \eta & 0 \\ 0 & \zeta \end{pmatrix},$$

we see that every two-cycle, and hence every permutation, is induced by an element of the Weyl group. □

A diagonal matrix $D = (\vartheta_1, \ldots, \vartheta_n)$ is in $\mathrm{S}\Delta(n)$ if and only if

$$\vartheta_1 + \cdots + \vartheta_{n-1} = -\vartheta_n \mod \mathbb{Z}.$$

Thus the torus $\mathrm{S}\Delta(n)$ has dimension $n - 1$, but the ϑ_ν of a generating element are still pairwise distinct. Since the transformation used in the proof of (3.2) has determinant 1, we get

(3.3) Theorem. *The Weyl group of* SU(*n*) *is also the symmetric group* S(*n*). □

Next we consider the special orthogonal group SO(*n*). Recall that

$$\mathrm{SO}(2) \cong \mathrm{U}(1) \cong S^1 \cong \left\{ \begin{pmatrix} \cos 2\pi\vartheta & -\sin 2\pi\vartheta \\ \sin 2\pi\vartheta & \cos 2\pi\vartheta \end{pmatrix} \middle| \vartheta \in \mathbb{R}/\mathbb{Z} \right\}.$$

Also recall that we may decompose

$$\mathbb{R}^{2n} = \mathbb{R}^2 \oplus \cdots \oplus \mathbb{R}^2 \quad \text{into } n \text{ summands, and}$$
$$\mathbb{R}^{2n+1} = \mathbb{R}^2 \oplus \cdots \oplus \mathbb{R}^2 \oplus \mathbb{R} \quad \text{into } (n+1) \text{ summands}$$

to obtain the inclusions

$$T(n) = \mathrm{SO}(2) \times \cdots \times \mathrm{SO}(2) \subset \mathrm{SO}(2n) \subset \mathrm{SO}(2n+1).$$

(3.4) Theorem. *$T(n)$ is a maximal torus in* SO(2*n*) *and* SO(2*n* + 1).

PROOF. A torus $T \subset \mathrm{SO}(2n)$ (or $\mathrm{SO}(2n+1)$) is conjugate to one contained in $T(n)$ by II, (8.5). □

The tori $\Delta(n)$ and $T(n)$ correspond under the canonical inclusions $\mathrm{U}(n) \subset \mathrm{SO}(2n) \subset \mathrm{SO}(2n+1)$. With this in mind we also represent the elements of $T(n)$ by n-tuples $(\vartheta_1, \ldots, \vartheta_n)$, $\vartheta_\nu \in \mathbb{R}/\mathbb{Z}$. The element represented by such an n-tuple has

$$\begin{pmatrix} \cos 2\pi\vartheta_\nu & -\sin 2\pi\vartheta_\nu \\ \sin 2\pi\vartheta_\nu & \cos 2\pi\vartheta_\nu \end{pmatrix} \in \mathrm{SO}(2)$$

as its νth component. The following will help us describe the Weyl group of SO(*n*):

(3.5) Notation. We let

$$\vartheta_{-\nu} = -\vartheta_\nu \quad \text{for } \nu = 1, \ldots, n$$

and let G(*n*) be the group of permutations φ of the set $\{-n, \ldots, -1, 1, \ldots, n\}$ for which $\varphi(-\nu) = -\varphi(\nu)$. There is a split exact sequence

$$1 \to (\mathbb{Z}/2)^n \to \mathrm{G}(n) \to \mathrm{S}(n) \to 1,$$

where S(*n*) operates on $(\mathbb{Z}/2)^n$ by permuting the factors. In other words, if you are familiar with the concept, you may identify G(*n*) with the wreath product $(\mathbb{Z}/2) \wr \mathrm{S}(n)$ of $\mathbb{Z}/2$ by S(*n*). The group G(*n*) operates on the torus $T(n)$, its action being given by

$$\varphi^{-1}(\vartheta_1, \ldots, \vartheta_n) = (\vartheta_{\varphi(1)}, \ldots, \vartheta_{\varphi(n)}).$$

(3.6) Theorem. *The Weyl group of* SO(2*n* + 1) *is* G(*n*). *Also, the Weyl group of* SO(2*n*) *is* SG(*n*), *the subgroup of* G(*n*) *consisting of even permutations.*

PROOF. We begin with $SO(2n+1)$. The eigenvalues of the matrix associated to the generating element $(\vartheta_1, \ldots, \vartheta_n)$ are the numbers $\exp(2\pi i(\pm\vartheta_\nu))$. (This familiar observation from linear algebra is nothing more than the general equation $e_{\mathbb{R}}^{\mathbb{C}} \circ r_{\mathbb{R}}^{\mathbb{C}} V \cong V \oplus \bar{V}$ for representations applied to the special case of the standard representation of $\Delta(n)$, see II, (6.1).)

Now, these eigenvalues remain fixed under conjugation, so the generating element above is mapped to an element $(\lambda_1, \ldots, \lambda_n)$ with

$$(-\lambda_n, \ldots, -\lambda_1, \lambda_1, \ldots, \lambda_n) = (\vartheta_{\varphi(-n)}, \ldots, \vartheta_{\varphi(-1)}, \vartheta_{\varphi(1)}, \ldots, \vartheta_{\varphi(n)})$$

for some permutation φ of $\{-n, \ldots, -1, 1, \ldots, n\}$. Since the $|\vartheta_\nu|$ are all distinct for a generating element, we certainly have $W \subset G(n)$. We need to show that $G(n) \subset W$.

Case 1. Let $\varphi \in G(n)$ and $\varphi\{1, \ldots, n\} = \{1, \ldots, n\}$. That is to say $\varphi \in S(n) \subset (\mathbb{Z}/2)^n \cdot S(n) = G(n)$ in the description of $G(n)$ as a semidirect product. Then we may use an equation like the one we had earlier, namely

$$\begin{pmatrix} 0 & 1 \\ 1 & 0 \end{pmatrix} \begin{pmatrix} A & 0 \\ 0 & B \end{pmatrix} \begin{pmatrix} 0 & 1 \\ 1 & 0 \end{pmatrix} = \begin{pmatrix} B & 0 \\ 0 & A \end{pmatrix}, \qquad 1, A, B \in SO(2).$$

Case 2. Let $\varphi \in G(n)$ be the element given by $\varphi(-1) = 1$ and $\varphi(\nu) = \nu$ for $|\nu| > 1$. This is the element $(-1, 1, \ldots, 1) \in (\mathbb{Z}/2)^n \subset G(n)$. Then conjugation by the matrix

$$h = \begin{bmatrix} -1 & & & & \\ & 1 & & & \\ & & \ddots & & \\ & & & 1 & \\ & & & & -1 \end{bmatrix}$$

yields φ.

Since $G(n) = (\mathbb{Z}/2)^n \cdot S(n)$ is clearly generated by the elements considered in Cases 1 and 2, we have $G(n) \subset W$.

Now we turn to the group $SO(2n)$. The subgroup $S(n)$ of $G(n)$ operates via even permutations and, as in Case 1 above, lies in the Weyl group. The parity of a permutation in $G(n) = (\mathbb{Z}/2)^n \cdot S(n)$ depends only on the factor in $(\mathbb{Z}/2)^n$, and even permutations in $(\mathbb{Z}/2)^n$ may be obtained via conjugation by a product of an even number of matrices of the type in Case 2. Since such a product is in $SO(2n)$, we conclude that $SG(n) \subset W$.

We now have to show that W is not bigger than $SG(n)$. Since $[G(n) : SG(n)] = 2$, it suffices to show that the permutation φ with $\varphi(1) = -1$ and $\varphi(\nu) = \nu$ for $|\nu| > 1$ is not in W. But if conjugation by some $g \in SO(2n)$ gave rise to this automorphism of $T(n)$, we would have $g^{-1}h$ acting as the identity on $T(n)$, where h is the matrix in Case 2 above. This leads to $g^{-1}h \in T(n) \subset SO(2n)$ by (2.4), and this, in turn, says that $h \in SO(2n)$, which is false. □

Next in line is the symplectic group $\mathrm{Sp}(n) \subset \mathrm{U}(2n)$ of unitary matrices of the form $\begin{pmatrix} A & -\bar{B} \\ B & \bar{A} \end{pmatrix}$. This is the group of $\mathbb{H}$-linear endomorphisms of $\mathbb{H}^n$ which are unitary when viewed as endomorphisms of $\mathbb{C}^{2n}$. Recall that we have a canonical inclusion $\mathrm{U}(n) \to \mathrm{Sp}(n)$, $A \mapsto \begin{pmatrix} A & 0 \\ 0 & \bar{A} \end{pmatrix} = A + 0j$. We let $T^n \subset \mathrm{Sp}(n)$ be the image of the torus $\Delta(n)$ under this inclusion; see I, (1.10).

(3.7) Theorem. *The torus T^n is maximal in* Sp(*n*).

PROOF. We show that $Z(T^n) = T^n$. Assume that $x \in Z(T^n)$ with $x = A + Bj$ for some $A, B \in \mathrm{GL}(n, \mathbb{C})$. Since x commutes with the element $iE \in \Delta(n) = T^n$, we have $iA + iBj = ix = xi = iA - iBj$. Thus $B = 0$ and $A \in \mathrm{U}(n)$. Since $\Delta(n)$ is maximal abelian in $\mathrm{U}(n)$, it follows that $A \in \Delta(n)$, which means $x \in T^n$. □

The group $\mathrm{G}(n)$ operates on $T^n = \Delta(n) = T(n)$ as in (3.5), where we identify these tori using the inclusions $\mathrm{Sp}(n) \supset \mathrm{U}(n) \subset \mathrm{SO}(2n)$.

(3.8) Theorem. *The Weyl group of* Sp(*n*) *is* G(*n*).

PROOF. The Weyl group W is certainly contained in $\mathrm{G}(n)$, since the eigenvalues of a generating element

$$t = \begin{bmatrix} z_1 & & & & & \\ & \ddots & & & & \\ & & z_n & & & \\ & & & \bar{z}_1 & & \\ & & & & \ddots & \\ & & & & & \bar{z}_n \end{bmatrix}$$

of T^n remain fixed under the operation of W, see the proof of (3.6). On the other hand, the subgroup $\mathrm{S}(n)$ of $\mathrm{G}(n)$ is contained in W since $\mathrm{S}(n)$ is the Weyl group of $\mathrm{U}(n)$. In other words, the transformations of the torus corresponding to the elements of $\mathrm{S}(n)$ are obtained via conjugation by unitary matrices. To get all of $\mathrm{G}(n)$, one applies the equation $j\lambda j^{-1} = \bar{\lambda}$ in $\mathbb{H}$ in order to exchange λ_ν with $\bar{\lambda}_\nu$ while leaving the other components fixed. □

As our final group we look at the spinor group, i.e., the universal cover $\rho\colon \mathrm{Spin}(n) \to \mathrm{SO}(n)$. The reader may wish to refer to the description of the spinor group and the projection ρ in I, §6.

(3.9) Theorem. *The elements of the form*

$$(\cos\eta_1 - e_1e_2\sin\eta_1)\cdot(\cos\eta_2 - e_3e_4\sin\eta_2)\cdot\ldots\cdot(\cos\eta_n - e_{2n-1}e_{2n}\sin\eta_n)$$

with $0 \le \eta_\nu \le 2\pi$ *form a maximal torus* $\tilde{T}(n)$ *in* Spin(k) *for* $k = 2n$ *and* $k = 2n + 1$.

Proof. It is easy to check that these elements form a connected abelian group: multiplying two such elements amounts to adding the corresponding "angles" η_ν. This group is clearly the continuous image of the torus with coordinates η_ν, so $\tilde{T}(n)$ is compact and, in particular, closed. It follows that $\tilde{T}(n)$ is a torus. To show that $\tilde{T}(n)$ is maximal, it suffices to show that $\rho\tilde{T}(n) = T(n)$. But this can be seen by computing, for example, that the transformation $\rho(\cos\eta - e_1e_2\sin\eta)$ maps the plane determined by e_1 and e_2 onto itself via

$$\begin{pmatrix} \cos 2\eta & -\sin 2\eta \\ \sin 2\eta & \cos 2\eta \end{pmatrix},$$

while acting as the identity on the orthogonal complement spanned by $e_3, \ldots, e_k$. The reader is referred to the calculation in the proof of I, (6.17). □

We actually have an explicit description of the double cover

$$\rho\colon \tilde{T}(n) \to T(n).$$

The element written out in the statement of (3.9) is mapped onto $(1/\pi)(\eta_1, \ldots, \eta_n) \in T(n)$. We remark by way of caution that, for $n > 1$, the parameters η_ν in (3.9) do not give a diffeomorphism between $\mathbb{R}^n/\mathbb{Z}^n$ and $\tilde{T}(n)$. Indeed, all the subgroups $\{(\cos\eta - e_{2k-1}e_{2k}\sin\eta) \mid 0 \le \eta \le 2\pi\}$ isomorphic to S^1 contain (-1). But we do have an isomorphism

(3.10)
$$\beta\colon \mathbb{R}^n/\mathbb{Z}^n \to \tilde{T}(n),$$
$$(\zeta_1, \zeta_2, \ldots, \zeta_n) \mapsto \beta_1 \cdot \ldots \cdot \beta_n, \qquad \zeta_\nu \in \mathbb{R}/\mathbb{Z},$$

with

$$\beta_1 = \cos 2\pi\zeta_1 - e_1e_2 \sin 2\pi\zeta_1,$$
$$\beta_j = (\cos\pi\zeta_j + e_1e_2\sin\pi\zeta_j)(\cos\pi\zeta_j - e_{2j-1}e_{2j}\sin\pi\zeta_j), \qquad j > 1.$$

The map $\rho \circ \beta\colon \mathbb{R}^n/\mathbb{Z}^n \to T(n)$ is given by

(3.11)
$$(\zeta_1, \ldots, \zeta_n) \mapsto (2\zeta_1 - \zeta_2 - \cdots - \zeta_n, \zeta_2, \ldots, \zeta_n).$$

This map $\rho\beta$ is clearly a double cover of one torus by another with kernel $\mathbb{Z}/2$. Since the same is true for ρ, the map β must be an isomorphism.

$$\begin{array}{ccc} \mathbb{R}^n/\mathbb{Z}^n & \xrightarrow{\ \beta\ } & \tilde{T}(n) \\ & {\scriptstyle \rho\beta}\searrow \quad \swarrow{\scriptstyle \rho} & \\ & T(n) & \end{array}$$

Therefore we denote the element $x \in \tilde{T}(n)$ by the corresponding $\beta^{-1}(x) = (\zeta_1, \ldots, \zeta_n)$, and in this notation the projection $\tilde{T}(n) \to T(n)$ is given by (3.11).

By (2.9) the Weyl group W of Spin($2n$) (or Spin($2n + 1$)) is isomorphic to that of SO($2n$) (or SO($2n + 1$)). Its operation on $\tilde{T}(n)$ is completely determined by the fact that ρ is W-equivariant. We may also interpret (3.11) as a formula for $L\rho\colon L\tilde{T}(n) \to LT(n)$. Since $L\rho$ is a W-equivariant isomorphism, an element φ in the Weyl group SG(n) of SO($2n$) or G(n) of SO($2n + 1$) operates on $L\tilde{T}(n)$ via $(L\rho)^{-1} \circ \varphi \circ (L\rho)$. If one so desired one could write this operation out explicitly in terms of the coordinates $\zeta_1, \ldots, \zeta_n$.

$$
\textbf{(3.12)}\qquad
\begin{array}{ccccc}
(\zeta_1, \ldots, \zeta_n) & L\tilde{T}(n) & \xrightarrow{\ \varphi\ } & L\tilde{T}(n) & (\tfrac{1}{2}\sum \vartheta_\nu, \vartheta_2, \ldots, \vartheta_n) \\
\downarrow L\rho & \downarrow\cong & & \cong\downarrow & \uparrow (L\rho)^{-1} \\
(2\zeta_1 - \sum_{\nu>1} \zeta_\nu, \zeta_2, \ldots, \zeta_n) & LT(n) & \xrightarrow[\ \varphi\ ]{} & LT(n) & (\vartheta_1, \ldots, \vartheta_n)
\end{array}
$$

Incidentally, for $n \geq 2$ there is no isomorphism $\tilde{T}(n) \to T(n)$ of the tori themselves which is compatible with this operation of the group SG(n) ((3.14), Ex. 5).

As an application of the explicit formulas we have just derived, and as a preview of some of the general results to come later in this book, we will compute the representation ring of U(n).

The maximal torus $\Delta(n)$ has representation ring

$$\mathbb{Z}[z_1, \ldots, z_n, z_1^{-1}, \ldots, z_n^{-1}],$$

see II, (8.2) ff. The homomorphism $\mathrm{RU}(n) \to \mathrm{R}\Delta(n)$ of representation rings is injective, and the image is contained in $\mathrm{R}\Delta(n)^{\mathrm{S}(n)}$ since S(n) is the Weyl group of U(n), see (2.7). We write $\mathrm{R}\Delta(n)$ in the form

$$\mathrm{R}\Delta(n) = \mathbb{Z}[z_1, \ldots, z_n, \sigma_n^{-1}], \qquad \sigma_n = z_1 \cdots z_n.$$

Since σ_n is already symmetric, the subring of symmetric functions in $\mathrm{R}\Delta(n)$—i.e., the functions invariant under the Weyl group S(n)—is simply given by

$$\mathrm{R}\Delta(n)^{\mathrm{S}(n)} = \mathbb{Z}[\sigma_1, \ldots, \sigma_n, \sigma_n^{-1}],$$

where σ_ν is the νth elementary symmetric function. Namely, if $f \in \mathrm{R}\Delta(n)$ is symmetric, then $f \cdot \sigma_n^k \in \mathbb{Z}[z_1, \ldots, z_n]^{\mathrm{S}(n)}$ for some k. Thus $f \cdot \sigma_n^k \in \mathbb{Z}[\sigma_1, \ldots, \sigma_n]$ and $f \in \mathbb{Z}[\sigma_1, \ldots, \sigma_n, \sigma_n^{-1}]$.

(3.13) Application. *The representation ring* $\mathrm{RU}(n) \subset \mathrm{R}\Delta(n)$ *is the ring* $\mathbb{Z}[\sigma_1, \ldots, \sigma_n, \sigma_n^{-1}]$.

Proof. Let V be the standard representation of U(n), so $V = \mathbb{C}^n$ with basis $e_1, \ldots, e_n$ as a complex vector space. If we restrict the character of the U(n)-module $\Lambda^\nu V$ to the torus $\Delta(n)$, then we get σ_ν. In fact, the elements

$$e_{i_1} \wedge \cdots \wedge e_{i_\nu}, \qquad i_1 < \cdots < i_\nu$$

form a complex basis for $\Lambda^\nu V$. Now $\Delta(n)$ fixes the corresponding one-dimensional space, and the action of $(z_1, \ldots, z_n) \in \Delta(n)$ on $e_{i_1} \wedge \cdots \wedge e_{i_\nu}$ is multiplication by $z_{i_1} \cdot \ldots \cdot z_{i_\nu}$. Therefore the value of the character of $\Lambda^\nu V$ at $(z_1, \ldots, z_n) \in \Delta(n)$ is σ_ν.

The representation $\Lambda^n V$ is one-dimensional and hence equal to its character. The conjugate representation has the conjugated and hence inverted character. Thus $\sigma_1, \ldots, \sigma_n$ and σ_n^{-1} are all in the image of $\mathrm{RU}(n) \to \mathrm{R}\Delta(n)$, which is all we had left to show. □

(3.14) Exercises

1. What is the center Z of $U(n)$? Show that the composition $\mathrm{SU}(n) \subset \mathrm{U}(n) \to \mathrm{U}(n)/Z$ is surjective and find a maximal torus and the Weyl group of $\mathrm{U}(n)/Z$.

2. Show that $\mathrm{SO}(2n + 1)$ and $\mathrm{Sp}(n)$ have isomorphic representation rings. Use $\mathrm{R}G = \mathrm{R}T^W$.

3. Compute the representation ring of $\mathrm{SO}(2n + 1)$ in a fashion analogous to (3.13). The result you should get is the following: Let V be the standard real representation of $\mathrm{SO}(2n + 1)$ and $W = \mathbb{C} \otimes_{\mathbb{R}} V$ its complexification. Let $\lambda^j = \Lambda^j W$. Then
$$\mathrm{RSO}(2n + 1) = \mathbb{Z}[\lambda^1, \ldots, \lambda^n].$$
Hint: Let σ_ν be the νth elementary symmetric polynomial and
$$\tau_\nu = \sigma_\nu(z_1, \ldots, z_n, \bar{z}_1, \ldots, \bar{z}_n).$$
Then
$$\mathrm{R}T(n)^{\mathrm{G}(n)} = \mathbb{Z}[\tau_1, \ldots, \tau_n].$$
To prove this, one defines an appropriate lexicographic ordering on monomials in $\mathrm{R}T(n)^{\mathrm{G}(n)}$ and uses induction with respect to this ordering. We will recover this result in VI, (5.4).

4. Compute the representation ring of $\mathrm{SU}(n)$. The result is: Let V be the standard representation of $\mathrm{SU}(n)$ and $\lambda^k = \Lambda^k V$. Then $\mathrm{RSU}(n) = \mathbb{Z}[\lambda^1, \ldots, \lambda^{n-1}]$. In this case the representation ring of the maximal torus has a natural description as $\mathrm{RS}\Delta(n) = \mathbb{Z}[z_1, \ldots, z_n]/(\sigma_n - 1)$. Again we will recover this result in VI, (5.1).

5. Show that for $n \geq 2$ there is no isomorphism $\tilde{T}(n) \to T(n)$ which is compatible with the operation of $\mathrm{SG}(n)$ (given in (3.11)–(3.12)) on these tori. *Hint*: Show that there is no $\mathrm{SG}(n)$-equivariant endomorphism $T(n) \to T(n)$ whose kernel contains exactly two elements.

4. Cartan Subgroups of Nonconnected Compact Groups

Our aim in this section is to generalize some of the results on maximal tori to the case where G is compact but not necessarily connected. The reader who is primarily interested in the classical theory for connected groups will therefore suffer no loss by skipping this section. We will follow the treatment given by Segal [1].

Throughout this section G will only be assumed to be compact. We will let G_0 be the connected component of the identity and $\Gamma = G/G_0$ the group of components. A group is called ***topologically cyclic*** if it contains an element, called a ***generator***, whose powers are dense in the group. Recall that a compact Lie group is topologically cyclic precisely if it is isomorphic to the product of a torus and a finite cyclic group (I, (4.14)).

(4.1) Definition. A closed subgroup $S \subset G$ is called a ***Cartan subgroup*** if it is topologically cyclic and of finite index in its normalizer $\mathrm{N}(S)$. The finite group $\mathrm{W}(S) = \mathrm{N}(S)/S$ is called the ***Weyl group*** of S.

If G is in fact connected, then the Cartan subgroups are easily seen to be precisely the maximal tori of G ((4.10), Ex. 1).

(4.2) Proposition. *Each element $g \in G$ is contained in a Cartan subgroup S of G such that S/S_0 is generated by gS_0. Here S_0 is the connected component of the identity of S.*

Proof. Let T be a maximal torus in the centralizer $Z(g)$ of g and let S be the closed subgroup generated by T and g. Then S is compact and abelian, so S_0 is a torus. Since T is connected, $T \subset S_0$, and since T is maximal in S, we have $S_0 = T$. The finite group S/S_0 is generated by gS_0, so it is also cyclic. Thus S is topologically cyclic by I, (4.14).

It remains to show that $[\mathrm{N}(S):S]$ is finite. We have

$$[\mathrm{N}(S):S] = [\mathrm{N}(S):Z(S)] \cdot [Z(S):S].$$

Now $Z(S)/S$ is certainly smaller than $Z(S)/S_0$. Since $Z(S) \subset Z(g)$ and $T \subset Z(S)$, we see that T is a maximal torus in $Z(S)$ and hence in $Z(S)_0$. By (2.4), T is its own centralizer in $Z(S)_0$, so $S_0 = T = Z(S)_0$. Hence $Z(S)/S_0 = Z(S)/Z(S)_0$ is finite.

For the other factor, observe that the group $\mathrm{N}(S)/Z(S)$ can be viewed as being a subgroup of the group of automorphisms of S. Thus $gZ(S)$ operates on S via conjugation by g. Now S is isomorphic to a product $\mathbb{Z}/k \times S^1 \times \cdots \times S^1$, and since $\mathrm{Hom}(S^1, S^1) \cong \mathbb{Z}$ and $\mathrm{Hom}(S^1, \mathbb{Z}/k) = 0$, we see that the automorphisms of S form a countable discrete group. Since $\mathrm{N}(S)/Z(S)$ is compact, it is finite. □

(4.3) Proposition. *If S is a Cartan subgroup of G generated by z, then any $g \in G_0 z$ is conjugate to an element of $S_0 z$ via conjugation by an element of G_0. In other words, the mapping*

$$\kappa\colon G_0 \times S_0 z \to G_0 z, \qquad (g, sz) \mapsto gszg^{-1}$$

is surjective.

PROOF. We use a modified version of the proof of (1.7). Since S is abelian and hence leaves $S_0 z$ pointwise fixed under conjugation, the map κ induces a map

$$q\colon G_0/(G_0 \cap S) \times S_0 z \to G_0 z.$$

The inverse image of z under q consists of those pairs $(g(G_0 \cap S), sz)$ such that $gszg^{-1} = z$. Since $gszg^{-1} = z$ implies that $g^{-1}zg = sz \in S_0 z \subset S$, and z generates S, the equation for g implies that $g \in \mathrm{N}(S)$. Letting $B = \{b \in G_0 \cap \mathrm{N}(S) \mid b^{-1}zb \in S_0 z\}$ (so B is the closed subgroup of elements which induce the identity on S/S_0 when acting via conjugation), we see that $q^{-1}\{z\}$ is the finite set

$$\{(b(G_0 \cap S), b^{-1}zb) \mid b \in B\}.$$

The finite group $F = B/(G_0 \cap S) \subset \mathrm{N}(S)/S$ operates on the right on $G_0/(G_0 \cap S)$ via the translation $(g(G_0 \cap S), b(G_0 \cap S)) \mapsto gb(G_0 \cap S)$ and on the right on $S_0 z$ via conjugation $(sz, b(G_0 \cap S)) \mapsto b^{-1}szb$. This yields a diagonal operation of F on $G_0/(G_0 \cap S) \times S_0 z$ for which the map q is equivariant if F acts trivially on $G_0 z$. Since the operation of F on $G_0/(G_0 \cap S)$ is free—i.e., the only $f \in F$ with a fixed point is the identity—the quotient $M = (G_0/(G_0 \cap S) \times S_0 z)/F$ possesses a canonical differentiable structure such that the projection $\pi\colon G_0/(G_0 \cap S) \times S_0 z \to M$ is a submersion of differentiable manifolds (I, (4.15), Ex. 3). This induces a differentiable map

$$\tilde{q}\colon M \to G_0 z$$

between manifolds such that $\tilde{q} \circ \pi = q$ with $\tilde{q}^{-1}\{z\}$ consisting of a single point, the class of (e, z) in M. Note that M and $G_0 z$ are compact, connected, and have the same dimension. Thus the proposition follows from the following two lemmas.

(4.4) Lemma. *The differential of $\tilde{q}$ at the point represented by (e, z) in M is regular.*

(4.5) Lemma. *Let $f\colon M \to N$ be a differentiable map between compact connected manifolds of the same dimension. Suppose there is a point $p \in N$ such that $f^{-1}\{p\}$ consists of a single x at which the differential of f is bijective. Then f is surjective.*

PROOF OF LEMMA (4.4). It suffices to show that the differential of $\kappa\colon G_0 \times S_0 z \to G_0 z$ is surjective at the point (e, z). Consider the map $G_0 \times S_0 \to G_0$ given by the composition

$$(g, s) \underset{r_z}{\mapsto} (g, sz) \underset{\kappa}{\mapsto} gszg^{-1} \underset{r_{z^{-1}}}{\longrightarrow} gszg^{-1}z^{-1} = g \cdot s \cdot c(z)(g^{-1}).$$

Since r_z and $r_{z^{-1}}$ are diffeomorphisms, we may achieve our goal by showing that the differential of this map is surjective at (e, e). Splitting $\mathrm{L}G_0$ into $\mathrm{L}G_0 = \mathrm{L}(G_0/S_0) \oplus \mathrm{L}(S_0)$ as in the proof of (1.8), the differential under consideration is given by

$$(X, Y) \mapsto X + Y - \mathrm{Ad}_{G_0/S_0}(z)X.$$

Thus we need only show that 1 is not an eigenvalue of $\mathrm{Ad}_{G_0/S_0}(z)$. But this follows as before from the observation that if 1 were an eigenvalue we could find a one-parameter group commuting with z but not contained in S_0. This would contradict the finiteness of $N(S)/S$. □

Proof of Lemma (4.5). We once again use an argument concerning mapping degrees, but we are unable to refer directly to I, (5.19) because our manifolds need not be orientable. However, the lemma is a consequence of the following more general result.

Statement. *If $f_0, f_1: M \to N$ are differentiably homotopic, and $p \in N$ is a regular value of both f_0 and f_1, then $f_0^{-1}\{p\}$ and $f_1^{-1}\{p\}$ contain the same number of points modulo 2.*

Once we have established this statement, we may apply it as follows: Suppose p and q are two regular values of f. Then, as in the proof of I, (5.19), there is a diffeomorphism $\varphi: N \to N$ which is homotopic to the identity and maps q to p. Letting $f_0 = f$ and $f_1 = \varphi \circ f$, we see that if $f^{-1}\{p\}$ contains exactly one point, then $f^{-1}\{q\}$ contains an odd number of points and hence is nonempty.

Proof of the Statement. Let $F: M \times [0, 1] \to N$ be a differentiable homotopy between f_0 and f_1. By Sard's theorem (Bröcker and Jänich [1], §6 or Milnor [1]) we may find points $q \in N$ arbitrarily close to p which are simultaneously regular values for F, f_0, and f_1. If we take such a q, then $F^{-1}\{q\}$ is a one-dimensional manifold with boundary $f_0^{-1}\{q\} \cup f_1^{-1}\{q\}$.

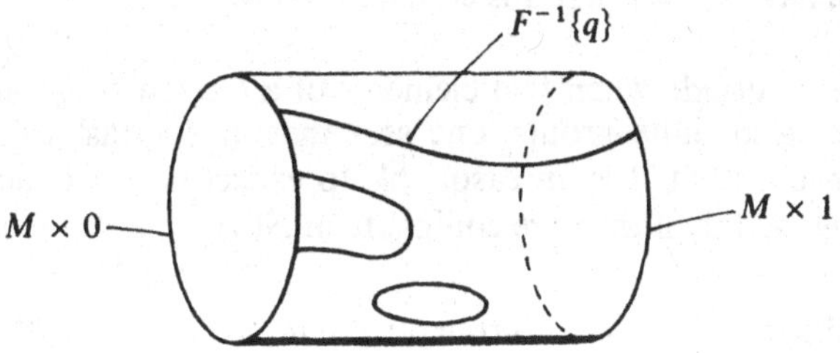

Figure 18

Since a compact one-dimensional manifold with boundary is easily seen to be diffeomorphic to a disjoint union of finitely many circles and intervals ((4.10), Ex. 2), the number of points in $f_0^{-1}\{q\} \cup f_1^{-1}\{q\}$ must be even. Since M is compact and p is a regular value of both f_0 and f_1, we have $f_0^{-1}\{q\} \cong f_0^{-1}\{p\}$ and $f_1^{-1}\{q\} \cong f_1^{-1}\{p\}$ for q close enough to p. □

The inclusion of a Cartan subgroup $S \subset G$ induces a homomorphism $S/S_0 \to \Gamma = G/G_0$. The image of this homomorphism is a cyclic subgroup of Γ denoted by $p(S)$. Conjugate Cartan subgroups have conjugate images under p.

(4.6) Proposition. *The projection*

$$p\colon \{\text{Cartan subgroups of } G\} \to \{\text{cyclic subgroups of } \Gamma\}$$

induces a bijection of conjugacy classes.

PROOF. A cyclic subgroup of Γ is generated by an element gG_0, and (4.2) tells us that g is contained in a Cartan subgroup S such that gS_0 generates S/S_0. Thus p is surjective.

Next, if S is a Cartan subgroup and $\gamma \in \Gamma$ is a generator of $p(S)$, then there is a generator $z \in S$ which is mapped to γ by $S \to S/(G_0 \cap S) \subset \Gamma$. In fact, $S = S_0 \times \mathbb{Z}/k$, and the generators of S are those elements decomposing into $z = (s, \alpha)$, where s is a generator of the torus S_0 and α generates $\mathbb{Z}/k$. Since S_0 projects to the unit in Γ, one need only verify that the inverse image of a generator of $\mathbb{Z}/l$ under a surjective homomorphism $\mathbb{Z}/k \to \mathbb{Z}/l$ always contains a generator of $\mathbb{Z}/k$ ((4.10), Ex. 3).

Now, if S is a Cartan subgroup generated by z and R is another Cartan subgroup with $p(S)$ conjugate to $p(R)$, then there is a $g \in G$ such that gzg^{-1} represents a generator of $p(R)$. Hence $gzg^{-1} \in rG_0$ for a generator r of R. But (4.3) then implies that there is a $g_0 \in G_0$ such that $g_0gzg^{-1}g_0^{-1} \in rR_0$. Thus S is conjugate to a subgroup of R, and vice versa. But compact Lie groups cannot be conjugate to proper subgroups—more generally, a compact manifold is never diffeomorphic to a proper submanifold ((4.10), Ex. 4). Therefore we have shown that S is conjugate to R. □

It remains to decide when two elements of a Cartan subgroup are conjugate. Looking at finite groups one sees that in contrast to the case of connected groups (2.5), it is unreasonable to expect that two elements of S will be conjugate only if they are conjugate in $N(S)$.

(4.7) Proposition. *Let S be a Cartan subgroup of G and let S^* denote the subset of elements which project onto a generator of S/S_0. Then two elements of S^* are conjugate in G if and only if they are conjugate in* $N(S)$.

PROOF. If x and gxg^{-1} are both in S^*, then S and $g^{-1}Sg$ are Cartan subgroups of G with $x \in S^* \cap (g^{-1}Sg)^*$. In particular, both S and $g^{-1}Sg$ are Cartan subgroups of $Z(x)$. Since x projects to a generator of both S/S_0 and $(g^{-1}Sg)/(g^{-1}Sg)_0$, we know by (4.6) that the two subgroups are conjugate in $Z(x)$. This means there is a $z \in Z(x)$ such that $zg^{-1}Sgz^{-1} = S$. Hence $zg^{-1} \in N(S)$ and $(zg^{-1})^{-1}x(zg^{-1}) = gxg^{-1}$ as required. □

(4.8) Corollary. *Let $[G]$ be the space of conjugacy classes of G with the induced quotient topology, and $[\Gamma]$ be the same for Γ. Then the inverse image* *of a class $\gamma \in [\Gamma]$ under the projection $[G] \to [\Gamma]$ is homeomorphic to $S^*/W(S)$, where S is a Cartan subgroup of G with a generator in γ.* □

(4.9) Proposition. *An element $g \in G$ is called* **general** *if it generates a Cartan subgroup and* **special** *otherwise. Suppose $H \subset G$ is a closed subgroup. Then there are only finitely many fixed points under the action of a general element on G/H via left translation.*

PROOF. We have $gxH = xH$ if and only if $g(xHx^{-1}) = xHx^{-1}$ and we have a G-equivariant diffeomorphism $G/H \to G/(xHx^{-1})$, $yH \mapsto yHx^{-1} = yx^{-1}(xHx^{-1})$. Thus if there are any fixed points we might as well assume that $gH = H$, i.e., $g \in H$. Now, let S be the Cartan subgroup generated by g. Then we have $gyH = yH \Leftrightarrow y^{-1}gy \in H \Leftrightarrow y^{-1}Sy \subset H$. Since by (4.6) there are only finitely many conjugacy classes of Cartan subgroups in H, we may choose a finite subset B of G such that $y^{-1}Sy \subset H$ if and only if $y^{-1}Sy = h^{-1}b^{-1}Sbh$ for some $h \in H$ and $b \in B$. That is to say, the coset yH is fixed by g if and only if $y \in N(S) \cdot b \cdot H$ for some $b \in B$. In the bijection

$$N(S)/(bHb^{-1} \cap N(S)) \to (N(S) \cdot b \cdot H)/H, \qquad [n] \mapsto [nb],$$

the first set is finite since it is a quotient of $N(S)/S$. Hence the second set and therefore the fixed-point set of g is finite. □

(4.10) Exercises

1. Show that a subgroup of a compact connected Lie group is a Cartan subgroup if and only if it is a maximal torus.
2. Show that up to diffeomorphism there are precisely four nonempty connected one-dimensional manifolds with boundary (the boundary may be empty). One possible approach involves first constructing a nonvanishing vector field.
3. Let $\varphi: \mathbb{Z}/k \to \mathbb{Z}/l$ be a surjective homomorphism, and let $g \in \mathbb{Z}/l$ be any generator. Show that $\varphi^{-1}\{g\}$ contains a generator of $\mathbb{Z}/k$.
4. Show that a compact smooth manifold cannot be diffeomorphic to a proper submanifold. *Hint*: Argue by dimension and number of components.
5. Consider the nontrivial semidirect product $D = S^1 \cdot (\mathbb{Z}/4)$ of S^1 and $\mathbb{Z}/4$. Elements of D are pairs (ϑ, n) with $\vartheta \in S^1$ and $n \in \mathbb{Z}/4$, and multiplication is given by

$$(\zeta, n) \cdot (\vartheta, m) = (\zeta \cdot \vartheta^{\varepsilon}, n + m), \qquad \varepsilon = (-1)^n.$$

Let $G = D/(\mathbb{Z}/2)$, where $\mathbb{Z}/2 \in D$ is the subgroup generated by $(e^{\pi i}, 2)$. Show that a Cartan subgroup of G is conjugate to either S^1 or $\mathbb{Z}/4$, where these two groups are considered as subgroups of G in the obvious fashion. This points out that Cartan subgroups may have different dimensions and that S_0 need not be equal to $S \cap G_0$. Show that $Z(S) = S$ and $W(S) = \mathbb{Z}/2$ for both Cartan subgroups S of D.

6. Show that the set of general elements is dense in any compact Lie group.

7. (Generalization of (2.9)). Let $K \subset G$ be a closed subgroup of finite index in its normalizer, and let $H \subset G$ be any closed subgroup. Show that $(G/H)^K$ is finite. *Hint*: In general, if K and H are closed, then $(G/H)^K$ has finitely many $N(K)/K$-orbits, cf. Bredon [1], II.57. The essential tool needed is quoted in II, (4.17), Ex. 2.

8. Show that the groups O(2) and SO(3) have infinitely many conjugacy classes of closed subgroups K with $N(K)/K$ finite. See also V, (2.15), Ex. 5.

9. Show that the homomorphism

$$\mathrm{res}_S^G\colon R(G) \to \prod_S R(S)$$

is injective, where S runs through the family of all Cartan subgroups $S \subset G$.

CHAPTER V
Root Systems

As we saw in the last chapter, the space of conjugacy classes of a compact connected Lie group G is the orbit space T/W of a maximal torus T under the action of the Weyl group W. Characters are really functions on T/W, and the Weyl integral formula transforms integrals of class functions on G into integrals of W-invariant functions on T. We therefore wish to investigate the structure of T as a W-manifold and of $\mathrm{L}T$ as a W-module. To do this we study the restriction of the adjoint representation to the maximal torus. In other words, we study $\mathrm{L}(G/T)$ as a T-module. It turns out that this representation is really the fundamental object of the theory—it contains all the information needed to specify both the structure of Lie groups and the arithmetic of their characters.

In the first section we compute all groups of rank 1. This is used in §2, where the root system of a Lie group first appears. Sections 3, 4, and 5 deal with general root systems and their classification, after which the root systems of the classical matrix groups are computed in §6. The fundamental group is the focus of §7, and §8 concerns itself with the structure of compact connected Lie groups in general. Sections 1–4 and 6 are the ones primarily needed for representation theory.

1. The Adjoint Representation and Groups of Rank 1

In this chapter G will always be a nontrivial compact connected Lie group and T a fixed maximal torus in G. Note that $\dim T > 0$ since G possesses nontrivial connected abelian subgroups, for example, one-parameter groups.

The infinitesimal structure of conjugation by elements of G in a neighborhood of the identity is described by the adjoint representation I, (2.10)

$$\mathrm{Ad}: G \times \mathrm{L}(G) \to \mathrm{L}(G).$$

We therefore wish to investigate the adjoint representation, and this turns out to require precise knowledge of groups of rank 1. Our plan, then, is to make a few general remarks concerning the weights of Ad and to proceed to classify all groups of rank 1. We will give two proofs of this classification: One due to H. Hopf [2], which is topological in nature and one which utilizes computations in the Lie algebra.

The infinitesimal form II, (2.9) of the adjoint representation is simply multiplication in the Lie algebra $\mathrm{L}(G)$

$$\mathrm{L}(G) \times \mathrm{L}(G) \to \mathrm{L}(G), \qquad (X, Y) \mapsto [X, Y]$$

as can be seen directly from the definition of $\mathrm{ad}(X)$ in I, (2.11). Since a representation of G is determined up to isomorphism by its restriction to the maximal torus T (IV, (2.8)), we will study $\mathrm{L}(G)$ as a T-module. As we know from II, (8.2), a T-module is determined by its weights and weight spaces. We will use the three different formulations of infinitesimal weights given in II, §9.

(1.1) Definition. The (infinitesimal) ***real weights*** of a G-module V are the (infinitesimal) real weights of V viewed as a T-module through restriction.

We remind the reader that, in the case of a real module V, one first has to pass to the complexification $V_{\mathbb{C}} = \mathbb{C} \otimes V$ and then consider the weights there (II, (8.2), (9.2)). We therefore need to investigate the weights of the adjoint representation $LG_{\mathbb{C}}$.

If T' is a different maximal torus we have a commutative diagram

$$\begin{array}{ccc} T \times V & \longrightarrow & V \\ \Big\downarrow{\scriptstyle c(g) \times l_g} & & \Big\downarrow{\scriptstyle l_g} \\ T' \times V & \longrightarrow & V \end{array}$$

with $c(g)t = gtg^{-1}$. It follows that the map $\alpha \mapsto \alpha \circ c(g)^{-1}$ induces a bijection between weights of V as a T-module and as a T'-module. The analogous statement is true for infinitesimal weights with $c(g)$ replaced by its differential. Applying these observations to the case $T = T'$ yields:

(1.2) Proposition. *Let $P(V)$ be the set of weights of V. Then the Weyl group $W = \mathrm{W}(G, T)$ acts on $P(V)$ via*

$$W \times P(V) \to P(V), \qquad (w, \alpha) \mapsto w(\alpha) = \alpha \circ w^{-1}.$$

If V is real, then $w(\bar{\alpha}) = \overline{w(\alpha)}$. In other words, the operation of W commutes with conjugation.

Conjugation is replaced by the map $\theta \mapsto -\theta$ in the corresponding statements for infinitesimal and real weights.

(1.3) Definition. The nontrivial weights of the adjoint representation are called the ***roots*** of G. In order to differentiate precisely between types of weights, we call the global weights $\vartheta: T \to U(1)$ the ***global*** roots of G, the corresponding linear forms $\Theta = L\vartheta: LT \to i\mathbb{R}$ the ***infinitesimal*** roots of G, the induced linear forms $\Phi: LT_{\mathbb{C}} \to \mathbb{C}$ the ***complex*** roots of G, and the linear forms $\alpha = \Theta/2\pi i$ the ***real*** roots of G.

In the theory of Lie algebras, it is the complex roots of G which are called the roots of $LG_{\mathbb{C}}$. Note that the complex roots are not the complexified real but rather the complexified infinitesimal roots, see II, (9.6), (9.7).

The real roots are more convenient for the purposes of investigating the elementary geometry of the corresponding vectors in the dual space LT^* (root systems, §3). The complex roots are more convenient for computations in the classical groups (§6) and for describing the structure of $LG_{\mathbb{C}}$.

Let $R\langle\mathbb{C}\rangle$ be the set of complex roots and $R = R\langle\mathbb{R}\rangle$ be the set of real roots of G. The decomposition into weight spaces then has the form

$$LG_{\mathbb{C}} = \bigoplus_{\alpha} L_\alpha, \qquad \alpha \in R\langle\mathbb{C}\rangle \cup \{0\}$$

with weight spaces

$$L_\alpha = \{X \in LG_{\mathbb{C}} \mid [H, X] = \alpha(H)X \text{ for all } H \in LT_{\mathbb{C}}\}.$$

Of course L_α is defined for an arbitrary linear form $\alpha: LT_{\mathbb{C}} \to \mathbb{C}$. In fact, we have

(1.4) Lemma. $[L_\alpha, L_\beta] \subset L_{\alpha+\beta}$.

Proof. By $[L_\alpha, L_\beta]$ we mean of course the set of all $[X_\alpha, X_\beta]$ with $X_\alpha \in L_\alpha$ and $X_\beta \in L_\beta$. If $H \subset LT_{\mathbb{C}}$, the Jacobi identity I, (2.12)(ii) gives us

$$\begin{aligned}[H, [X_\alpha, X_\beta]] &= -[X_\alpha, [X_\beta, H]] - [X_\beta, [H, X_\alpha]] \\ &= [X_\alpha, \beta(H)X_\beta] - [X_\beta, \alpha(H)X_\alpha] \\ &= (\alpha(H) + \beta(H)) \cdot [X_\alpha, X_\beta].\end{aligned}$$ □

(1.5) Theorem. *Let G be a compact connected Lie group of rank* 1 *with* $\dim G > 1$. *Then* $\dim G = 3$ *and the Weyl group of G has order* 2.

From this theorem, of which we give two proofs, we obtain the following

(1.6) Corollary. *With the hypotheses of the theorem either* $G \cong SO(3)$ *or* $G \cong SU(2) \cong Sp(1) \cong Spin(3)$.

We will first prove (1.6) from (1.5), but in doing so we will only use the conclusion that $\dim G = 3$. Since the Weyl groups of the groups listed in (1.6) all have order 2, it will only be necessary to prove that $\dim G = 3$ when proving theorem (1.5).

PROOF OF COROLLARY (1.6). Choose an Ad-invariant metric on LG. Then the adjoint representation gives us a homomorphism

$$\mathrm{Ad}\colon G \to \mathrm{SO}(3)$$

which we claim is immersive. Were this not the case, there would be a non-zero $X \in LG$ in the kernel of $\mathrm{ad} = L\mathrm{Ad}$. The one-parameter group $\{\exp(tX) \mid t \in \mathbb{R}\}$ would then be invariant under conjugation. And since $\dim T = 1$, the maximal torus T would have to lie in the center of G and hence would be normal. But this contradicts the conjugation theorem, since $\dim G > 1$.

It follows that $\mathrm{Ad}\colon G \to \mathrm{SO}(3)$ is a covering. But there are only two coverings of SO(3)—the trivial and the universal (double) cover $\mathrm{Spin}(3) \to \mathrm{SO}(3)$. □

ALGEBRAIC PROOF OF THEOREM (1.5). Fix an $H \in LT\backslash\{0\}$. Taking the structure of the infinitesimal roots (II, §9) into account, there are numbers $n_\alpha \in \mathbb{R}$ such that

$$[H, X_\alpha] = i n_\alpha X_\alpha$$

for all $X_\alpha \in L_\alpha$. Furthermore, $n_{-\alpha} = -n_\alpha$.

Let $c\colon LG_{\mathbb{C}} \to LG_{\mathbb{C}}$ denote complex conjugation. Suppose $0 \neq X_\alpha \in L_\alpha$. Then $cX_\alpha \in L_{-\alpha}$. If $[X_\alpha, cX_\alpha]$ were zero, then the subalgebra of LG generated by $X_\alpha + cX_\alpha$ and $i(X_\alpha - cX_\alpha)$ would be abelian. But since G has rank 1 there are no abelian subalgebras of dimension greater than 1, see IV, (1.12), Ex. 3. Thus $[X_\alpha, cX_\alpha] \neq 0$. However $[X_\alpha, cX_\alpha] \in L_0$, and therefore the subalgebra of LG generated by H and $i[X_\alpha, cX_\alpha]$ is abelian. So we end up with $[X_\alpha, cX_\alpha] = \lambda \cdot H$, $\lambda \in i\mathbb{R}\backslash\{0\}$.

Now set $R_+ = \{\alpha \mid n_\alpha > 0\}$ and let n_β be the smallest of the numbers n_α, $\alpha \in R_+$. Choose $X_\beta \in L_\beta\backslash\{0\}$ and set $X_{-\beta} = cX_\beta$. Then the subspace V of $LG_{\mathbb{C}}$ generated by $LT_{\mathbb{C}}$, $\bigoplus_{\alpha \in R_+} L_\alpha$, and $X_{-\beta}$ is mapped into itself by $\mathrm{ad}\, H\colon X \mapsto [H, X]$. On V the endomorphism $\mathrm{ad}\, H$ has trace

$$i \sum_{\alpha \in R_+\backslash\{\beta\}} n_\alpha \dim L_\alpha + i n_\beta(\dim L_\beta - 1).$$

By lemma (1.4), $\mathrm{ad}\, X_\beta$ and $\mathrm{ad}\, X_{-\beta}$ also map V into V. We have seen that $[X_\beta, X_{-\beta}] = \lambda H$, $\lambda \neq 0$, so ((1.7), Ex. 1.5)

$$\mathrm{ad}\, H = -\frac{1}{\lambda}(\mathrm{ad}\, X_\beta\, \mathrm{ad}\, X_{-\beta} - \mathrm{ad}\, X_{-\beta}\, \mathrm{ad}\, X_\beta).$$

Therefore we have that ad H has trace zero on V. This is only possible if $L_\alpha = 0$ for $\alpha \in R_+ \setminus \{\beta\}$ and dim $L_\beta = 1$. This implies that

$$LG_{\mathbb{C}} = LT_{\mathbb{C}} \oplus L_\beta \oplus L_{-\beta},$$

and this space has dimension 3. □

Topological Proof of Theorem (1.5). We start by choosing a Euclidean metric on LG which is invariant under the adjoint representation as in the proof of (1.6). Let $T \cong S^1$ be a maximal torus in G and let $H \in LT$ a vector of norm 1. Consider the map

$$f\colon G \to S^{n-1} \subset LG, \qquad n = \dim G,$$
$$g \mapsto \mathrm{Ad}(g)H.$$

For $t \in T$ we have $f(gt) = \mathrm{Ad}(gt)H = \mathrm{Ad}(g)\mathrm{Ad}(t)H = \mathrm{Ad}(g)H = f(g)$, so f factors through a map $\varphi\colon G/T \to S^{n-1}$:

$$\begin{array}{ccc} G & \xrightarrow{\ f\ } & S^{n-1} \\ & {\scriptstyle\pi}\searrow \quad \nearrow{\scriptstyle\varphi} & \\ & G/T & \end{array}$$

(i) **Claim.** *The map φ is injective.*

Proof of (i). Suppose that $\mathrm{Ad}(g)H = \mathrm{Ad}(b)H$. Then $\mathrm{Ad}(bg^{-1})|LT = \mathrm{id}_{LT}$, and so (naturality of the exponential map I, (3.2)) T is fixed under conjugation by bg^{-1}. Thus $bg^{-1} \in Z(T) = T$, and $b \in gT$.

(ii) **Claim.** *The map φ is a diffeomorphism.*

Proof of (ii). Since both manifolds are compact the image of φ is closed. Moreover G/T and S^{n-1} have equal dimension. The group G acts on $S^{n-1} \subset$ LG via Ad and acts transitively on G/T via left translation. Since φ is equivariant for these operations, φ has constant rank. Since φ is injective, it must have rank $n - 1$ by the rank theorem of calculus (see Bröcker and Jänich [1], 5.4). Therefore φ is open. Hence $\varphi(G/T) \subset S^{n-1}$ is both open and closed, so φ is surjective. Since φ is locally invertible it is a diffeomorphism.

Now the projection

$$\pi\colon G \to G/T \cong S^{n-1}$$

is a locally trivial fibration with fiber $T \cong S^1$. Such a fibration induces an exact sequence of homotopy groups (Hu [1], V, 6, p. 152 or G. W. Whitehead [1])

$$\ldots \to \pi_2(S^{n-1}) \to \pi_1(T) \xrightarrow{i_*} \pi_1(G) \xrightarrow{\pi_*} \pi_1(S^{n-1}).$$

Since $n - 1 = \dim(G/T)$ must be even (see IV, (1.10)), we know that $n \neq 2$. Assume, then, that $n > 3$. In this case $\pi_2(S^{n-1}) = \pi_1(S^{n-1}) = 0$, so $i_*: \pi_1(T) \to \pi_1(G)$ is an isomorphism. We show that this is impossible.

Indeed by (ii) there is an element $g \in G$ such that $\mathrm{Ad}(g)H = -H$. The induced automorphism $\mathrm{c}(g)_*$ of $\pi_1(T) \cong \mathbb{Z}$ is the map $-\mathrm{id}$. But in G we may find a path $s \mapsto g_s$, $0 \leq s \leq 1$, with $g_0 = e$ and $g_1 = g$. This path gives us a homotopy $\mathrm{c}(g_s)$ between the automorphisms id and $\mathrm{c}(g)$ of G. Thus $\mathrm{c}(g)_*: \pi_1(G) \to \pi_1(G)$ is the identity, a contradiction. □

Note. The few facts about homotopy groups which were used in this proof are developed in (1.7), Ex. 2–4 in an elementary fashion.

(1.7) Exercises

1. Let G be a compact connected Lie group and $K \subset LG$ an abelian subalgebra of maximal dimension. Show that $K = LT$ for some maximal torus T in G.

2. A closed Euclidean cell D is a space homeomorphic to $[0, 1]^k$ for some k. Show that every locally trivial bundle

$$F \to E \xrightarrow{\pi} D$$

over a closed Euclidean cell is trivial. In other words, there is always a commutative diagram

$$\begin{array}{ccc} E & \xrightarrow{\cong} & D \times F \\ & {\scriptstyle\pi}\searrow \quad \swarrow{\scriptstyle\mathrm{pr}_1} & \\ & D & \end{array}$$

Hint: Start with a decomposition of $[0, 1]^k$ into sufficiently small cubes.

3. Show that every S^1-principal bundle

$$S^1 \to G \xrightarrow[\pi]{} S^n$$

is trivial if $n > 2$. *Hint*: Let D^- and D^+ be the lower and upper hemispheres of S^n. We are looking for a section $\sigma: S^n \to G$. Since the bundle over D^- is trivial (Exercise 2), there already is a section σ^- over D^-. And since the bundle over D^+ is also trivial, a section over D^+ is given by a map $D^+ \to S^1$ into the fiber. Restricting σ^- to $S^{n-1} = \partial D^- = \partial D^+$ defines a map $\partial D^+ \to S^1$ which must be extended to all of D^+. To do this, show that every map $S^{n-1} \to S^1$ is null-homotopic as it factors through $\exp: \mathbb{R} \to S^1$.

4. Use Exercise 3 to argue that the map $i_*: \pi_1(T) \to \pi_1(G)$ in the topological proof of (1.5) must be an isomorphism.

5. (a) Show that

$$\mathrm{ad}[X, Y] = \mathrm{ad}\, X\, \mathrm{ad}\, Y - \mathrm{ad}\, Y\, \mathrm{ad}\, X$$

in any Lie algebra L. In other words, $\mathrm{ad}: L \to \mathrm{End}\, L$ is always a homomorphism of Lie algebras.

(b) Let X and Y be endomorphisms of a finite-dimensional vector space. Show that $\mathrm{Tr}[X, Y] = 0$.

2. Roots and Weyl Chambers

Let R be the set of real roots of the Lie algebra LG associated to the maximal torus T and let

$$k = \operatorname{rank}(G) = \dim T.$$

The global root corresponding to the real root $\alpha\colon \mathrm{L}T \to \mathbb{R}$ is the homomorphism

$$\vartheta_\alpha\colon T \to \mathrm{U}(1), \qquad \exp(H) \mapsto e^{2\pi i\alpha(H)}.$$

Notice that $\vartheta_{-\alpha} = \bar{\vartheta}_\alpha$. The adjoint representation decomposes into the direct sum of its weight spaces

$$\text{(2.1)} \qquad \begin{aligned} \mathrm{L}G_{\mathbb{C}} &= L_0 \oplus \bigoplus_{\alpha\in R} L_\alpha, \\ \mathrm{L}G &= M_0 \oplus \bigoplus_{\alpha\in R^+} M_\alpha. \end{aligned}$$

Here $M_\alpha = (L_\alpha \oplus L_{-\alpha}) \cap \mathrm{L}G = M_{-\alpha}$, and R^+ is supposed to contain exactly one element from each pair $\{\alpha, -\alpha\}$ of real roots. Later we will fix R^+ more precisely.

We also use the notation

$$\text{(2.2)} \qquad U_\alpha = \ker \vartheta_\alpha = \ker \vartheta_{-\alpha}.$$

Thus U_α is a closed subgroup of T with codimension one and L_α is the isotypical summand of L$G_{\mathbb{C}}$ corresponding to the irreducible character ϑ_α. Similarly, the M_α are the real isotypical summands of the real T-module LG.

(2.3) Proposition.

(i) $M_0 = \mathrm{L}T$, $L_0 = \mathrm{L}T_{\mathbb{C}} = \mathbb{C} \otimes \mathrm{L}T$.
(ii) *If* $t \in T$ *and* $Z(t)$ *is the centralizer of* t *in* G, *then* $\mathrm{L}Z(t) = M_0 \oplus \bigoplus_{\alpha\in N} M_\alpha$, *where* $N = \{\alpha \in R^+ \mid t \in U_\alpha\}$.
(iii) $\bigcap_{\alpha\in R} U_\alpha = Z(G)$ *is the center of* G.
(iv) $\bigcup_{\alpha\in R} U_\alpha$ *is the set of singular elements of* G *in* T (*see* IV, (2.10)).

Proof. (i) If we decompose the T-module LG as $\mathrm{L}T \oplus \mathrm{L}(G/T)$, then IV, (1.10) says that a generator t of T acts on $\mathrm{L}(G/T)$ without any real eigenvalue. Therefore $\mathrm{L}T = M_0$ is the trivial summand of the T-module LG and $L_0 = \mathbb{C} \otimes M_0 = \mathbb{C} \otimes \mathrm{L}T$, see II, (8.7)(ii).

(ii) Recall that $\mathrm{Ad}(t)X = X \Leftrightarrow \mathrm{c}(t)\exp(sX) = \exp(sX)$ for all $s \Leftrightarrow \exp(sX) \in Z(t)$ for all $s \Leftrightarrow X \in \mathrm{L}Z(t)$. Thus $\mathrm{L}Z(t)$ is the eigenspace associated to the eigenvalue 1 of $\mathrm{Ad}(t)$. By the definition of U_α we have

$$\mathrm{Ad}(t)X = X \Leftrightarrow t \in U_\alpha$$

for all nonzero X in M_α.

(iii) $\bigcap_{\alpha\in R} U_\alpha = \{t \in T \mid \mathrm{Ad}(t) = \mathrm{id}\} = \{t \in T \mid \mathrm{c}(t) = \mathrm{id}\}$.

(iv) $\bigcup_{\alpha\in R} U_\alpha$ is just the set of those $t \in T$ such that $\mathrm{c}(t)$ fixes a one-parameter group not contained in T. □

From the proposition we see once again that

$$\dim G/T = \sum_{\alpha \in R^+} \dim_{\mathbb{R}} M_\alpha = 2 \sum_{\alpha \in R^+} \dim_{\mathbb{C}} L_\alpha$$

is even. The nontrivial irreducible real representations of the torus are two-dimensional.

The subgroups $U_\alpha \subset T$ need not be connected.

(2.4) Example. Let $G = \mathrm{Sp}(1) = \mathrm{SU}(2)$ and, as usual, let

$$T = S^1 = \left\{ \begin{pmatrix} \lambda & 0 \\ 0 & \bar{\lambda} \end{pmatrix} \middle| \, |\lambda| = 1 \right\}.$$

We then have

$$\begin{pmatrix} \lambda & 0 \\ 0 & \bar{\lambda} \end{pmatrix} \begin{pmatrix} 0 & x \\ -\bar{x} & 0 \end{pmatrix} \begin{pmatrix} \bar{\lambda} & 0 \\ 0 & \lambda \end{pmatrix} = \begin{pmatrix} 0 & \lambda^2 x \\ -\bar{\lambda}^2 \bar{x} & 0 \end{pmatrix}.$$

Thus SU(2) has a root α with $\vartheta_\alpha(\lambda) = \lambda^2$, and so $U_\alpha = \{1, -1\}$.

(2.5) Remark. If U is a closed subgroup of a torus T with codimension 1, then U is topologically cyclic.

Proof. We have $U = U_0 \times A$ where U_0 is a torus and A is finite (I, (3.7)). Also, the homomorphism $A \subset T \to T/U_0 \cong S^1$ is injective. Therefore A is cyclic and the statement follows from I, (4.14). □

Incidentally, A can have at most order two for any U_α as in (2.2). This will be shown in (2.10).

(2.6) Proposition. *The singular elements of G form a subset of codimension ≥ 3. More precisely:*

There exist a compact manifold M with $\dim M + 3 \leq \dim G$ and a smooth map $M \to G$ whose image contains all the singular elements of G.

Proof. Let u_α be a generator of U_α. By (2.3)(ii) we have $\dim Z(u_\alpha) \geq k + 2$, $k = \mathrm{rank}(G)$, and therefore $\dim Z(U_\alpha) \geq k + 2$. The elements conjugate to elements of U_α lie in the image of

$$G/Z(U_\alpha) \times U_\alpha \to G, \qquad (gZ, u) \mapsto gug^{-1}.$$

The dimension of the left-hand side is at most $\dim G - (k + 2) + k - 1 = \dim G - 3$. The proposition now follows from the fact that every singular element of G is conjugate to an element of some U_α. □

(2.7) Proposition. *Let U_α^0 denote the connected component of the identity in U_α.*

(i) *If $U_\alpha^0 = U_\beta^0$, then either $\alpha = \beta$ or $\alpha = -\beta$.*
(ii) *The real weight-spaces M_α of the ϑ_α have dimension 2.*
(iii) $\dim G = \operatorname{rank}(G) + 2m$, *where $2m = |R|$ is the number of roots of G.*

PROOF. (i) and (ii) follow directly from (2.3)(ii) and the following lemma (2.8), and (iii) then follows from (2.1).

(2.8) Lemma. *Let u be a generator of U_α^0 and let $k = \operatorname{rank}(G) = \dim T$. Then* $\dim Z(u) = k + 2$.

PROOF. $Z(u) = Z(U_\alpha^0)$ is connected by IV, (2.3)(ii). Consider the diagram

$$\begin{array}{ccc} Z(u) & \longrightarrow & Z(u)/U_\alpha^0 \\ \cup & & \cup \\ T & \longrightarrow & T/U_\alpha^0. \end{array}$$

Since T is a maximal torus in G and hence in $Z(u)$, the group T/U_α^0 is a maximal torus in $Z(u)/U_\alpha^0$, see IV, (2.9). But $\dim T/U_\alpha^0 = 1$ and $\dim Z(u) \geq k + 2$ by (2.3)(ii). Therefore theorem (1.5) concerning groups of rank one tells us that $\dim Z(u)/U_\alpha^0 = 3$. Thus $\dim Z(u) = k + 2$. □

(2.9) Theorem. *Let u_α be a generator of the subgroup U_α^0 of the maximal torus (see (2.7)).*

(i) *The Weyl group of $Z(u_\alpha) = Z(U_\alpha^0)$ has order 2.*
(ii) *If α and β are proportional real roots, then either $\alpha = \beta$ or $\alpha = -\beta$.*
(iii) *We have $Z(U_\alpha) = Z(U_\alpha^0)$. Thus if w_α is a generator of the Weyl group of $Z(u_\alpha)$, then $w_\alpha(u) = u$ for all $u \in U_\alpha$.*

PROOF. (i) By IV, (2.9) the Weyl group of $Z(u_\alpha)$ with respect to T is isomorphic to the Weyl group of $Z(u_\alpha)/U_\alpha^0$ with respect to T/U_α^0. The statement now follows from (1.5).

(ii) Let $\alpha = c \cdot \beta$, $c \in \mathbb{R}$. Then $\ker(\alpha) = \ker(\beta) = LU_\alpha^0 = LU_\beta^0$. Thus $U_\alpha^0 = U_\beta^0$ and by (2.7)(i) we have $\alpha = \pm\beta$.

(iii) By (2.5) U_α is topologically cyclic. If u is a generator then $Z(u) = Z(U_\alpha) \subset Z(U_\alpha^0) = Z(u_\alpha)$. But $\dim Z(u) \geq k + 2$ by (2.3)(ii) and $Z(U_\alpha^0)$ is connected by IV, (2.3)(ii). Thus $Z(u) = Z(u_\alpha)$. Since the automorphism w_α of the torus is induced by conjugation by an element of $Z(u_\alpha) = Z(U_\alpha)$, it leaves U_α pointwise fixed. □

(2.10) Corollary. *The subgroups U_α have at most two components.*

PROOF. The automorphism w_α of T in (2.9)(iii) induces an automorphism

$$\varphi\colon S^1 \cong T/U_\alpha^0 \to T/U_\alpha^0 \cong S^1.$$

This is not the identity since, as the proof of (2.8) shows, φ is the nontrivial element of the Weyl group of $Z(u)/U_\alpha^0$ and therefore operates nontrivially on the maximal torus T/U_α^0 of this group. Thus φ has exactly two fixed points and, since every component of U_α determines a fixed point, there are at most two components. □

In (2.18), Ex. 1, we explicitly describe the automorphism w_α of T.

Of course, the elements w_α lie in the Weyl group of G with respect to T. We are now in a position to give a more precise geometric description of the Weyl group and its action on LT.

(2.11) Definitions and Notation. The hyperplanes

$$\mathscr{H}_\alpha = LU_\alpha = \ker \alpha, \qquad \alpha \in R^+$$

decompose LT into finitely many convex regions, namely the nonempty sets of the form

$$\{v \in LT \mid \varepsilon_\alpha \cdot \alpha(v) > 0 \text{ for } \alpha \in R^+\},$$

with $\varepsilon_\alpha = \pm 1$. These regions are called ***Weyl chambers***. We choose a Euclidean inner product on LG which is invariant for the adjoint representation. We then consider W to be a group of orthogonal transformations of LT. Since W permutes the roots, W also operates on $LT \setminus \bigcup_{\alpha \in R} \mathscr{H}_\alpha$ and thereby on the set of Weyl chambers. The ***walls*** of a Weyl chamber K are those subsets $\bar{K} \cap \mathscr{H}_\alpha$ of LT which have dimension $k - 1$, where $k = \operatorname{rank}(G)$.

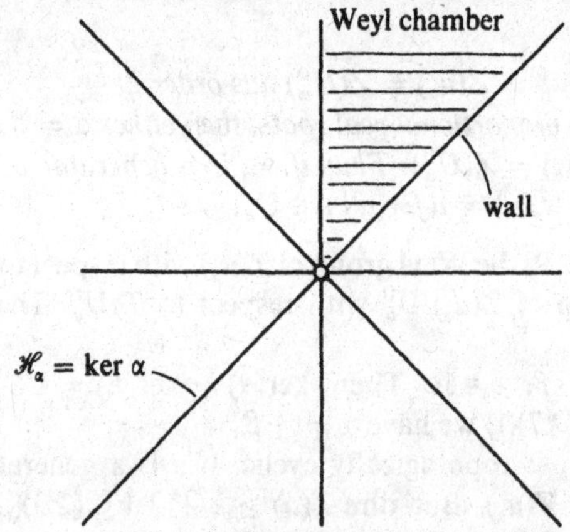

Figure 19

The Weyl chambers and their walls form the elementary geometric configuration which will be used to elucidate the action of the Weyl group.

Each hyperplane $\mathscr{H}_\alpha$ determines a ***reflection*** s_α of LT. Thus s_α is the nontrivial orthogonal transformation of LT which leaves $\mathscr{H}_\alpha$ pointwise fixed. If K is a Weyl chamber and $\bar{K} \cap \mathscr{H}_\alpha$ a wall of K, we also call s_α a reflection in this wall.

(2.12) Theorem.

(i) *The reflections* s_α *in the hyperplanes* $\mathscr{H}_\alpha = \ker \alpha$ *are elements of the Weyl group* W.
(ii) *The reflections in the walls of any given Weyl chamber* K *generate all of* W.
(iii) *The Weyl group acts simply transitively on the set of Weyl chambers (i.e., for every pair of Weyl chambers* K, K' *there is exactly one* $w \in W$ *with* $wK = K'$*).*

PROOF. (i) The inclusion $Z(U_\alpha) \subset G$ induces an inclusion of the Weyl groups with respect to T. Therefore the elements w_α from (2.9)(iii) are also in W. The transformation w_α of LT is nontrivial, orthogonal, and fixes $\mathscr{H}_\alpha = LU_\alpha$. Thus $w_\alpha = s_\alpha$.

As a first step leading to (ii) and (iii) we prove the

Claim. *Let* K *be a Weyl chamber. The subgroup* $W' \subset W$ *generated by the reflections in the walls of* K *acts transitively on the set of Weyl chambers.*

PROOF. Let L be another Weyl chamber and let $x \in K$, $y \in L$ be fixed. We have to show that there is a $w \in W'$ with $wy \in K$. Choose w so that the distance $|wy - x|$ is minimal (here we use the fact that W, and hence W', is finite). Now suppose that $wy \notin K$. Then wy and x are on different sides of a hyperplane $\mathscr{H}_\alpha$, $\alpha \in R$ which supports a wall of K.

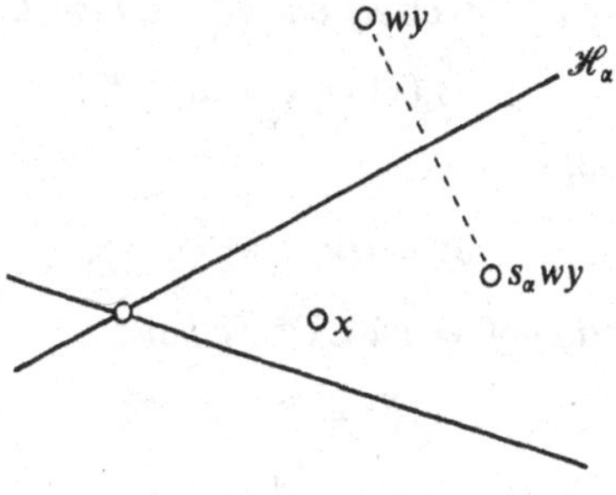

Figure 20

But then $|s_\alpha wy - x| < |wy - x|$, and this contradicts the choice of w.

Now (ii) and (iii) follow immediately once we have established the following

Claim. *If* K *is a Weyl chamber,* $w \in W$ *and* $wK \subset K$*, then* $w = 1$.

PROOF. Let n be the order of w as an element of W and let $x \in K$ be arbitrary. Then the point

$$y = \frac{1}{n} \sum_{j=1}^{n} w^j x$$

lies in K since K is convex. Note that, in particular, $y \neq 0$. Furthermore, $wy = y$ and therefore w has a fixed point in K. Since w operates linearly on LT, the entire line $\{s \cdot y \mid s \in \mathbb{R}\}$ is left fixed by w. If w is induced via conjugation by $g \in N(T) \subset G$, then conjugation by g fixes the one-parameter group $S = \{\exp(sy) \mid s \in \mathbb{R}\}$. By IV, (2.3)(ii) we can find a maximal torus which contains both g and S. But S is not contained in the singular set $\bigcup_{\alpha \in R} U_\alpha$ because $y \notin \bigcup_\alpha \mathscr{H}_\alpha$, see (2.3)(iv). Thus S is not in the intersection of two distinct maximal tori. Since $S \subset T$ we see that $g \in T$ and thus $w = 1$. □

In summary, the Weyl group acts freely on the complement of the hyperplanes $\mathscr{H}_\alpha$ and nowhere else. That is to say, the isotropy group is $\{1\}$ for each point in this complement, whereas $\mathscr{H}_\alpha$ is fixed by s_α. In (4.1) we will give a more detailed description of the action of the Weyl group on points of the walls.

The W-invariant inner product on LT determines an isomorphism

$$\kappa\colon LT \to LT^*, \qquad x \mapsto \langle x, \cdot \rangle.$$

The Weyl group acts on LT and dually on LT^*, and κ is a W-equivariant isomorphism. In what follows we will often identify LT with LT^* as W-modules via κ. The transformations s_α of LT and LT^* may be described as follows:

(2.13) Proposition.

(i) *For every $\alpha \in R$ there is precisely one $\alpha^* \in LT$ such that*

$$s_\alpha(x) = x - \alpha(x)\alpha^*$$

for all $x \in LT$, namely

$$\alpha^* = 2\kappa^{-1}(\alpha)/\langle \alpha, \alpha \rangle.$$

(ii) *For the dual operation of W on LT^* we have*

$$s_\alpha(\beta) = \beta - \beta(\alpha^*)\alpha$$

for all $\beta \in LT^*$. This, too, uniquely determines α^*.

Proof. We identify LT with LT^* via κ. The defining equation (i) of α^* is then

$$s_\alpha(x) = x - \langle \alpha, x \rangle \alpha^*.$$

If x is orthogonal to α, then both sides are just x. If $x = \alpha$, then the equation becomes $-\alpha = \alpha - \langle \alpha, \alpha \rangle \alpha^*$. Thus the equation holds precisely if

$$\alpha^* = 2\alpha/\langle \alpha, \alpha \rangle.$$

This proves (i).

For (ii), consider $\beta \in LT^*$. We then have $\beta s_\alpha(x) = \beta(x) - \beta(\alpha^*)\alpha(x)$ and, since $s_\alpha = s_\alpha^{-1}$, $s_\alpha(\beta) = \beta - \beta(\alpha^*)\alpha$. This is (ii), and if this equation holds for a particular α^* and all $\beta \in LT^*$, then (i) follows. □

With LT and LT^* identified via κ, the equations (2.13) take on the following form:

$$\textbf{(2.14)}\qquad \begin{aligned} s_\alpha(x) &= x - \langle \alpha, x\rangle\alpha^* = x - \langle \alpha^*, x\rangle\alpha, \\ \alpha^* &= 2\alpha/\langle \alpha, \alpha\rangle. \end{aligned}$$

(2.15) Notation. The elements $\alpha^* \in LT$ for $\alpha \in R$ are called ***inverse roots***. We denote the set of inverse roots by $R^* = \{\alpha^* \mid \alpha \in R\}$. The kernel I of the exponential map $\exp\colon LT \to T$ is called the ***integral lattice*** and the group $I^* = \{\alpha \in LT^* \mid \alpha I \subset \mathbb{Z}\}$ is called the ***lattice of integral forms***.

Observe that $\alpha^* \in LT$ is independent of the choice of inner product. If one wants to distinguish between LT and LT^*, one writes

$$\alpha^\vee = \kappa(\alpha^*) = 2\alpha/\langle \alpha, \alpha\rangle.$$

A real root α corresponds to a global root ϑ_α and the diagram

$$\begin{array}{ccccccccc} 0 & \longrightarrow & I & \longrightarrow & LT & \xrightarrow{\exp} & T & \longrightarrow & 1 \\ & & \downarrow & & \downarrow{\scriptstyle \alpha} & & \downarrow{\scriptstyle \vartheta_\alpha} & & \\ 0 & \longrightarrow & \mathbb{Z} & \longrightarrow & \mathbb{R} & \xrightarrow[e]{} & U(1) & \longrightarrow & 1, \end{array} \qquad e(t) = e^{2\pi i t},$$

shows that R is contained in I^*. In other words, the real roots are integral forms.

(2.16) Proposition. $R^* \subset I$, *that is to say, the inverse roots are integral.*

Proof. Let $\alpha \in R$ and $2x = \alpha^* \in LT$. By (2.13), $\alpha(\alpha^*) = 2$, and hence $\alpha(x) = 1$. Looking at the diagram above we see that $\exp(x) \in U_\alpha = \ker \vartheta_\alpha$. Next observe that $s_\alpha \in W$ also acts on T and that $\exp(s_\alpha x) = s_\alpha \exp(x) = \exp(x)$ since s_α is the identity on U_α by (2.9)(iii). But $s_\alpha x = -x$, so $\exp(-x) = \exp(x)$. Thus $\exp(\alpha^*) = \exp(2x) = 1$, and therefore $\alpha^* \in I$. □

(2.17) Corollary. *If* $\alpha, \beta \in R$ *then*

$$s_\alpha(\beta) = \beta - \beta(\alpha^*)\alpha, \qquad \beta(\alpha^*) \in \mathbb{Z}.$$

Note.

$$\alpha^* \in R^* \subset I \subset LT,$$
$$\alpha \in R \subset I^* \subset LT^*.$$

Observe, however, that even if we do identify LT with LT^* via κ, then α, α^*, and R, R^*, and I, I^* in general do *not* correspond under this identification.

(2.18) Exercises

1. Let w be a nontrivial automorphism of the torus $\mathbb{R}^k/\mathbb{Z}^k$ of finite order. Suppose further that the set of points left fixed by w has dimension $(k-1)$. Show that:

   (i) $w^2 = 1$.
   (ii) There is an automorphism γ of T such that $L(\gamma w \gamma^{-1})\colon \mathbb{R}^k \to \mathbb{R}^k$ sends $(x_1, \ldots, x_k)$ to $(-x_1, x_2 + vx_1, x_3, \ldots, x_k)$ for some $v \in \mathbb{Z}$.
   (iii) The fixed-point set of w has two components for v even and one for v odd.

N.B. For Exercises 2, 3, 4, 7 and 8 we assume that G is a compact connected Lie group, T a maximal torus in G, and W the associated Weyl group.

2. Suppose that $\operatorname{rank}(G) = 2$ and $\dim G = 2(n+1)$. Show that there is a split exact sequence

$$0 \to \mathbb{Z}/n \to W \to \mathbb{Z}/2 \to 0$$

   such that the generator of $\mathbb{Z}/2$ acts on $\mathbb{Z}/n$ by sending v to $-v$. Describe the decomposition into Weyl chambers. For more information, see (3.11) and (3.15), Ex. 3.

3. Let $\dim G = 4$ and $\operatorname{rank}(G) = 2$. Show that $Z(G)$ is isomorphic to S^1 or $S^1 \times \mathbb{Z}/2$ and that $G/Z(G) \cong SO(3)$.

4. Suppose that $\operatorname{rank}(G) = k$ and that $\dim Z(G) = 0$. Show that if $\dim G = 3k$, then $W \cong (\mathbb{Z}/2)^k$.

5. Let $\mathscr{H}_1, \ldots, \mathscr{H}_m$ be hyperplanes in $\mathbb{R}^k$ and let $s_1, \ldots, s_m$ be the reflections of $\mathbb{R}^k$ in the $\mathscr{H}_i$. Assume further that each s_i permutes the set of hyperplanes $\{\mathscr{H}_1, \ldots, \mathscr{H}_m\}$. Let $W \subseteq O(k)$ be the group generated by the s_i, and let K be a component of $\mathbb{R}^k \setminus \bigcup_i \mathscr{H}_i$. Reflections in the walls of K are defined as in (2.11). Show that:

   (i) W is finite.
   (ii) W is generated by reflections in the walls of K.
   (iii) W acts simply transitively on the set of components of $\mathbb{R}^k \setminus \bigcup_i \mathscr{H}_i$.

   Hint: In this case the fact that W is generated by the reflections $s_1, \ldots, s_m$ appears as part of the hypothesis. In the text we cannot assume this. The solution can be taken from the proof of (4.1).

6. Let $W \subseteq O(k)$ be a finite group generated by reflections as in Exercise 5 with $\dim(\mathscr{H}_1 \cap \cdots \cap \mathscr{H}_m) \leq 1$. Show that the normalizer of W in $O(k)$ is finite. Show that $O(k)$ contains infinitely many conjugacy classes of such subgroups if $k \geq 2$.

7. Let $H \subseteq G$ be a closed connected subgroup and suppose that the respective maximal tori T_H and T_G are chosen so that $T_H \subseteq T_G$. Show that every root of H is the restriction $\vartheta | T_H$ of a root $\vartheta\colon T_G \to U(1)$ of G.

8. With the hypotheses from Exercise 7, suppose that $\operatorname{rank}(H) = \operatorname{rank}(G)$. Show that every root of H is also a root of G, and thus the Weyl group of H is a subgroup of the Weyl group of G.

9. Show that for every $\alpha \in R$ there are elements

$$X_\alpha \in L_\alpha, \qquad X_{-\alpha} \in L_{-\alpha}, \quad \text{and} \quad H_\alpha \in LT_{\mathbb{C}}$$

such that

$$[H_\alpha, X_\alpha] = 2X_\alpha,$$

$$[H_\alpha, X_{-\alpha}] = -2X_{-\alpha},$$

$$[X_\alpha, X_{-\alpha}] = -H_\alpha,$$

Hint: Use the description of the structure of sl(2, ℂ) in II, (10.1), $LSO(3)_{\mathbb{C}} \cong LSU(2)_{\mathbb{C}} \cong$ sl(2, ℂ), and the proof of (2.8). Thus for each $\alpha \in R$ there is an sl(2, ℂ)-module structure on $LG_{\mathbb{C}}$, i.e., X_α, $X_{-\alpha}$, H_α generate a Lie algebra isomorphic to sl(2, ℂ).

3. Root Systems

The investigation of the T-module $LG_{\mathbb{C}}$ in the previous section displayed an elementary geometric object which governs the structure theory of Lie groups: the root system. In this section we look at the notion of a root system in general. Our exposition is based on that of Serre [3]—also see Humphreys [1].

In what follows V is a k-dimensional real vector space.

(3.1) Definition. A ***symmetry*** associated to a nonzero vector $\alpha \in V$ is an automorphism s of V which leaves some hyperplane $\mathscr{H} \subset V$ pointwise fixed and maps α to $-\alpha$.

(3.2) Lemma. *Let $R \subset V$ be a finite set which spans V. Then there is at most one symmetry associated to a vector α which maps R into itself.*

Proof. Let S be the group of automorphisms of V which map R into itself. Observe that S is finite since R is finite. We choose an S-invariant inner product on V. With respect to this metric, the symmetries in the lemma are orthogonal. Thus the only choice is reflection in the hyperplane orthogonal to α. □

(3.3) Definition. A finite subset $R \subset V$ satisfying the following is called a (reduced) ***root system in*** V:

(i) R spans V and $0 \notin R$.
(ii) If $\alpha, \beta \in R$ are proportional, then $\alpha = \beta$ or $\alpha = -\beta$.
(iii) To each vector $\alpha \in R$ there is an associated symmetry s_α mapping R into itself.
(iv) $s_\alpha(\beta) - \beta$ is an integer multiple of α for all $\alpha, \beta \in R$.

The dimension of V is called the ***rank*** of the root system and the elements of R are called ***roots***.

Property (ii) is often omitted in the literature. A root system with this property is then called ***reduced***. We, however, only consider reduced root systems.

The symmetry s_α in (iii) is uniquely determined due to (3.2). To every root α we associate the hyperplane $\mathscr{H}_\alpha$ left fixed by s_α and the element α^* in the dual V^* of V such that:

(3.4)
$$\alpha^* \mid \mathscr{H}_\alpha = 0, \qquad \alpha^*(\alpha) = 2 \quad \text{and hence}$$
$$s_\alpha(v) = v - \alpha^*(v) \cdot \alpha.$$

The element α^* is called the ***inverse root*** of α. Condition (iv) is then equivalent to

(iv)′
$$\alpha^*(\beta) \in \mathbb{Z} \quad \text{for all} \quad \alpha, \beta \in R.$$

From the symmetry of this formulation one gathers that the system R^* of inverse roots is a root system in V^* ((3.15), Ex. 2). This is called the ***inverse root system*** of R.

(3.5) Definition. The ***Weyl group*** of a root system R in V is the subgroup W of Aut(V) generated by the symmetries s_α in (3.3)(iii).

The root system R is mapped to itself by elements of W. Since R generates V and is finite, W is also finite. We choose a W-invariant Euclidean inner product $\langle\ ,\ \rangle$ on the W-module V. This allows us to identify V with its dual V^* via the isomorphism

$$\kappa\colon V \to V^*, \qquad v \mapsto \langle v, \cdot \rangle.$$

With respect to this metric, the symmetries $s_\alpha \in W$ are reflections of V in hyperplanes $\mathscr{H}_\alpha$ orthogonal to the α's. Such a reflection is given by

$$s_\alpha\colon v \mapsto v - 2\frac{\langle \alpha, v\rangle}{\langle \alpha, \alpha\rangle} \cdot \alpha.$$

Comparing this formula with the definition (3.4) of α^* and identifying V with V^* via κ, we get

(3.6)
$$\alpha^* = 2/|\alpha|^2 \cdot \alpha$$

as a vector in V and condition (iv) or (iv)′ in (3.3) says that

$$\langle \alpha^*, \beta \rangle = 2/|\alpha|^2 \cdot \langle \alpha, \beta \rangle \in \mathbb{Z} \quad \text{for all} \quad \alpha, \beta \in R.$$

(3.7) Definition. The integers

$$n_{\alpha\beta} = \langle \alpha^*, \beta \rangle = 2\frac{\langle \alpha, \beta\rangle}{\langle \alpha, \alpha\rangle}, \qquad \alpha, \beta \in R,$$

are called the ***Cartan numbers*** of the root system R.

These numbers are subject to rather strong restrictions, since

$$\langle \alpha, \beta \rangle = |\alpha| \cdot |\beta| \cdot \cos(\alpha, \beta),$$

where $\cos(\alpha, \beta)$ is the cosine of the angle $\measuredangle(\alpha, \beta)$ between α and β. Thus

$$n_{\alpha\beta} = 2\frac{|\beta|}{|\alpha|}\cos(\alpha, \beta), \tag{3.8}$$

$$n_{\alpha\beta} \cdot n_{\beta\alpha} = 4\cos^2(\alpha, \beta).$$

Since $n_{\alpha\beta}$ is an integer, $4\cos^2(\alpha, \beta)$ must be equal to 0, 1, 2, 3, or 4 and in the last case $\alpha = \pm\beta$. From this we immediately obtain:

(3.9) Proposition. *Let α, β be nonproportional roots and $\langle \alpha, \beta \rangle > 0$ (this means $n_{\alpha\beta} > 0$ or, equivalently, the roots α and β form an acute angle). Then $\alpha - \beta$ is also a root.*

PROOF. We have $n_{\alpha\beta} > 0$ if and only if $n_{\beta\alpha} > 0$. Thus if $n_{\alpha\beta} \cdot n_{\beta\alpha} < 4$, either $n_{\alpha\beta} = 1$ or $n_{\beta\alpha} = 1$. In the second case, $\alpha - \beta = \alpha - n_{\beta\alpha} \cdot \beta = \alpha - \langle \beta^*, \alpha \rangle \beta = s_\beta \alpha$. Similarly, in the first case we get $\beta - \alpha = s_\alpha \beta$. □

Using (3.8) we find that, possibly after switching α and β, there are only the following seven possibilities for nonproportional roots:

(3.10)

	$n_{\alpha\beta}$	$n_{\beta\alpha}$	$\measuredangle(\alpha, \beta)$	$\|\alpha\|^2/\|\beta\|^2$
(i)	0	0	$\pi/2$	?
(ii)	1	1	$\pi/3$	1
(iii)	-1	-1	$2\pi/3$	1
(iv)	1	2	$\pi/4$	2
(v)	-1	-2	$3\pi/4$	2
(vi)	1	3	$\pi/6$	3
(vii)	-1	-3	$5\pi/6$	3

Up to similarity transformation there is only one root system of rank 1 and, with the aid of the list above, the reader may easily verify that up to similarity the only root systems of rank 2 are the following ((3.15), Ex. 3). The names $A_1 \times A_1, \ldots$ of these diagrams are standard, see (5.6).

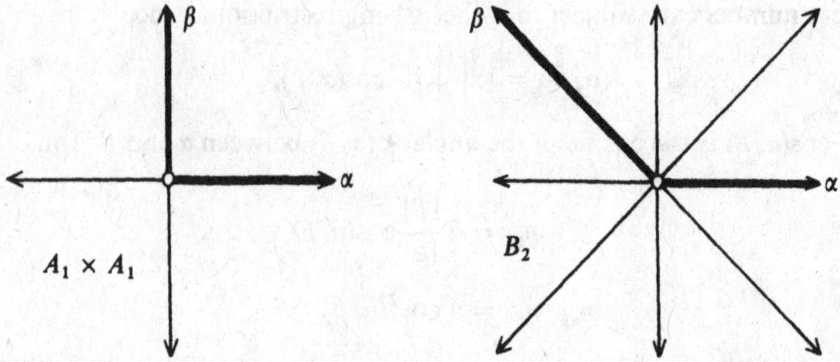

(3.11)

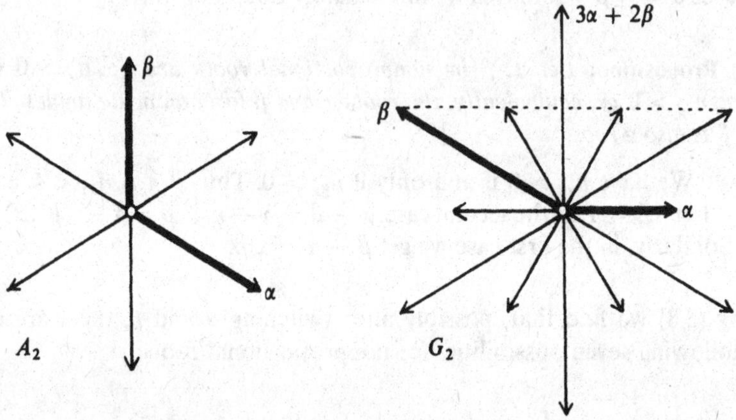

Figure 21

Finally, before we say anything more about root systems in general, we point out how what we said in the last section about the roots of a Lie group fits into our current scheme.

(3.12) Theorem. *Let G be a compact connected Lie group with maximal torus T and let $V \subset \mathrm{L}T^*$ be the subspace generated by the set R of real roots of G. Then R is a root system in V and the Weyl group W of G, viewed as a subgroup of* $\mathrm{Aut}(V)$, *is the same as the Weyl group of this root system defined in* (3.5).

Proof. We consider the individual parts of definition (3.3): (i) is trivial, (ii) is in (2.9)(ii), (iii) follows from (2.12), and (iv) follows from (2.17). □

Let Z_0 be the connected component of the identity of the center $Z(G)$. The W-module $\mathrm{L}T$ decomposes into the direct sum

$$\mathrm{L}T = \mathrm{L}Z_0 \oplus \mathrm{L}(T/Z_0).$$

Furthermore, $S = T/Z_0$ is the maximal torus of G/Z_0 and $\mathrm{L}Z_0 = \bigcap_{\alpha \in R} \mathscr{H}_\alpha$ is trivial as a W-module, see (2.3) and IV, (2.3). We have the corresponding decomposition of the W-module

$$\mathrm{L}T^* = \mathrm{L}Z_0^* \oplus V.$$

Thus $R \subset V$ is also the root system of G/Z_0.

This root system and the system of real roots of G in LT^* differ only by the trivial W-module LZ_0^*. For the purpose of the present discussion it does not hurt to simply forget about this difference.

(3.13) Definition. A Lie group is called ***semisimple*** if it possesses no abelian connected normal subgroup other than $\{1\}$.

(3.14) Remark. A compact connected Lie group is semisimple if and only if its center is finite.

PROOF. A connected abelian normal subgroup is a torus which is contained in every maximal torus and therefore in the center. □

Thus, for a semisimple compact connected Lie group, the root system R in LT^* is a root system in the sense of (3.3). We will give a more precise description of the structure of compact Lie groups in §8 of this chapter.

(3.15) Exercises

1. Let V be a finite-dimensional real vector space. Show that:
   (i) If s is an automorphism of V with finite order which leaves some hyperplane pointwise fixed, then either $s = \mathrm{id}$ or s is the symmetry associated to some vector α.
   (ii) If s is a symmetry associated to a vector α in V, then $s^2 = \mathrm{id}$.
2. Let R be a root system in V. Show that R^* is a root system in the dual V^* of V, that $\alpha^{**} = \alpha$ for $\alpha \in R$, and that R and R^* have isomorphic Weyl groups.
3. Verify that, up to similarity, the only root systems of rank 2 are those given in (3.11).
4. Show that the Weyl group of a root system in $\mathbb{R}^k$ has finite normalizer in $O(k)$.
5. Let R be a root system and α, β nonproportional roots. Let p be the smallest and q be the largest integer n with $\beta + n\alpha \in R$. Show that $\beta + n\alpha \in R$ for $p \leq n \leq q$.

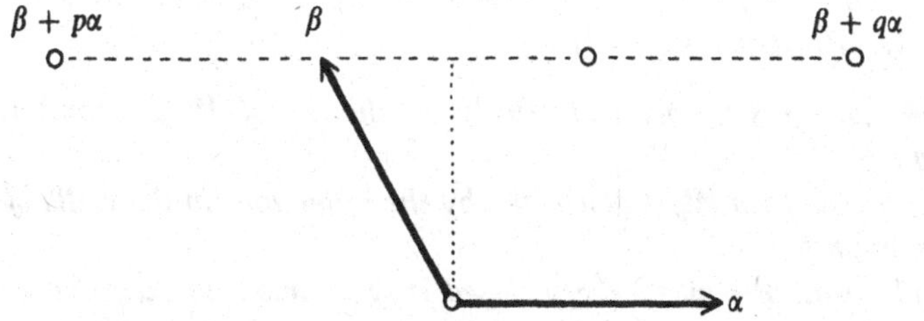

Figure 22

The set of roots of the form $\beta + n\alpha$, $n \in \mathbb{Z}$, is called ***the α-string through β***. Show that this string is invariant under s_α and, in particular, that $s_\alpha(\beta + p\alpha) = \beta + q\alpha$. Show that $p + q = -n_{\alpha\beta}$, that this string contains at most four elements, and that strings of length 2, 3, 4 actually exist. *Hint*: (3.9).

6. Let G be a compact connected Lie group and α, β real roots of G. Let

$$L(\alpha, \beta) = \bigoplus_{n=p}^{q} L_{\beta+n\alpha}$$

with p, q as in Exercise 5. Then $L(\alpha, \beta)$ is an $sl(2, \mathbb{C})$-submodule of $LG_{\mathbb{C}}$ by (2.18), Ex. 9, $L_{\beta+q\alpha}$ is primitive in $L(\alpha, \beta)$, see II, (10.3), and $L(\alpha, \beta)$ is a simple $sl(2, \mathbb{C})$-module. In particular, $[L_\alpha, L_\beta] = L_{\alpha+\beta}$ if $\alpha + \beta \in R$. Compare Bourbaki [1], VIII, §2, No. 2, Prop 4.

7. Show that $\alpha - \beta$ may also be a root when $\langle \alpha, \beta \rangle < 0$.

8. Let $\mathrm{Aut}(R)$ be the group of orthogonal automorphisms of V which map the root system R into itself. Show that the Weyl group W of R is normal in $\mathrm{Aut}(R)$.

9. Let R be a root system in V and $R' \subset R$ a subset satisfying: If $\alpha, \beta \in R'$, $m, n \in \mathbb{Z}$, and $m\alpha + n\beta \in R$, then $m\alpha + n\beta \in R'$. Show that R' is a root system in the subspace V' of V generated by R'.

10. Suppose the root system R contains a root of length c. Show that the set of all roots of length c forms a root system in the subspace they span.

4. Bases and Weyl Chambers

Let R be a root system in V with Weyl group W. To every root α we have associated the hyperplane $\mathscr{H}_\alpha \subset V$ which is fixed under the corresponding symmetry $s_\alpha \in W$. We introduce a W-invariant metric on V, so $\mathscr{H}_\alpha$ is the hyperplane orthogonal to α and s_α is the reflection in $\mathscr{H}_\alpha$. The hyperplanes $\mathscr{H}_\alpha$ divide V into finitely many convex regions—these are called the ***Weyl chambers*** of the root system. In other words, the Weyl chambers are the connected components of $V \setminus \bigcup_{\alpha \in R} \mathscr{H}_\alpha$. They are open by definition. We will denote the closure of a Weyl chamber K by $\bar{K}$. The ***walls*** of a Weyl chamber are defined as in (2.11). The Weyl group operates on the set of Weyl chambers and, as a generalization of (2.12), we have

(4.1) Theorem. *Let W_p be the isotropy group of the point $p \in V$ in the Weyl group W of a root system R in V.*

(i) *W_p operates simply transitively on the set of Weyl chambers K with $p \in \bar{K}$.*
(ii) *If $p \in \bar{K}$, then W_p is generated by the reflections in the walls of K which contain p.*
(iii) *The orbit of p meets every closed Weyl chamber in exactly one point.*

PROOF. We start by considering a special case.

(4.2) Lemma. *W is generated by reflections in the walls of any given Weyl chamber K and operates transitively on the set of all Weyl chambers.*

Proof. Let W' be the group generated by the reflections in the walls of K. As in the proof of (2.12), it follows that W' acts transitively on the set of Weyl chambers. Therefore we need to show that $W = W'$. Let s_α be the reflection in a hyperplane $\mathscr{H}_\alpha$. Since $\mathscr{H}_\alpha$ belongs to a wall of some Weyl chamber K_1, and $K_1 = wK$ for some $w \in W'$, we see that $\mathscr{H}_\alpha = w\mathscr{H}_\beta$ for a wall $\mathscr{H}_\beta \cap \bar{K}$ of K. Therefore $s_\alpha = ws_\beta w^{-1} \in W'$. Since, by definition, W is generated by the reflections s_α, it follows that $W = W'$. The theorem now follows easily from the following somewhat technical lemma.

(4.3) Lemma. *Let K be a Weyl chamber, $x, y \in \bar{K}$, and $w \in W$. If $wx = y$, then $x = y$ and w is a product of reflections in walls of K which contain x.*

Proof of Theorem (4.1). (i) Let K and L be Weyl chambers and $p \in \bar{K} \cap \bar{L}$. By (4.2) there is a $w \in W$ with $w\bar{L} = \bar{K}$. Since $p \in \bar{L}$ we have $wp \in \bar{K}$. But p is also in $\bar{K}$, so $p = wp$ by (4.3) and hence $w \in W_p$. This shows transitivity of W_p as claimed. If we choose for x in (4.3) a point of K, then x is contained in no wall. Thus the lemma says that the only element of W mapping x into K is the identity. Thus all of W acts simply transitively on the set of Weyl chambers.

(ii) follows immediately from (4.3) with $x = p$.

(iii) Since W acts transitively on the Weyl chambers, the orbit of p meets every closed Weyl chamber—in precisely one point by (4.3). □

Proof of Lemma (4.3). We express w as a product of reflections in walls of K and proceed by induction on the number of reflections in this expression. So let $w = s_{\mathscr{H}} \cdot w'$, where $s_{\mathscr{H}}$ is the reflection in the wall $\bar{K} \cap \mathscr{H}$ of K.

Case 1. The Weyl chambers K and wK lie on different sides of $\mathscr{H}$. Then $\bar{K} \cap w\bar{K} \subset \mathscr{H}$, and hence $y \in \mathscr{H}$. Therefore $y = s_{\mathscr{H}}y = s_{\mathscr{H}}wx = s_{\mathscr{H}}s_{\mathscr{H}}w'x = w'x$ and the statement follows by induction.

Case 2. The Weyl chambers K and wK lie on the same side of $\mathscr{H}$. Let $w' = s_1 \cdot \ldots \cdot s_k$ with each s_ν a reflection in a wall of K. There is a first l such that $s_{\mathscr{H}}s_1 \ldots s_l K$ lies on the same side of $\mathscr{H}$ as K. Let $u = s_{\mathscr{H}}s_1 \ldots s_{l-1}$. Then uK and us_lK lie on different sides of $\mathscr{H}$. Thus K and s_lK lie on different sides of $u^{-1}\mathscr{H}$, and so $\bar{K} \cap s_l\bar{K} \subset u^{-1}\mathscr{H}$. Thus s_l is the reflection in the hyperplane $u^{-1}\mathscr{H}$ and $s_l = u^{-1}s_{\mathscr{H}}u$. Plugging this into the expression $w = us_l \ldots s_k$, we can shorten this product. □

The hypotheses here are slightly different from those in the proof of (2.12). There are two points involved: First, that W is generated by the reflections s_α, $\alpha \in R$ and, second, that W operates freely on the set of Weyl chambers. Here we assume the first and derive the second, whereas in (2.12) we do the opposite.

Among other things, the theorem says: The Weyl group operates simply transitively on the set of Weyl chambers, every closed Weyl chamber is a fundamental domain of this operation, and every closed Weyl chamber is mapped homeomorphically onto the orbit space of the operation of W by the canonical projection.

A given Weyl chamber may be described by specifying its walls and on which side of these walls the chamber lies. But the side of a wall is given by a root. Thus the Weyl chambers correspond to certain subsets of R which we now describe more algebraically.

(4.4) Definition. A subset S of the root system R in V is called a ***basis*** or a ***system of simple roots*** if S is linearly independent in V and every root $\beta \in R$ may be written as

$$\beta = \sum_{\alpha \in S} m_\alpha \cdot \alpha$$

with integers m_α such that either all $m_\alpha \geq 0$ or all $m_\alpha \leq 0$. The elements of S are called ***simple roots*** with respect to S.

We will soon see that bases actually exist. Suppose, then, that S is a fixed basis. Then R splits into two disjoint parts

$$R = R_+ \cup R_-, \qquad R_- = -R_+,$$

where R_+ consists of those roots β whose coefficients m_α in (4.4) are all nonnegative. These are called ***positive roots***, and the elements of R_- are called ***negative roots***. In general, an element in a subset $R' \subset R$ is called ***decomposable in*** R' if it can be expressed as the sum of at least two elements of R'. Otherwise it is called ***indecomposable***. The basis obviously consists of the elements indecomposable in R_+ (in fact, every decomposable positive root is the sum of two positive roots ((4.15), Ex. 2)). In this way the basis S is determined by the corresponding set of positive roots R_+.

The next theorem says that bases exist and are in one-to-one correspondence with Weyl chambers.

(4.5) Proposition. *The following assignment defines a bijection between the set of Weyl chambers and the set of bases of the root system R in V.*

We assign to every Weyl chamber K the set of positive roots $R_+(K) = \{\alpha \in R \mid \langle \alpha, t\rangle > 0$ for one and therefore all $t \in K\}$ and the basis $S(K)$ consisting of the elements indecomposable in $R_+(K)$.

We assign to each basis S the Weyl chamber

$$K(S) = \{t \in V \mid \langle \alpha, t\rangle > 0 \textit{ for all } \alpha \in S\}.$$

*It has walls $\mathscr{H}_\alpha \cap \bar{K}(S)$, $\alpha \in S$, where $\mathscr{H}_\alpha$ is the hyperplane orthogonal to α. We call $K(S)$ the **fundamental** Weyl chamber corresponding to S.*

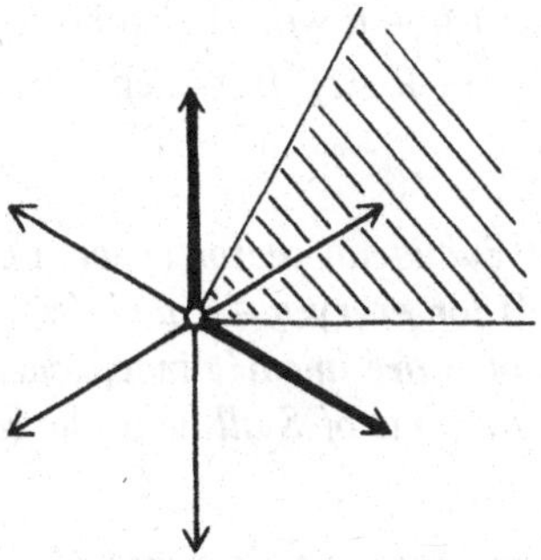

Figure 23

(4.6) Corollary.

(i) *If S is a basis of R, then the Weyl group is generated by the reflections s_α, $\alpha \in S$. These are called* **simple** *reflections.*
(ii) *The Weyl group operates simply transitively on the set of bases of R, and every root is an element of some basis.*

PROOF. The corollary follows from (4.1) with $p = 0$. Clearly every root α is orthogonal to a wall of some Weyl chamber K with $\langle \alpha, t \rangle > 0$ for $t \in K$.

We will prove the proposition in four steps:

Step 1. Let K be a Weyl chamber, and $t \in K$. Let

$$R_+(t) = \{\alpha \in R \mid \langle \alpha, t \rangle > 0\}$$

and let $S(t)$ be the set of elements indecomposible in $R_+(t)$. Then

$$R_+(K) = R_+(t), \qquad S(K) = S(t) \tag{4.7}$$

since the sign of $\langle \alpha, t \rangle$ as a function of t is constant on the connected set K. We show:

Every element of $R_+(t)$ is the sum of elements of $S(t)$.

PROOF. If not, choose a counterexample $\alpha \in R_+(t)$ with $\langle \alpha, t \rangle$ minimal. Since $\alpha \notin S(t)$, we have $\alpha = \alpha_1 + \cdots + \alpha_k$, $\alpha_\nu \in R_+(t)$, $k \geq 2$. But then $\langle \alpha_1, t \rangle < \langle \alpha, t \rangle$, a contradiction.

Step 2. $S(t)$ is a basis.

PROOF. We have to show that the elements of $S(t)$ are linearly independent. This follows from the next two lemmas (4.8) and (4.9) which, incidentally, are interesting in their own right:

(4.8) Lemma. *Let $\alpha, \beta \in S(t)$, $\alpha \neq \beta$. Then α and β form an obtuse angle, that is $\langle \alpha, \beta \rangle \leq 0$.*

Proof. Otherwise $\alpha - \beta$ and $\beta - \alpha$ would also be roots by (3.9) and one of these, say $\alpha - \beta$, would lie in $R_+(t)$. But then $\alpha = (\alpha - \beta) + \beta$ would be decomposable. □

(4.9) Lemma. *Let V be a Euclidean vector space, $t \in V$, and S contained in V. Suppose that $\langle \alpha, t \rangle > 0$ for every $\alpha \in S$ and $\langle \alpha, \beta \rangle \leq 0$ for all $\alpha, \beta \in S$, $\alpha \neq \beta$. Then the elements of S are linearly independent. Note that the hypotheses simply say that the elements of S all lie in the same half-space and that no two form an acute angle.*

Proof. A linear dependence between elements of S may be written in the form

$$\sum_\beta m_\beta \cdot \beta = \sum_\gamma n_\gamma \cdot \gamma, \qquad m_\beta \geq 0, \quad n_\gamma \geq 0,$$

where the β and γ run through disjoint subsets of S. Denoting one side of the equation by λ, we have

$$|\lambda|^2 = \sum_{\beta, \gamma} m_\beta \cdot n_\gamma \cdot \langle \beta, \gamma \rangle \leq 0,$$

and so $\lambda = 0$. Thus

$$0 = \langle \lambda, t \rangle = \sum_\beta m_\beta \langle \beta, t \rangle = \sum_\gamma n_\gamma \langle \gamma, t \rangle.$$

But by assumption we have $\langle \beta, t \rangle > 0$ and $\langle \gamma, t \rangle > 0$. Hence $m_\beta = n_\gamma = 0$ for all β, γ. □

Thus we have completed the second step and, in particular, proved the existence of bases.

Step 3. If S is a basis then $K(S)$ is a Weyl chamber with walls $\bar{K}(S) \cap \mathscr{H}_\alpha$, $\alpha \in S$.

Proof. The basis S determines a set R_+ of positive roots and $K(S) = \{t \mid \langle \alpha, t \rangle > 0 \text{ for all } \alpha \in R_+\}$. Since $\langle \alpha, t \rangle < 0$ for the negative roots α, all $t \in K(S)$ lie on the same side of every hyperplane $\mathscr{H}_\alpha$, $\alpha \in R$. Therefore $K(S)$ lies completely in a Weyl chamber K'. On the other hand, $\langle \alpha, t \rangle$ has constant sign on K' as a function of t for every root α. Thus $K' = K(S)$ and $K(S)$ is a Weyl chamber. The linear forms $\langle \alpha, \cdot \rangle$, $\alpha \in S$, form a basis of the dual V^* of V. Let B be the dual basis of $V = V^{**}$. Then $K(S)$ is the set of positive linear combinations of elements of B. The claim concerning the walls follows.

Step 4. $K = K(S(K))$ and $S = S(K(S))$.

Proof. Clearly $K \subset K(S(K))$ and, since $K(S(K))$ is a Weyl chamber, both are equal. Let R_+ be the set of positive roots corresponding to S and let $K(R_+) = K(S)$. Then it is equally clear that $R_+ \subset R_+(K(R_+))$, from which we again conclude equality. Thus the second equality follows.

This finishes the proof of (4.5). □

We now choose a fixed basis S of the root system R in V and let R_+ be the corresponding set of positive roots.

(4.10) Lemma. *If $\alpha \in S$ then the reflection s_α permutes the positive roots different from α.*

Proof. Let β be a positive root different from α. Then $\beta = \sum_{\gamma \in S} m_\gamma \cdot \gamma, m_\gamma \geq 0$. Since $\beta \neq \alpha$, we have $m_\gamma > 0$ for some $\gamma \neq \alpha$. But $s_\alpha(\beta) = \beta - n_{\alpha\beta} \cdot \alpha$. Thus, in the representation of $s_\alpha(\beta)$ as a linear combination of elements in the basis S, only the coefficient of α changes. In particular, the coefficient of γ is m_γ. Since this is positive, all coefficients are positive by definition of a basis. □

In applications to representation theory we will have to consider the half sum of the positive roots at key junctures. It is denoted by

$$\varrho = \tfrac{1}{2} \sum_{\alpha \in R_+} \alpha. \tag{4.11}$$

We collect a few simple facts about this element.

(4.12) Proposition. *Let S be a basis of the root system R in V. For every $\alpha \in S$ we have*

(i) $s_\alpha(\varrho) = \varrho - \alpha$.
(ii) $\langle \varrho, \alpha^* \rangle = 1$ *(see (3.6)).*
(iii) *If W is the Weyl group of R and $w \in W$, then*

$$w(\varrho) = \varrho - \sum_S n_\alpha \cdot \alpha, \qquad n_\alpha \in \mathbb{Z}.$$

(iv) $\langle \varrho, \beta^* \rangle$ *is an integer for every root β.*

Proof. (i) Let $\varrho_\alpha = \varrho - \alpha/2$. Then by (4.10) we have $s_\alpha(\varrho_\alpha) = \varrho_\alpha$. Since $s_\alpha(\alpha/2) = -\alpha/2$, we get $s_\alpha(\varrho) = \varrho - \alpha$.

(ii) We have $\langle \varrho, \alpha^* \rangle = \langle \varrho_\alpha, \alpha^* \rangle + \frac{1}{2}\langle \alpha, \alpha^* \rangle = \langle \varrho_\alpha, \alpha^* \rangle + 1$. Furthermore, $\langle \varrho_\alpha, \alpha^* \rangle = \langle s_\alpha(\varrho_\alpha), \alpha^* \rangle = \langle \varrho_\alpha, s_\alpha(\alpha^*) \rangle = -\langle \varrho_\alpha, \alpha^* \rangle$. Thus $\langle \varrho_\alpha, \alpha^* \rangle = 0$.

(iii) This follows from (i) by writing w as a product of reflections $s_\alpha, \alpha \in S$.

(iv) Let β be a root. By (4.6)(ii), there is a $w \in W$ such that $w^{-1}\beta \in S$. Thus $\langle \varrho, \beta^* \rangle = \langle \varrho - w\varrho, \beta^* \rangle + \langle w\varrho, \beta^* \rangle = \langle \varrho - w\varrho, \beta^* \rangle + \langle \varrho, (w^{-1}\beta)^* \rangle \in \mathbb{Z}$ by (ii) and (iii). □

Looking again at the case of a Lie group, the element $\varrho \in LT^*$ need not be an integral form. The element $\varrho - w\varrho$, however, is an integral form for every w in the Weyl group since it is a linear combination of roots with integer coefficients.

The inverse roots are integral elements. They generate a free abelian subgroup Γ of I. This subgroup has a geometric interpretation. In (7.1) we will show that I/Γ is the fundamental group of the Lie group G.

For the sake of convenience we identify LT and LT^* in this section by means of a W-invariant metric. This isn't really essential (see (4.15), Ex. 1). Nevertheless, we have to distinguish the integral lattice I and its dual $I^* = \{\alpha | \langle \alpha, \gamma \rangle \in \mathbb{Z} \text{ for } \gamma \in I\}$.

(4.13) Proposition. *If S is a basis of the root system R in V, then $S^* = \{\alpha^* | \alpha \in S\}$ is a basis of the inverse root system R^*. Thus S^* is a basis of the abelian group Γ generated by the inverse roots.*

PROOF. The second statement follows from the first. Since α^* and α differ only by a positive constant, the elements of S^* are linearly independent. Now given $\beta^* \in R^*$, we may find a composition of reflections s_α, with $\alpha \in S$, sending β^* to an element of S^*. Hence from the equation

$$s_\alpha(\beta^*) = \beta^* - \langle \alpha, \beta^* \rangle \alpha^* = \beta^* - n_{\beta\alpha} \cdot \alpha^*,$$

with $\alpha \in S$, we see that every inverse root is an integral linear combination of elements of S^*. This uniquely determines an expression for β^* of the form

$$\beta^* = \sum_{\gamma \in S} m_\gamma \cdot \gamma^*.$$

But since roots differ from their inverses by a positive factor, all the m_γ must have the same sign. □

The half sum ϱ of the positive roots lies in the fundamental chamber K since $\langle \varrho, \alpha^* \rangle = 1$ for all $\alpha \in S$. The roots of a compact connected Lie group lie in the lattice I^* of integral forms, and we note the following for later applications:

(4.14) Note. Let β be an integral form in LT^* and let ϱ be the half sum of the positive roots. Then

$$\beta + \varrho \in K \quad \text{if and only if} \quad \beta \in \bar{K}.$$

PROOF. For $\alpha \in S$ we have $\langle \varrho, \alpha^* \rangle = 1$ by (4.12) and $\langle \beta, \alpha^* \rangle \in \mathbb{Z}$ by (2.16). Thus $\langle \beta + \varrho, \alpha^* \rangle > 0$ for all $\alpha \in S$ precisely when $\langle \beta, \alpha^* \rangle \geq 0$ for all $\alpha \in S$. □

(4.15) Exercises

1. In this section we have identified the real vector space V with its dual V^* by choosing a W-invariant Euclidean metric. Show that the following do not depend upon the choice of metric: The decomposition of V into Weyl chambers, the hyperplanes $\mathscr{H}_\alpha$, the angles between the roots, and the bijection between Weyl chambers and bases. Show that the vectors α^*, $\alpha \in R$, correspond to well-defined elements of the dual V^*.

2. Show that every decomposable positive root is the sum of a positive and a simple positive root.

3. Show that there is precisely one element w in the Weyl group of a root system R which sends R_+ to R_-. Furthermore, if w is represented as a product of reflections s_α, $\alpha \in S$, then every s_α appears as a factor and the number of factors is at least as large as the cardinality of R_+.

4. If R is a root system and $A \subset R$, we set $A^* = \{\alpha^* \mid \alpha \in A\}$. Show that

$$(R_+)^* = (R^*)_+.$$

5. Show that the Weyl group of a root system R contains no reflections other than the s_α, $\alpha \in R$.

6. Show that there is precisely one homomorphism of the Weyl group to $\mathbb{Z}/2$ sending every reflection onto the generator of $\mathbb{Z}/2$.

7. Define Aut R as in (3.15), Ex. 7 and let $E = \text{Aut}\, R/W$. Show that Aut R is the semidirect product of W and E.

8. Let α and β be roots in a basis and let s_α and s_β be the associated reflections. Show that $s_\alpha s_\beta$ has order $n = 2, 3, 4$, or 6 according to whether the angle between α and β is $\pi/2$, $2\pi/3$, $3\pi/4$, or $5\pi/6$.

9. Let S be a basis of a root system in V. Suppose $\beta \in V$ is a vector such that for each w in the Weyl group all the nonzero coefficients of $w\beta$ with respect to S have the same sign. Show that β is a multiple of some root.

5. Dynkin Diagrams

In this paragraph we will give a short overview of Dynkin diagrams. Despite the fact that we will hardly need them, they belong to a systematic treatment of Lie theory and serve as a guide through what follows: A simply connected Lie group is determined by its Lie algebra and in this way the so-called compact semisimple Lie algebras are in one-to-one correspondence (up to isomorphism, of course) with the compact Lie groups. The compact semisimple Lie algebras are the real Lie algebras with negative definite Killing form (see (5.10) below). Continuing, complexification (tensoring with $\mathbb{C}$) yields a one-to-one correspondence between these and the complex semisimple Lie algebras. But the complex semisimple Lie algebras are classified by their root systems and the root systems are classified by their bases. Now, Dynkin diagrams give a quick and easy description of the bases. As a result, we have a bijective correspondence:

$$\text{Simply connected compact Lie groups} \leftrightarrow \text{Dynkin diagrams.}$$

We now clarify the Dynkin diagram of a root system and related topics.

Let R be a root system in V with basis S and Weyl group W. We assume that V is equipped with a W-invariant Euclidean inner product. The Cartan numbers $n_{\alpha\beta} = \langle \alpha^*, \beta \rangle = 2/|\alpha|^2 \cdot \langle \alpha, \beta \rangle$ form a matrix

(5.1) $$(n_{\alpha\beta}), \qquad \alpha, \beta \in S$$

called the ***Cartan matrix*** of the root system. We know that

$$n_{\alpha\alpha} = 2, \qquad n_{\alpha\beta} \in \{0, -1, -2, -3\} \quad \text{for } \alpha \neq \beta.$$

For example, the root system G_2 gives rise to the matrix $\begin{pmatrix} 2 & -1 \\ -3 & 2 \end{pmatrix}$. Of course, the order of the rows and columns depends upon an ordering of the roots in S. Up to this, the matrix is independent of the choice of basis because the Weyl group acts transitively on the set of all bases. It is an invariant of the root system. Cartan matrices are regular ((5.12), Ex. 2) and a root system is determined by its Cartan matrix. More precisely:

(5.2) Proposition. *Let R' be another root system in a vector space V' with basis S' and let $\varphi: S \to S'$ be a bijection such that*

$$n_{\alpha\beta} = n_{\varphi(\alpha)\varphi(\beta)}, \qquad \alpha, \beta \in S.$$

Then φ extends to exactly one linear isomorphism $\varphi: V \to V'$ which sends R to R'. Furthermore, $n_{\alpha\beta} = n_{\varphi(\alpha)\varphi(\beta)}$ for all $\alpha, \beta \in R$.

PROOF. The bijection φ has a unique linear extension $\varphi: V \to V'$ because S and S' are vector space bases. The formula for a reflection

$$s_\alpha(\beta) = \beta - n_{\alpha\beta} \cdot \alpha$$

for $\alpha, \beta \in S$ (or S') shows

$$\begin{aligned} \varphi \circ s_\alpha(\beta) = \varphi(\beta) - n_{\alpha\beta} \cdot \varphi(\alpha) &= \varphi(\beta) - n_{\varphi(\alpha)\varphi(\beta)} \cdot \varphi(\alpha) \\ &= s_{\varphi(\alpha)}(\varphi(\beta)). \end{aligned}$$

Therefore $\varphi \circ s_\alpha = s_{\varphi(\alpha)} \circ \varphi$. Thus, if W and W' are the Weyl groups of R and R', then φ induces an isomorphism

$$W \to W', \qquad w \mapsto \varphi w \varphi^{-1},$$

since the simple reflections generate the Weyl groups (see (4.6)). But the Weyl groups act transitively on the sets of bases and every root is in some basis, and so $\varphi R = R'$. The above formula $s_\alpha(\beta) = \beta - n_{\alpha\beta}\alpha$ then implies that all Cartan numbers are invariant under φ. □

Choosing $R = R'$, we obtain the automorphism group of the root system

(5.3) $$\operatorname{Aut}(R) = \{\varphi \in \operatorname{Aut}(V) \mid \varphi(R) = R\}.$$

Given a finite family of root systems R_ν in V_ν, $\nu = 1, \ldots, k$, we may form the ***sum*** of the root systems. It consists of the disjoint union

$$R = \bigcup_{\nu=1}^{k} R_\nu \quad \text{canonically embedded in} \quad V = \bigoplus_{\nu=1}^{k} V_\nu$$

with $\langle \alpha^*, \beta \rangle = 0$ for $\alpha \in V_\nu$, $\beta \in V_\mu$ if $\nu \neq \mu$ and $\langle \alpha^*, \beta \rangle$ equal to its value on V_ν if $\nu = \mu$.

Conversely, if the root system R in V is given, one can ask if R is the sum of two subsystems.

(5.4) Proposition. *Let $V = V_1 \oplus V_2$ and suppose that $R = R_1 \cup R_2$ with $R_\nu \subset V_\nu$, $\nu = 1, 2$. Then R_ν is a root system in V_ν and R is the sum of these root systems. In particular, V_1 is orthogonal to V_2.*

PROOF. It is obviously enough to check the last statement. So let $\alpha \in R_1$ and $\beta \in R_2$. Then $\alpha - \beta \notin V_1 \cup V_2$ and hence by assumption $\alpha - \beta \notin R$. By (3.9) we have $\langle \alpha, \beta \rangle \leq 0$ and analogously $\langle \alpha, -\beta \rangle \leq 0$. Thus $\langle \alpha, \beta \rangle = 0$. □

(5.5) Definition. A nonempty root system is called ***reducible*** if it decomposes into the sum of two nonempty root systems. Otherwise it is called ***irreducible***.

Every root system may clearly be written as a sum of irreducible root systems, and this decomposition is unique ((5.10), Ex. 3). Thus we should attempt to find all possible bases of irreducible root systems. As a first step, we look at the angles between the roots.

(5.6) Definition. Let R be a root system in V with basis S. The associated ***Coxeter graph*** has one vertex for each element of S, and every pair α, β of distinct vertices is connected by $n_{\alpha\beta} \cdot n_{\beta\alpha}$ edges.

By (3.8), the only possible values of $n_{\alpha\beta} \cdot n_{\beta\alpha} = 4\cos^2(\alpha, \beta)$ for $\alpha \neq \pm\beta$ are 0, 1, 2 and 3. Moreover, these actually occur. The Coxeter graphs for the root systems of rank two in (3.11) are

$A_1 \times A_1$

A_2

B_2

G_2

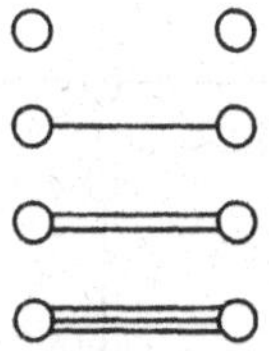

Since the Cartan numbers are invariant under the operation of the Weyl group, and since the Weyl group transitively permutes the bases, the Coxeter graph is an invariant of the root system.

The Coxeter graph is connected if and only if the root system is irreducible: For suppose the basis splits into two orthogonal subsets S_1 and S_2. Then they generate two orthogonal subspaces V_1 and V_2 which are invariant under the Weyl group. Since every root is sent to a simple root by some element of the Weyl group, the set of roots decomposes accordingly as $R = R_1 \cup R_2$ with $R_v \subset V_v$.

The Coxeter graph characterizes the root system when all the roots have the same length. If, however, two different lengths occur, then the graph only gives the angle between two roots α and β. In this case one requires the additional information of which root is longer.

(5.7) Definition. The *Dynkin diagram* consists of the Coxeter graph together with the symbol $>$ or $<$ attached to each doubled or tripled edge pointing to the shorter root.

$$B_2 \quad \text{○═════○}, \qquad G_2 \quad \text{○═════○} \quad \text{(see (3.11))}.$$
$$\qquad \beta \qquad \alpha \qquad\qquad \alpha \qquad \beta$$

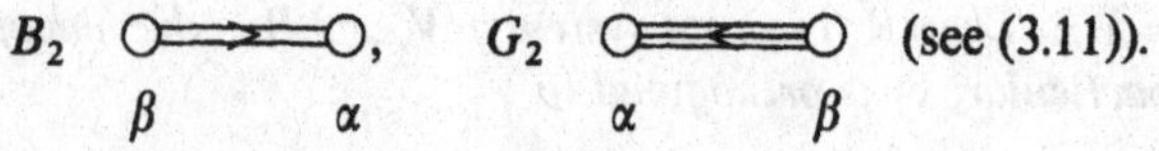

Dynkin diagrams are classified as follows:

(5.8) Theorem. *The Dynkin diagram of an irreducible root system is isomorphic to exactly one of the diagrams in the following list having n vertices:*

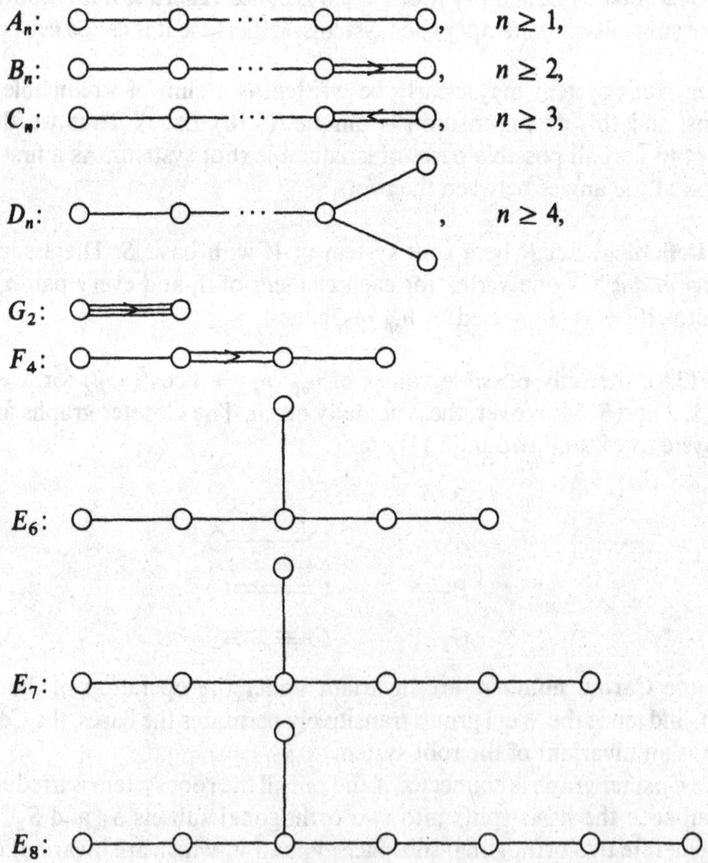

The restrictions on the ranks in the series B_n, C_n, and D_n only serve to avoid repetitions.

For the proof, which is an ingenious argument involving only elementary geometry, we refer the reader to the literature on Lie algebras, see, e.g., Demazure [1], Th. 1, or Humphreys [1], p. 57 ff. There it is also shown that every Dynkin diagram on the list actually corresponds to a root system.

(5.9) Proposition. *A root system is determined up to isomorphism by its Dynkin diagram.*

PROOF. We show how to reconstruct the Cartan matrix from the Dynkin diagram. The information necessary for the following statements is contained in table (3.10). The roots of the basis correspond to the vertices of the diagram and we have:

If $\alpha = \beta$, then $n_{\alpha\beta} = 2$.
If $\alpha \neq \beta$, then $n_{\alpha\beta} \leq 0$ and moreover:
If α and β are not connected by an edge, then $n_{\alpha\beta} = 0$.
If α and β are connected by at least one edge and $|\alpha| > |\beta|$, then $n_{\alpha\beta} = -1$, see (3.10)(v), (vii).
If α and β are connected by j edges and $|\alpha| \leq |\beta|$, then $n_{\alpha\beta} = -j$.

To see this, compare the columns $n_{\alpha\beta}$ and $|\alpha|^2/|\beta|^2$ in (3.10)(iii), (v), (vii). □

Once we assume the classification (5.6), we can read off general statements about root systems simply by inspecting the list. For example:

(5.10) Remark. There are at most two different lengths of roots in an irreducible root system. We correspondingly speak of the ***long*** and the ***short*** roots in R. If α is long and β short, then $|\alpha|^2/|\beta|^2 = 3$ in the case of G_2 and $|\alpha|^2/|\beta|^2 = 2$ in the cases B_n, C_n, and F_4.

It is not difficult to show this directly ((5.12), Ex. 7).

Thus the lengths of the roots are determined up to a common factor by the Dynkin diagram—one simply reads them off. Together with the Cartan matrix, this determines all inner products $\langle \alpha, \beta \rangle$ with $\alpha, \beta \in S$. In other words, if R is an irreducible root system in V, then there is only one Euclidean W-invariant inner product on V—of course, up to a constant. This may also be seen directly ((5.12), Ex. 6). There is a canonical choice of such a metric given by the normalization

(5.11)
$$\langle u, v \rangle = \sum_{\alpha \in R} \langle \alpha, u \rangle \cdot \langle \alpha, v \rangle.$$

This metric has a more natural explanation.

(5.12) Definition. The ***Killing form*** ψ of a Lie algebra L is the symmetric bilinear form

$$\psi(X, Y) = \mathrm{Tr}(\mathrm{ad}\, X \circ \mathrm{ad}\, Y) \quad \text{for } X, Y \in LG.$$

This form is invariant under the adjoint representation because the trace is invariant under conjugation.

Now let $L = LG$ be the Lie algebra of a compact Lie group and let X be in the Lie algebra of the maximal torus (this is no real restriction since, given X, one may choose the torus accordingly). Let $(2\pi)^{-1}R$ be the set of real roots of G. Then we can split $LG_{\mathbb{C}}$ into the weight spaces of the roots and choose a corresponding basis of weight vectors. With respect to this basis, ad X is in diagonal form with the diagonal entries $i\alpha(X)$ with $\alpha \in R$ and 0. The number of zeros equals the rank of G. In this case we consequently have

$$\psi(X, X) = -\sum_{\alpha \in R} \alpha(X)^2.$$

This immediately yields

(5.13) Proposition. *The Killing form ψ of a compact connected Lie group is negative definite if and only if the center of G is finite, i.e., when G is semi-simple (see* (3.14)).

PROOF. The center of G is finite (discrete) if and only if $\bigcap_{\alpha \in R} \ker(\alpha) = 0$. □

Thus, if G is a semisimple compact connected Lie group, the Killing form ψ yields a distinguished Ad-invariant metric on LG and thereby a W-invariant metric $\langle u, v\rangle = -\psi(u, v)$ on LT. Now, for a basis of weight vectors as above, ad(u) and ad(v) are given by diagonal matrices and

$$\psi(u, v) = \sum_{\alpha \in R} i\alpha(u) \cdot i\alpha(v).$$

Thus if we identify the roots α with elements of LT using the chosen metric $-\psi$, then $\alpha(u) = \langle \alpha, u\rangle$ and $-\psi(u, v) = \langle u, v\rangle$. Hence

$$\langle u, v\rangle = \sum_{\alpha \in R} \langle \alpha, u\rangle \cdot \langle \alpha, v\rangle.$$

With these identifications one can therefore say: The negative Killing form $-\psi$ is the canonical inner product on the root system of a semisimple compact Lie group. In order to compute the Killing form, one can start with an arbitrary Ad-invariant metric and normalize so that (5.11) is valid.

The classification of connected Dynkin diagrams is simultaneously a classification of the simple complex Lie algebras. This appears as a central result in books on Lie algebras, see, e.g., Humphreys [1], 14.2, p. 75 ff. In the next chapter we will show how to read off the irreducible characters of a compact connected Lie group from the position of the root system in the lattice of integral forms.

(5.14) Exercises

1. Compute the Cartan matrices of the root systems of rank two (3.11).

2. Show that the Cartan matrices are regular.

3. Show that a root system splits uniquely into a sum of irreducible root systems. The subspaces $V_\nu \subset V$ and the subsets $R_\nu \subset R$ of such a splitting are uniquely determined by $R \subset V$. Show that the Weyl group of R is the direct product of the Weyl groups of the irreducible components of R.

4. Let R be a root system in V with Weyl group W. Show that R is irreducible if and only if V is irreducible as a W-module. In this case every W-orbit of R spans V. Show that the isotypical summands of the W-module V are irreducible. *Hint*: If $V = V' \oplus V''$ is an s_α-invariant splitting, then either $\alpha \in V'$ or $\alpha \in V''$.

5. Show that the Coxeter graph determines the Weyl group and its operation on V up to isomorphism.

6. Let R be an irreducible root system in V. Show that up to a constant factor there is exactly one W-invariant Euclidean metric on V (Ex. 4 and II, (1.16), Ex. 7).

7. Let α and β be roots of an irreducible root system with $|\alpha| > |\beta|$. Let W be the Weyl group. Show that:

   (i) There is a $w \in W$ such that $\langle w\alpha, \beta\rangle \neq 0$. *Hint*: Exercise 4.
   (ii) $|\alpha|^2/|\beta|^2$ is either 2 or 3.
   (iii) There are at most two lengths of roots in R.

8. Let R be a root system in V with Weyl group W. Show that there is precisely one W-invariant positive-definite inner product on V for which

$$\langle u, v\rangle = \sum_{\alpha \in R} \langle \alpha, u\rangle \cdot \langle \alpha, v\rangle.$$

   This is called the ***canonical*** inner product of the root system. *Hint*: Consider an irreducible root system. If $\langle\langle\ ,\ \rangle\rangle$ is any W-invariant inner product on V, then so is

$$(u, v) \mapsto \sum_{\alpha \in R} \langle\langle \alpha, u\rangle\rangle \cdot \langle\langle \alpha, v\rangle\rangle.$$

   Now apply Exercise 6.

9. Show that

$$\sum_{\alpha \in R} \langle \alpha, \alpha\rangle = \dim V$$

   for the canonical inner product (Exercise 8). *Hint*: If A is the matrix of the inner products $\langle \alpha, \beta\rangle$ with $\alpha, \beta \in R$, then $A^2 = A$ and $\operatorname{rank}(A) = \dim V$.

10. Show directly that there are no cycles in a Coxeter graph, i.e., that a basis contains no roots $\alpha_1, \ldots, \alpha_k$ such that α_j is connected to α_{j+1} and α_k is connected to α_1. *Hint*: Let $\beta = \alpha_1 + \cdots + \alpha_k$. Then we would have $\langle\beta, \beta\rangle \leq 0$.

11. Compute the determinant of the Cartan matrix for every type in the list (5.6). The answer is

type	A_n	B_n	C_n	D_n	E_6	E_7	E_8	F_4	G_2
determinant	$n+1$	2	2	4	3	2	1	1	1

12. Let R be a root system of type B_n. Show that its inverse root system R^* is of type C_n and that, for every other type of root system on the list (5.6), R is isomorphic to R^*.

13. Let A, B, and C be endomorphisms of a finite-dimensional vector space and let $[A, B] = AB - BA$. Show that

$$\operatorname{Tr}([A, B] \circ C) = \operatorname{Tr}(A \circ [B, C]).$$

Now let L be a Lie algebra and ad X the endomorphism $Z \mapsto [X, Z]$ of L. Show that $\operatorname{ad}[X, Y] = [\operatorname{ad} X, \operatorname{ad} Y]$. Let $\psi(X, Y) = \operatorname{Tr}(\operatorname{ad} X \circ \operatorname{ad} Y)$ be the Killing form of L. Show that

$$\psi([X, Y], Z) = \psi(X, [Y, Z]).$$

14. Let G be semisimple and $H \subset LG$ be an ideal (i.e., a subalgebra such that $[X, Y] \in H$ whenever $X \in L$ and $Y \in H$). Let $H^\perp$ be the orthogonal complement of H with respect to the Killing form. Show that $H^\perp$ is an ideal (Exercise 13). Show that LG is the direct sum of simple Lie algebras (i.e., algebras with no nontrivial proper ideals).

6. The Roots of the Classical Groups

In the abstract theory of Lie algebras it is most natural to work with the infinitesimal roots in the sense of (1.3) and II, (9.3). But with our approach, which best elucidates the connection with the integral lattice, we prefer to consider the real roots. We will give these for the various classical groups. This leads to the factor $2\pi i$ appearing at many places in our formulas. This factor can be avoided in abstract Lie theory. But we don't bother to do that since we only use Lie algebras as a means to an end.

The reader should refer to IV, §3 for the description of the maximal tori and the Weyl groups, as well as for notational conventions.

We start with the unitary group $\mathrm{U}(n)$ of complex $(n \times n)$-matrices A such that $^*AA = E$ with $^*A = {}^t\bar{A}$. The other compact linear Lie groups appear as closed subgroups of the unitary groups. This gives us certain information about the other groups for free.

The Lie algebra consists of the skew-Hermitian matrices (but see I, (2.21) for physicists' notation)

$$\mathrm{u}(n) = \{A \in \mathrm{End}(\mathbb{C}^n) \mid A + {}^*A = 0\}.$$

The Lie bracket is

$$[A, B] = AB - BA$$

and the exponential map is given by

$$\exp(A) = \sum_{k=0}^{\infty} A^k/k!.$$

A convenient inner product on $\mathrm{u}(n)$ which is invariant for the adjoint representation is given by

$$\text{(6.1)} \qquad \langle A, B\rangle = (2\pi)^{-2}\,\mathrm{Tr}({}^*AB) = -(2\pi)^{-2}\,\mathrm{Tr}(AB).$$

The maximal torus $\Delta(n)$ consists of the diagonal matrices, see VI, §3:

$$D = \begin{bmatrix} z_1 & & \\ & \ddots & \\ & & z_n \end{bmatrix}, \qquad z_\nu = \exp(2\pi i \vartheta_\nu).$$

We use the n-tuple $(\vartheta_1, \ldots, \vartheta_n)$, $\vartheta_\nu \in \mathbb{R}/\mathbb{Z}$, to denote D and have corresponding coordinates $(\vartheta_1, \ldots, \vartheta_n)$, $\vartheta_\nu \in \mathbb{R}$, for the Lie algebra $\mathrm{L}\Delta(n) \cong \mathbb{R}^n$. The exponential map sends $(\vartheta_1, \ldots, \vartheta_n) \in \mathrm{L}\Delta(n)$ to D above and, thanks to the factor $(2\pi)^{-2}$, the invariant metric above restricts to the standard metric on $\mathrm{L}\Delta(n) \cong \mathbb{R}^n$.

As usual, we also denote the projection of $\mathrm{L}\Delta(n)$ onto the νth coordinate by

$$\vartheta_\nu\colon \mathrm{L}\Delta(n) \to \mathbb{R}, \qquad (\vartheta_1, \ldots, \vartheta_n) \mapsto \vartheta_\nu.$$

Thus, in the following notation for roots, ϑ_ν is not an element of but rather a linear form on the Lie algebra of the torus. If we identify $\mathrm{L}\Delta(n)$ with its dual space as before, then ϑ_ν corresponds to the νth standard unit vector in $\mathrm{L}\Delta(n)$.

We wish to determine the (real) roots of $\mathrm{U}(n)$. In other words, we are looking for the linear forms $\alpha\colon \mathrm{L}\Delta(n) \to \mathbb{R}$, $\alpha \neq 0$, for which there is an element $X \neq 0$ in $\mathbb{C} \otimes \mathrm{LU}(n)$ such that

$$[H, X] = 2\pi i \alpha(H) \cdot X$$

for all $H \in \mathrm{L}\Delta(n) \subset \mathrm{u}(n)$; see (2.3)(f) and I, (2.11).

So much for the preparations which are also relevant for the various subgroups of $\mathrm{U}(n)$.

Now, the complexification of $\mathrm{u}(n)$ is $\mathrm{gl}(n, \mathbb{C})$ by the identification

$$\mathbb{C} \otimes \mathrm{u}(n) \to \mathrm{gl}(n, \mathbb{C}), \qquad z \otimes A \mapsto z \cdot A.$$

Here $\mathbb{C} \otimes \mathrm{L}(\mathrm{U}(n)/\Delta(n))$ is the space of matrices with zero diagonal. The condition on α now explicitly says

$$HX - XH = 2\pi i\alpha(H) \cdot X$$

where $X \neq 0$ is a complex matrix with zero diagonal and H runs through the diagonal matrices

$$H = 2\pi i \begin{bmatrix} \vartheta_1 & & \\ & \ddots & \\ & & \vartheta_n \end{bmatrix}, \qquad \vartheta_\nu \in \mathbb{R}.$$

(6.2) Proposition. *For $n \geq 2$ the group* $\mathrm{U}(n)$ *has the following root system*:

(i) *Roots*: $\vartheta_\mu - \vartheta_\nu$, $\mu \neq \nu$, $1 \leq \mu, \nu \leq n$.
(ii) *Positive roots*: $\vartheta_\mu - \vartheta_\nu$, $\mu < \nu$.
(iii) *Basis*: $\vartheta_\nu - \vartheta_{\nu+1}$, $1 \leq \nu < n$.
(iv) *Type*: A_{n-1} .
(v) *Fundamental Weyl chamber*: $\bar{K} = \{H \mid \vartheta_\nu \geq \vartheta_{\nu+1}, 1 \leq \nu < n\}$ *with* $H = (\vartheta_1, \ldots, \vartheta_n) \in \mathrm{L}\Delta(n)$.
(vi) *Sum of the positive roots*: $2\varrho = \sum_{\nu=1}^{n} (n - 2\nu + 1)\vartheta_\nu$.

PROOF. The statements (ii)–(vi) follow from (i). To see (i), let $E_{\mu\nu} \in \mathrm{gl}(n, \mathbb{C})$ be the matrix whose only nonzero entry is a 1 in the (μ, ν)th position. These matrices for $\mu \neq \nu$ form a basis for $\mathbb{C} \otimes \mathrm{L}(\mathrm{U}(n)/\Delta(n))$ and, for H as above, we have

$$[H, E_{\mu\nu}] = 2\pi i(\vartheta_\mu - \vartheta_\nu) \cdot E_{\mu\nu}. \qquad \square$$

The Weyl group is the symmetric group operating by permutations of the roots, see IV, (3.2).

Notice that the root system of $\mathrm{U}(n)$ only has rank $n - 1$; the center of $\mathrm{U}(n)$ is one-dimensional. Since $\mathrm{U}(n)$ and $\mathrm{SU}(n)$ are the same modulo their centers, their root systems agree up to a trivial summand coming from the Lie algebra of the center (see end of §3).

(6.3) Proposition.

(i) *The root system of* $\mathrm{SU}(n)$ *is that of* $\mathrm{U}(n)$ *as described in* (6.2).
(ii) *The Lie algebra of the maximal torus is given by*

$$\mathrm{LS}\Delta(n) = \left\{(\vartheta_1, \ldots, \vartheta_n) \in \mathrm{L}\Delta(n) \,\middle|\, \sum_{\nu=1}^{n} \vartheta_\nu = 0\right\}.$$

(iii) *The integral lattice is given by*

$$I = \mathbb{Z}^n \cap \mathrm{LS}\Delta(n) = \left\{(\vartheta_1, \ldots, \vartheta_n) \,\middle|\, \vartheta_\nu \in \mathbb{Z}, \sum_\nu \vartheta_\nu = 0\right\}.$$

(iv) *Let e_ν be the νth standard unit vector of* $\mathrm{L}\Delta(n) \cong \mathbb{R}^n$. *The inverse roots* $(\vartheta_\nu - \vartheta_{\nu+1})^* = e_\nu - e_{\nu+1}$, $1 \leq \nu < n$, *form a basis of the integral lattice of* $\mathrm{SU}(n)$.

Proof. Use the preceding remark and compare IV, (3.1). □

Note that the element ϱ in (6.2)(vi) is an integral form for SU(n) but not for U(n).

We now move on to the group SO($2n$) for $n \geq 2$. Once again we use coordinates $(\vartheta_1, \dots, \vartheta_n)$ to describe the maximal torus. Each ϑ_ν determines a (2×2)-block

$$D_\nu = \begin{bmatrix} \cos 2\pi\vartheta_\nu & -\sin 2\pi\vartheta_\nu \\ \sin 2\pi\vartheta_\nu & \cos 2\pi\vartheta_\nu \end{bmatrix}$$

in the matrix D which is an element of the torus. There is also a corresponding block

$$H_\nu = 2\pi \begin{bmatrix} 0 & -\vartheta_\nu \\ \vartheta_\nu & 0 \end{bmatrix}$$

of the matrix H in the Lie algebra of the torus.

The complexified Lie algebra $\mathbb{C} \otimes \mathrm{so}(2n)$ consists of the skew-symmetric complex $(2n \times 2n)$-matrices. Consider the two matrices

$$M = \begin{bmatrix} 1 & i \\ -i & 1 \end{bmatrix}, \qquad N = \begin{bmatrix} 1 & i \\ i & -1 \end{bmatrix}.$$

One easily computes that

$$H_\nu \cdot M = M \cdot H_\nu = 2\pi i \vartheta_\nu \cdot M, \qquad N \cdot H_\nu = -H_\nu \cdot N = 2\pi i \vartheta_\nu \cdot N.$$

Now we let

$$H_{\mu\nu} = \begin{bmatrix} H_\mu & 0 \\ 0 & H_\nu \end{bmatrix}, \quad X = \begin{bmatrix} 0 & M \\ -{}^tM & 0 \end{bmatrix}, \quad Y = \begin{bmatrix} 0 & N \\ -{}^tN & 0 \end{bmatrix}.$$

Using ${}^tH = -H$ we obtain

$$[H_{\mu\nu}, X] = 2\pi i(\vartheta_\mu - \vartheta_\nu)X, \qquad [H_{\mu\nu}, Y] = 2\pi i(\vartheta_\mu + \vartheta_\nu)Y.$$

From this one can read off the roots of SO($2n$):

(6.4) Proposition. *For $n \geq 2$ the group* SO($2n$) *has the following root system:*

(i) *Roots:* $\pm\vartheta_\mu \pm \vartheta_\nu$, $1 \leq \mu < \nu \leq n$.
(ii) *Positive roots:* $\vartheta_\mu - \vartheta_\nu$, $\vartheta_\mu + \vartheta_\nu$, $1 \leq \mu < \nu \leq n$.
(iii) *Basis:* $\alpha_\nu = \vartheta_\nu - \vartheta_{\nu+1}$, $1 \leq \nu < n$, $\alpha_n = \vartheta_{n-1} + \vartheta_n$.
(iv) *Type:* D_n 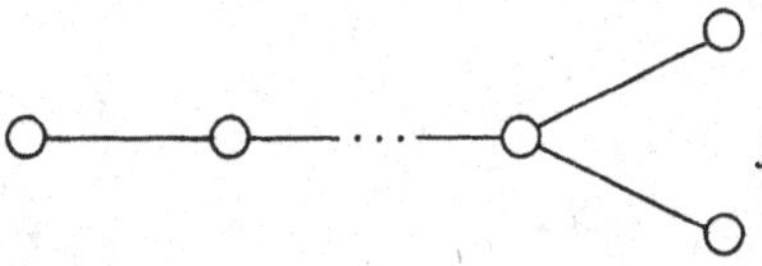 *for* $n \geq 4$ ((6.8), *Ex.* 1).
(v) *Fundamental Weyl chamber:* $\bar{K} = \{H \mid \vartheta_\nu \geq \vartheta_{\nu+1},\ 1 \leq \nu < n,$ *and* $\vartheta_{n-1} \geq |\vartheta_n|\}$ *for* $H = (\vartheta_1, \dots, \vartheta_n) \in LT(n)$.
(vi) *Sum of the positive roots:* $2\varrho = 2\sum_{\nu=1}^{n} (n - \nu)\vartheta_\nu$.

PROOF. (i) The preceding discussion shows that these are actually roots. There can't be more since we have already exhibited $2n(n-1) = \dim L(SO(2n)/T(n))$ roots. As before, the rest is easily verified. □

For $n \geq 2$, the group $SO(2n+1)$ has the same maximal torus and the roots of $SO(2n)$ are also roots of $SO(2n+1)$. With H_ν as above, we define

$$\tilde{H}_\nu = \left[\begin{array}{c|c} H_\nu & \\ \hline & 0 \end{array}\right], \qquad Z = \left[\begin{array}{c|c} 0 & \begin{matrix} 1 \\ -i \end{matrix} \\ \hline \begin{matrix} -1 & i \end{matrix} & 0 \end{array}\right].$$

One computes that

$$[\tilde{H}_\nu, Z] = 2\pi i \vartheta_\nu \cdot Z.$$

Thus $\pm\vartheta_\nu$ is a root of $SO(2n+1)$ and by comparing numbers and dimensions as above we obtain:

(6.5) Proposition. *For $n \geq 2$ the group $SO(2n+1)$ has the following root system:*

(i) *Roots*: $\pm\vartheta_\mu \pm \vartheta_\nu$, $1 \leq \mu < \nu \leq n$, *and* $\pm\vartheta_\nu$, $1 \leq \nu \leq n$.
(ii) *Positive roots*: $\vartheta_\mu \pm \vartheta_\nu$ *for* $\mu < \nu$, *and* ϑ_ν *for* $1 \leq \nu \leq n$.
(iii) *Basis*: $\alpha_\nu = \vartheta_\nu - \vartheta_{\nu+1}$ *for* $1 \leq \nu < n$ *and* $\alpha_n = \vartheta_n$.
(iv) *Type*: B_n .
(v) *Fundamental Weyl chamber*: $\bar{K} = \{H \mid \vartheta_\nu \geq \vartheta_{\nu+1},\ \vartheta_n \geq 0\}$ for $H = (\vartheta_1, \ldots, \vartheta_n) \in LT(n)$.
(vi) *Sum of the positive roots*: $2\varrho = \sum_{\nu=1}^{n} (2n - 2\nu + 1)\vartheta_\nu$. □

The root systems of the symplectic groups $Sp(n)$, $n \geq 2$, are similar to, but not exactly the same as these. We describe the maximal torus T^n and its Lie algebra by coordinates $(\vartheta_1, \ldots, \vartheta_n)$ which in this case correspond to the matrices

$$\begin{bmatrix} z_1 & & & & & \\ & \ddots & & & & \\ & & z_n & & & \\ & & & \bar{z}_1 & & \\ & & & & \ddots & \\ & & & & & \bar{z}_n \end{bmatrix} \in T^n,$$

$$H = 2\pi i \begin{bmatrix} \vartheta_1 & & & & & \\ & \ddots & & & & \\ & & \vartheta_n & & & \\ & & & -\vartheta_1 & & \\ & & & & \ddots & \\ & & & & & -\vartheta_n \end{bmatrix} \in LT^n$$

with $z_\nu \in \exp(2\pi i \vartheta_\nu)$. The complexified Lie algebra $\mathbb{C} \otimes \mathrm{sp}(n)$ is the same as the Lie algebra of $\mathrm{Sp}(n, \mathbb{C})$ and therefore consists of those complex $(2n \times 2n)$-matrices A such that

$$ {}^tAJ + JA = 0, \qquad J = \begin{bmatrix} 0 & -E \\ E & 0 \end{bmatrix}, $$

where E is the identity matrix in $\mathrm{GL}(n, \mathbb{C})$; see I, (1.12) and III, (8.7). Explicitly this means that $\mathrm{LSp}(n, \mathbb{C})$ consists of the complex matrices of the form

$$ \begin{bmatrix} U & W \\ V & -{}^tU \end{bmatrix} \qquad \begin{array}{l} U, V, W \in \mathrm{gl}(n, \mathbb{C}), \\ {}^tV = V, \quad {}^tW = W. \end{array} $$

Again, let $E_{\mu\nu}$ denote the matrix with a 1 at position (μ, ν) and zeros everywhere else. One computes that

$$ [H, E_{n+\mu,\nu} + E_{n+\nu,\mu}] = -2\pi i(\vartheta_\mu + \vartheta_\nu)\cdot(E_{n+\mu,\nu} + E_{n+\nu,\mu}) $$

for $1 \le \mu \le \nu \le n$, and

$$ [H, E_{\mu\nu} - E_{n+\nu,n+\mu}] = 2\pi i(\vartheta_\mu - \vartheta_\nu)\cdot(E_{\mu\nu} - E_{n+\nu,n+\mu}) $$

for $1 \le \mu < \nu \le n$.

From this we obtain:

(6.6) Proposition. *For $n \ge 2$ the group $\mathrm{Sp}(n)$ has the following root system:*

(i) *Roots:* $\pm\vartheta_\mu \pm \vartheta_\nu$, $1 \le \mu < \nu \le n$, *and* $\pm 2\vartheta_\nu$, $1 \le \nu \le n$.
(ii) *Positive roots:* $\vartheta_\mu \pm \vartheta_\nu$ *for* $\mu < \nu$ *and* $2\vartheta_\nu$ *for* $1 \le \nu \le n$.
(iii) *Basis:* $\alpha_\nu = \vartheta_\nu - \vartheta_{\nu+1}$ *for* $1 \le \nu < n$ *and* $\alpha_n = 2\vartheta_n$.
(iv) *Type:* C_n O——O—···—O⇐O ($C_n = B_n$ *for* $n = 2$).
(v) *Fundamental Weyl chamber:* $\bar{K} = \{H \mid \vartheta_\nu \ge \vartheta_{\nu+1},\ \vartheta_n \ge 0\}$ *with* $H = (\vartheta_1, \ldots, \vartheta_n) \in LT^n$.
(vi) *Sum of the positive roots:* $2\varrho = \sum_{\nu=1}^{n} 2(n - \nu + 1)\vartheta_\nu$. □

In all the cases above—with the exception of $\mathrm{SU}(n)$—we have chosen the coordinates of the torus so that the integral lattice $I \subset LT$ is simply the lattice $\mathbb{Z}^n$. Thus the lattice I^* of integral forms is given by

$$ I^* = \sum_\nu n_\nu \cdot \vartheta_\nu, \qquad n_\nu \in \mathbb{Z}. $$

With the chosen coordinates, the invariant metric on LT is the standard Euclidean metric. If we use this metric to identify LT and LT^*, then $I^* = \mathbb{Z}^n$ too.

Of course, the root system depends only on the Lie algebra. Indeed even the Lie algebra of a maximal torus may be characterized as a maximal abelian subalgebra. Thus, for example, the root system of $\mathrm{Spin}(m)$ is isomorphic to that of $\mathrm{SO}(m)$ since $\mathrm{Spin}(m)$ is a cover of $\mathrm{SO}(m)$ and has the same Lie algebra. But this is not the whole picture. The calculations above also yield information concerning the integral lattice and the lattice of integral forms. This will prove to be important for representation theory. One can read off how the root system of Spin(m) sits in the lattice of integral forms from IV, (3.12).

(6.7) Proposition. *Let $m = 2n$ or $m = 2n + 1$. The root system of* Spin(m) *is that of* SO(m). *Let ρ*: Spin(m) $\to$ SO(m) *be the covering map. Choose coordinates for* $L\tilde{T}(n) \cong LT(n)$ *in which* $L\rho$ *is the identity. Then the integral lattice of* Spin(m) *is given by*

$$\tilde{I} = \left\{(\vartheta_1, \dots, \vartheta_n) \mid \vartheta_\nu \in \mathbb{Z} \text{ for } 1 \le \nu \le n \text{ and } \sum_\nu \vartheta_\nu \in 2\mathbb{Z}\right\}.$$

The lattice $\tilde{I}^*$ of integral forms of Spin(m) *consists of the linear forms* $(\vartheta_1, \dots, \vartheta_n) \mapsto x_1\vartheta_1 + \cdots + x_n\vartheta_n$ *with either $x_\nu \in \mathbb{Z}$ for all ν or $x_\nu + \frac{1}{2} \in \mathbb{Z}$ for all ν. If we identify* $LT(n)$ *with* $(LT(n))^*$ *by means of the standard metric (which is invariant under the Weyl group), we therefore have*

$$\tilde{I}^* = \{\zeta + \varepsilon(1, \dots, 1) \mid \zeta \in \mathbb{Z}^n \text{ and } \varepsilon = 0 \text{ or } \varepsilon = \tfrac{1}{2}\}. \qquad \square$$

Note that the coordinates in this proposition are not the same as those in IV, (3.10), for which $\tilde{I}$ would be $\mathbb{Z}^n$ of course. However, the essential thing is to clarify how $\tilde{I}$ of Spin(m) is related to I of SO(m) in $L\tilde{T}(n) = LT(n)$.

(6.8) Exercises

1. Show that the root system of SO(6) is of type A_3 and that the root system of SO(4) splits into two summands of Type A_1. See also VI, (6.20), Ex. 12, 13.
2. Show that the root systems of SO(3) and Sp(1) are both of type A_1. See also I, (6.18).
3. For which irreducible root systems R in V is $-\mathrm{id}_V$ an element of the Weyl group?
4. Let G be a compact connected Lie group whose adjoint representation factors through Spin(n):

$$\begin{array}{ccc} & & \mathrm{Spin}(n) \\ & \nearrow & \downarrow \\ G & \xrightarrow{\mathrm{Ad}} & \mathrm{SO}(n) \end{array}$$

Show that the half sum of the positive roots of G is an integral form on LT. *Hint*: Let T be the maximal torus of G and $T(n)$ that of SO(n). Without loss of generality there is a restriction Ad: $T \to T(n)$. Consider the compositions

$$\alpha_\nu: LT \xrightarrow[L\,\mathrm{Ad}]{} LT(n) \xrightarrow[\vartheta_\nu]{} \mathbb{R}.$$

Show: If $\beta_1, \dots, \beta_m$ are the positive roots of G, then $\sum_{\nu=1}^m \varepsilon_\nu \beta_\nu = \sum_{\nu=1}^n \alpha_\nu$ with $\varepsilon_\nu = \pm 1$.

5. Does the adjoint representation factor through Spin? Show that the answer is no for U(n) and SO($2n + 1$) and yes for SU(n), Sp(n), and Spin(n). *Hint*: The last three groups are simply connected.

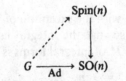

7. The Fundamental Group, the Center and the Stiefel Diagram

In this section we stick to the following notation: G is a compact connected Lie group, $j: T \to G$ is the inclusion of a maximal torus, W is the Weyl group, R is the root system, and we assume we have fixed a W-invariant inner product as well as a basis and thereby positive roots. We will show how to read off the fundamental group $\pi_1(G)$ from the position of the root system in the lattice of integral forms. In the process we will see that the second homotopy group of G is always zero. As a prerequisite we require a little elementary knowledge about fundamental groups and covering spaces.

Unless otherwise indicated, we will choose the unit element of G as base point.

First we consider the torus. The exponential map $\exp: LT \to T$ is the universal covering, since $LT \cong \mathbb{R}^k$ is contractible and the integral lattice $I = \ker(\exp)$ is discrete. Thus I is the group of covering transformations of the universal cover and hence

$$I \cong \pi_1(T).$$

Describing elements of the fundamental group as homotopy classes of paths, this isomorphism sends the element $z \in I$ to the homotopy class of the path $t \mapsto \exp(tz)$. From now on we identify I with $\pi_1(T)$ in this way. The main result of this section is:

(7.1) Theorem. *The inclusion of the maximal torus induces a surjection* $j_*: I = \pi_1(T) \to \pi_1(G)$.

The kernel of j_* *is the group* $\Gamma \subset I$ *generated by the inverse roots (see* (3.4)). *Thus*

$$\pi_1(G) \cong I/\Gamma.$$

The second homotopy group $\pi_2(G)$ *vanishes.*

The proof will be finished after (7.12).

(7.2) Lemma. $\Gamma \subset \ker(j_*)$.

PROOF. An inverse root $\alpha^* \in \Gamma$ corresponds to the path $t \mapsto \exp(t\alpha^*)$ in G. But for $0 \le t \le \frac{1}{2}$ we have

$$\exp(1 - t)\alpha^* = \exp(-t\alpha^*) = s_\alpha \circ \exp(t\alpha^*).$$

Thus it suffices to show that the two paths

$$t \mapsto j \circ \exp(t\alpha^*), \qquad 0 \le t \le \tfrac{1}{2},$$
$$t \mapsto j \circ s_\alpha \circ \exp(t\alpha^*),$$

are homotopic with fixed endpoints. For the difference of these two paths is the path $t \mapsto j \circ \exp(t\alpha^*)$.

Now, the endpoint $\exp(\frac{1}{2}\alpha^*)$ is in U_α, the kernel of the global root $\vartheta_\alpha\colon T \to \mathrm{U}(1)$ corresponding to α. This is because $\vartheta_\alpha \circ \exp(\frac{1}{2}\alpha^*) = \exp(2\pi i\alpha(\frac{1}{2}\alpha^*)) = \exp(2\pi i) = 1$. However, s_α is induced by conjugation with an element in the centralizer of U_α

$$g \in Z(U_\alpha) = Z(U_\alpha^0),$$

and this group is connected (IV, (2.3)(ii) and (2.9)(iii)). Any path $\tau \mapsto g(\tau)$ in $Z(U_\alpha)$ with $g(0) = 1$ and $g(1) = g$ induces a homotopy

$$(t, \tau) \mapsto g(\tau)\exp(t\alpha^*)g(\tau)^{-1}, \qquad 0 \le \tau \le 1,$$

between the two paths above leaving the endpoints fixed. □

Next, recall the notion of the regular elements IV, (2.10). We denote the set of regular elements of G by G_r.

(7.3) Lemma. *G_r is path-connected and the inclusion $G_r \overset{\subset}{\to} G$ induces an isomorphism*

$$\pi_1(G_r) \cong \pi_1(G)$$

and a surjection

$$\pi_2(G_r) \to \pi_2(G).$$

Here we must choose the basepoint in G_r.

Proof. The set of singular points $G_s = G \backslash G_r$ has codimension at least 3 in G in the sense of (2.7). Now in general, a continuous map between differentiable manifolds can be deformed into a differentiable map. And in this situation, a differentiable path in G joining two points of G_r can be deformed with fixed endpoints into a path in G_r. Similarly, a differentiable map $S^2 \to G$ sending the basepoint to G_r can be deformed with fixed basepoint into a map with image in G_r (see, e.g., Bröcker and Jänich [1], 12.9, 14.8, 14.9, Ex. 3). This shows that G_r is path-connected and the maps in the lemma are surjective. Applying the same argument to a contraction of a loop

$$f\colon D^2 \to G, \qquad f(\partial D^2) \subset G_r,$$

we see that $\pi_1(G_r) \to \pi_1(G)$ is injective. See also (7.16), Ex. 2. □

Thus our task is to compute the first homotopy groups of the regular part G_r. Here we can explicitly give suitable coverings including the universal covering.

(7.4) Lemma. *The conjugation map*

$$q\colon G/T \times T_r \to G_r, \qquad (gT, t) \mapsto gtg^{-1}$$

is a (not necessarily connected) $|W|$*-fold covering. Here* $|W|$ *is the order of the Weyl group and* T_r *is the regular part of* T.

PROOF. In a regular point t we have

$$\det(q) = \det(\mathrm{Ad}_{G/T}(t^{-1}) - E_{G/T}) \neq 0$$

(see IV, (1.8)). Thus q is a local diffeomorphism. The Weyl group W operates from the left on G/T via

$$w(gT) = gTn^{-1} = gn^{-1}T \quad \text{for} \quad w = nT \in \mathrm{N}(T)/T = W.$$

It also operates on T_r and hence diagonally on $G/T \times T_r$. Moreover, q is equivariant for this action:

$$qw(gT, t) = q(gn^{-1}T, ntn^{-1}) = gtg^{-1} = q(gT, t).$$

Let $G/T \times_W T_r$ be the orbit space of this operation. Then the map q factors as

$$q\colon G/T \times T_r \xrightarrow[\mathrm{pr}]{} G/T \times_W T_r \xrightarrow[\tilde{q}]{} G_r.$$

Since W acts freely on G/T, the projection pr is a covering of manifolds and $\tilde{q}$ is a local diffeomorphism. Since q is surjective, so is $\tilde{q}$. Finally, we show that $\tilde{q}$ is also injective and hence is a diffeomorphism. By definition, each $t \in T_r$ lies in no maximal torus other than T. Thus if $gtg^{-1} = g_1t_1g_1^{-1}$, then $(g_1^{-1}g)t(g_1^{-1}g)^{-1} = t_1 \in T_r$ and therefore $g_1^{-1}g = n \in \mathrm{N}(T)$. Hence if $w = nT$, then $w(gT, t) = (g_1T, t_1)$. □

(7.5) Proposition. $\pi_2(G) = 0$.

PROOF. A covering induces isomorphisms of the higher homotopy groups. Thus we have an isomorphism

$$q_*\colon \pi_2(G/T \times K) \to \pi_2(G_r)$$

for every connected component $K \subset T_r$. By (7.3), every element of $\pi_2(G)$ is therefore induced by a map $S^2 \to G$ in the following diagram:

$$\begin{array}{ccccc} S^2 \longrightarrow G/T \times K \subset & G/T \times T_r & \xrightarrow{\;q\;} & G_r \\ & \downarrow & & \cap\downarrow \\ \overset{(\varphi_1, \varphi_2)}{\cdots\cdots\cdots\cdots\to} & G/T \times T & \xrightarrow{\;q\;} & G \end{array}$$

Now, the second component $\varphi_2\colon S^2 \to T$ is null-homotopic since $\pi_2(T) \cong \pi_2(\mathrm{L}T) = 0$. Thus we may assume that $\varphi_2(S^2) = \{1\} \subset T$. But then $q(\varphi_1(x), \varphi_2(x)) = q(\varphi_1(x), 1) = 1$. □

(7.6) Proposition. $\pi_1(G/T) = 0$.

Proof. Let K be a component of T_r as above and let $p \in K$. Consider the diagram

$$\begin{array}{ccccc} \pi_1(G/T) = \pi_1(G/T \times \{p\}) & \xrightarrow{\subset} & \pi_1(G/T \times K) & \xrightarrow[q_*]{\subset} & \pi_1(G_r) \\ \downarrow \cong & & & & \downarrow \cong \\ \pi_1(G/T \times \{1\}) & & \xrightarrow{q_* = 0} & & \pi_1(G) \end{array}$$

The map q_* in the top row is injective since q is a covering there. Starting with a path joining p and 1, one constructs a homotopy which shows that the diagram commutes. □

(7.7) Corollary. *The map* $j_*\colon \pi_1(T) \to \pi_1(G)$ *is surjective.*

Proof. This follows from the exact sequence of homotopy groups for the fibration $T \subset G \to G/T$ (see Hu [1], V, 6, p. 152):

$$\ldots \to \pi_1(T) \to \pi_1(G) \to \pi_1(G/T) \to \ldots. \qquad \square$$

If we were willing to apply some facts from differential geometry we could show this result as follows: An element in $\pi_1(G)$ is represented by a closed curve which may be deformed into a geodesic through the unit element with respect to a translation invariant Riemannian metric. But the geodesics through the unit are simply the one parameter groups, that is, they are curves with constant covariant derivative. And each one-parameter group lies entirely in a torus. Thus, up to conjugation and hence homotopy, it lies in T.

We can now directly give a universal covering of G_r: Since $\pi_1(G/T) = 0$, we need only replace the factor K in $G/T \times K$ by its universal covering. This we will do somewhat systematically.

(7.8) Definition. The family of affine hyperplanes in LT

$$L_{\alpha n} = \alpha^{-1}\{n\}, \qquad \alpha \in R_+, \qquad n \in \mathbb{Z}$$

together with the union $LT_s = \bigcup_{\alpha, n} L_{\alpha n}$ is called the ***Stiefel diagram*** of G. The Stiefel diagram is the inverse image of the set of singular points of T under the exponential map. Therefore we call the elements of LT_s ***singular*** and the points in the complement ***regular***. The set LT_r of regular points in LT decomposes into convex connected components which are called the ***chambers of the Stiefel diagram*** or the ***alcoves*** of G. The group Ω of isometries of LT generated by the reflections $s_{\alpha n}$ in the hyperplanes $L_{\alpha n}$ of the Stiefel diagram is called the ***extended Weyl group*** of G. The ***walls*** of an alcove and the ***reflections in a wall*** are defined as for the Weyl group.

Figure 24 illustrates some two-dimensional Stiefel diagrams together with the integral lattices (see Adams [1], (5.2), pp. 102–104, Bourbaki [1], Ch. IV–VI, PL. X, p. 276).

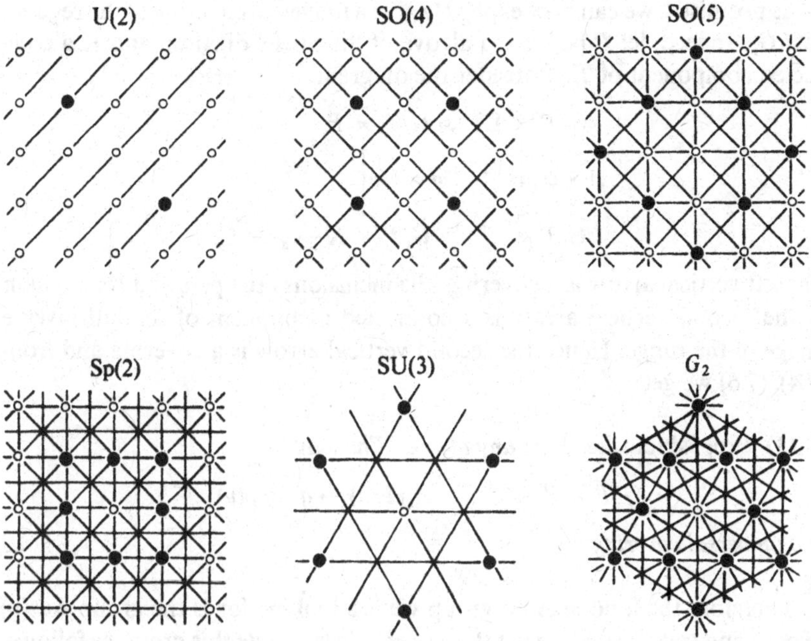

o integral elements • inverse roots

Figure 24

(7.9) Proposition. *Let Γ be the group of translations of LT generated by the inverse roots. Then Γ is normal in Ω and $\Omega = W \cdot \Gamma$ as a semidirect product.*

Proof. Let γ_α be the translation by α^*, $\alpha \in R_+$. From $\alpha(\alpha^*) = 2$ we obtain $\gamma_\alpha L_{\alpha n} = L_{\alpha, n+2}$ and hence the relations

$$s_{\alpha 1} s_{\alpha 0} = \gamma_\alpha, \qquad s_{\alpha, n+2} = \gamma_\alpha s_{\alpha n} \gamma_\alpha^{-1}.$$

From this we see that Ω is generated by Γ and W. We also have $w\gamma_\alpha w^{-1} = \gamma_{w(\alpha)}$ for $w \in W$, $\alpha \in R_+$. Thus Γ is normal in Ω. Also, $W \cap \Gamma = \{1\}$, since only the identity translation fixes the origin. □

In particular, this description shows that Ω acts discontinuously on LT: An orbit can't have any cluster point.

(7.10) Proposition. *The extended Weyl group Ω is generated by the reflections in the walls of any fixed alcove and acts simply transitively on the set of all alcoves.*

Proof. If one uses the fact that the Ω-orbits in LT have no cluster points, then this follows word for word as in the proof of the corresponding statement for the Weyl group (4.1) and (4.2), (2.12). □

As promised, we can now explicitly give a universal covering of the regular part G_r. Indeed, let P be a fixed alcove of the Stiefel diagram and K a connected component of T_r. Consider the diagram

$$\begin{array}{ccccc} G/T \times \mathrm{LT} & \supset & G/T \times P & & \\ \downarrow{\scriptstyle \mathrm{id} \times \exp} & & \downarrow{\scriptstyle \mathrm{id} \times \exp} & \searrow & \\ G/T \times T & \supset & G/T \times K & \xrightarrow{q} & G_r \end{array}$$

The left vertical arrow is a covering, the inclusions are open, and the domain of the second vertical arrow is a connected component of the full inverse image of the range. Hence the second vertical arrow is a covering and from (7.4), (7.6) we get:

(7.11) Proposition. *Let P be any alcove. The map*

$$p\colon G/T \times P \to G_r, \qquad (gT, v) \mapsto g \cdot \exp(v) \cdot g^{-1}$$

is a universal covering. □

Therefore the fundamental group we are looking for is the group $\mathrm{Aut}(p)$ of covering transformations of this cover. We compute this group as follows:

First, we describe a group Δ of covering transformations of

$$G/T \times \mathrm{LT}_r \to G_r, \qquad (gT, v) \mapsto g \cdot \exp(v) \cdot g^{-1}$$

which acts simply transitively on the fibers. This group Δ of isometries of LT is generated by the group I of translations together with W. Just as in the case of Ω, we see that I is normal in Δ and $\Delta = W \cdot I$ as a semidirect product. The group Δ acts ***properly*** and ***discontinuously*** on LT. That is to say that every point possesses a neighborhood U such that $\delta U \cap U = \varnothing$ for all but finitely many $\delta \in \Delta$. Furthermore, $\delta \in \Delta$ transforms the Stiefel diagram into itself and hence operates on LT_r. The group W operates on G/T via

$$w(gT) = gn^{-1}T \quad \text{for} \quad w = nT \in \mathrm{N}(T)/T = W$$

and the projection $\Delta \to \Delta/I = W$ induces an action of Δ on G/T. With these actions Δ operates diagonally on $G/T \times \mathrm{LT}_r$. This action is not only proper and discontinuous but also ***free***. In other words, the only element of Δ with a fixed point is the identity. To see this, let $w = nT$ as above and $\delta = wz$ with $z \in I$. If $(gn^{-1}T, n(t + z)n^{-1}) = (gT, t)$, then $n \in T$ and so w is the identity in W. Thus $z = 0$.

As in the proof of (7.3), it now follows that the map of the orbit space

$$G/T \times_\Delta \mathrm{LT}_r \to G_r, \qquad (gT, v)/\Delta \mapsto g \cdot \exp(v) \cdot g^{-1}$$

is a diffeomorphism. And this, in turn, means that the map

$$G/T \times \mathrm{LT}_r \to G_r$$

is a covering with Δ operating simply transitively on the fibers. In other words, $G/T \times LT_r \to G_r$ is the projection of a Δ-principal bundle, see I, (4.2). Now, Ω is a subgroup of Δ and, in fact:

(7.12) Lemma. *Ω is normal in Δ.*

PROOF. Conjugation by $w \in W$ sends Ω into itself and (if we think of I as a group of translations) conjugation by $\iota \in I$ fixes Γ. Thus we only need to check that $\iota w \iota^{-1} \in \Omega$. But $\iota w \iota^{-1} = w(w^{-1}\iota w)\iota^{-1} = w \cdot w^{-1}(\iota) \cdot \iota^{-1}$. Here $w^{-1}(\iota)$ is the result of $w^{-1} \in W$ acting on $\iota \in I \subset LT$. Thus $w^{-1}(\iota) = \gamma\iota$ for some $\gamma \in \Gamma$ since (in additive notation) $s_\alpha(\iota) = \iota - \langle \alpha, \iota \rangle \alpha^*$ for any reflection $s_\alpha \in W$. Putting this together, $\iota w \iota^{-1} = w\gamma$. □

PROOF OF THEOREM (7.1). After factoring out the action of Ω, the covering $G/T \times LT_r \to G_r$ induces the covering

$$G/T \times_\Omega LT_r \to G_r.$$

The factor group Δ/Ω operates as a group of covering transformations on this covering space—still simply transitively on fibers. But $\Delta/\Omega = WI/W\Gamma \cong I/\Gamma$ and $G/T \times_\Omega LT_r \cong G/T \times P$ because Ω acts simply transitively on the set of alcoves. Therefore $G/T \times_\Omega LT_r \to G_r$ is the universal cover and $\Delta/\Omega \cong I/\Gamma \cong \mathrm{Aut}(p)$ its group of covering transformations. Consequently, $I/\Gamma \cong \pi_1(G_r) \cong \pi_1(G)$. Admittedly, this proof doesn't clarify why the isomorphism $I/\Gamma \to \pi_1(G)$ is induced by the inclusion of the maximal torus as claimed. However, we already know that $j_*: I = \pi_1(T) \to \pi_1(G)$ is surjective and that Γ is in the kernel (7.2), (7.7). And since the image is isomorphic to I/Γ, one readily sees that the kernel must be equal to Γ. □

(7.13) Remark. *If G is a compact connected Lie group, then the following are equivalent*:

(i) *G is semisimple (see* (3.13)*).*
(ii) *The center $Z(G)$ is finite.*
(iii) *$\pi_1(G)$ is finite.*
(iv) *The universal cover $\tilde{G}$ of G is compact.*
(v) *The Killing form of G is negative-definite.*

PROOF. The equivalence of (i) and (ii) is in (3.14). But (ii) says the rank of the root system—and hence of Γ—is equal to dim LT. This is the same as saying that I/Γ is finite, so (ii) and (iii) are equivalent. The equivalence of (iii) and (iv) is clear from the theory of coverings and the equivalence of (ii) and (v) is in (5.11). □

We have already remarked that, in the case of a covering $p: \tilde{G} \to G$, the root systems of $\tilde{G}$ and G agree, whereas the integral lattices may be different. But for a simply connected Lie group $I = \Gamma$ and the lattice Γ is determined by the root system. Thus if G is semisimple, Γ is the integral lattice of the universal cover $\tilde{G}$. In a simply connected group, the half sum of the positive roots is always integral since it is integral on Γ; see (4.12)(iv).

The following proposition complements the main theorem (7.1) of this section and facilitates reading off $\pi_1(G) \cong I/\Gamma$ from the Stiefel diagram.

(7.14) Proposition. *Let A be an alcove of G. Then the set $I \cap \bar{A}$ forms a system of representatives of I/Γ. Hence the order of $\pi_1(G)$ is the number of integral elements in $\bar{A}$.*

PROOF. Every Ω-orbit and hence every Γ-orbit meets $\bar{A}$ in at most one point ((7.16), Ex. 3). Thus different elements of $I \cap \bar{A}$ lie in different Γ-cosets in I. Suppose $z \in I$. Since Ω acts transitively on the set of alcoves, there are $w \in W$ and $\gamma \in \Gamma$ such that $\gamma + w(z) \in \bar{A}$. But we have already seen that $w(z) - z \in \Gamma$ since $s_\alpha(z) = z - \langle \alpha, z \rangle \alpha^*$. Together we have $z + w(z) - z + \gamma \in \bar{A}$ and $w(z) - z + \gamma \in \Gamma$. □

The elements of $I \cap \bar{A}$ always lie on the boundary of A. If G is semisimple, they are vertices since they meet every family $\bigcup_{n \in \mathbb{Z}} L_{\alpha n}$ of parallel hyperplanes of the Stiefel diagram.

(7.15) Definition. We will call an integral element ***central*** if it meets every family $(L_{\alpha n} \mid n \in \mathbb{Z})$ of parallel hyperplanes of the Stiefel diagram. Thus the closed subgroup of LT of all central elements is

$$\bigcap_{\alpha \in R} \bigcup_{n \in \mathbb{Z}} L_{\alpha n}.$$

In other words, the central elements are precisely those which the exponential map sends to $\bigcap_{\alpha \in R} U_\alpha$, the center of G.

(7.16) Proposition. *Let $Z(G)$ be the center, Λ the set of central elements, and A an alcove of G.*

(i) *The exponential map induces an isomorphism*

$$\Lambda/I \cong Z(G).$$

(ii) *If G is simply connected then*

$$Z(G) \cong \Lambda/\Gamma.$$

(iii) *The set $\Lambda \cap \bar{A}$ forms a representative system for Λ/Γ.*

PROOF. (i) and (ii) are clear since $\Lambda = \exp^{-1} Z(G)$. For (iii) we may assume that G is semisimple and hence Λ is discrete since $LZ(G)$ is orthogonal to the subspace of LT generated by Γ. However Λ is the integral lattice for $G/Z(G)$ if G is semisimple and (iii) is a special case of (7.14). □

To summarize: If we are given the root system of a semisimple Lie group and therewith the discrete groups $\Gamma \subset \Lambda$ of translations, then there is precisely one simply connected compact Lie group $\tilde{G}$ with this root system and we have

$$I(\tilde{G}) = \Gamma, \qquad Z(\tilde{G}) = \Lambda/\Gamma,$$
$$I(\tilde{G}/Z(\tilde{G})) = \Lambda.$$

The other semisimple Lie groups with the same root system are obtained from $\tilde{G}$ by factoring out a central subgroup. Every central subgroup $C \subset \Lambda/\Gamma$ gives rise to a compact Lie group $G = \tilde{G}/C$ with $\pi_1(G) \cong C$ and integral lattice $I(G)$ equal to the inverse image of C under the projection $\Lambda \to \Lambda/\Gamma$. Thus Γ is the smallest possible integral lattice (the inverse roots are integral) and Λ is the largest possible (the integral elements are central). The other possible integral lattices lie inbetween. The reader will not be denied the pleasure of using the Stiefel diagrams displayed following (7.8) to read off the possible coverings and quotients of the given groups (see also (7.17), Ex. 11).

The geometric treatment of the fundamental group is due to Weyl [1]. For the Stiefel diagram, see Stiefel [1], [2], and for its structure, see also Bourbaki [1], V, §3.9.

(7.17) Exercises

1. Show that the fundamental group of any topological group is abelian. *Hint*: Group multiplication induces a homomorphism $\mu_*\colon \pi_1(G) \times \pi_1(G) \to \pi_1(G)$ which is multiplication in $\pi_1(G)$.

2. For the proof of (7.3) we need the following result about general position: Let L, M, N be (compact) differentiable manifolds with $\dim L + \dim N < \dim M$. Suppose we are given differentiable maps $L \xrightarrow{f} M \xleftarrow{g} N$. Let $A \subset N$ be closed and $f(L) \cap g(A) = \emptyset$. Then there is a differentiable map $g_1\colon N \to M$ homotopic to g with $f(L) \cap g_1(N) = \emptyset$ and $g_1|A = g|A$. Here is a way to prove this:

   Choose a closed neighborhood B of A such that $f(L) \cap g(B) = \emptyset$. First look at a local situation

$$\begin{array}{ccccc} L & & M & & N \\ \cup & & \cup & & \cup \\ f^{-1}U & \xrightarrow{f} & U & \xleftarrow{g} & D, \end{array}$$

   where D is compact, $D \cap A = \emptyset$ and $U \cong \mathbb{R}^n$ may be considered as a vector space. Choose a small $p \in U$ not in the image of the map $f^{-1}U \times g^{-1}U \to U$, $(x, y) \mapsto f(x) - g(y)$ and put $g_1 = g + \psi \cdot p$ for a function ψ on N with value 1 on D and 0 on A as well as outside $g^{-1}U$. Given a finite set of similar local situations, there is a p sufficiently small so that $f(L) \cap g_1(B \cup D) = \emptyset$ and the other situations are not thrown out of wack. Thus one may proceed inductively, covering N by B and (locally) finitely many D's as above.

3. Generalize (4.1) to the operation of the extended Weyl group Ω on LT. In particular, show that every Ω-orbit meets every closed alcove in exactly one point.

4. Use the following to show that Ω acts transitively on the set of alcoves of the Stiefel diagram: Let P, Q be alcoves with a common wall. An orbit containing P also contains Q, and every two alcoves may be connected by a finite chain in which every pair of consecutive alcoves has a wall in common.

5. The group of covering transformations Aut(p) of $G/T \times P \to G$ may be thought of as that subgroup of Δ consisting of elements mapping P into itself. Give an injective homomorphism $I/\Gamma \to \Delta$ whose image is this subgroup.

6. Show that $\pi_2(G/T) \cong \Gamma$.

7. Show that there is exactly one translation $\gamma \in \Gamma$ with $0 \in \gamma\bar{P}$ for a given alcove P.

8. Show that the rank of the abelian group $\pi_1(G)$ is $\dim Z(G)$ and that the map $\pi_1(Z(G)) \to \pi_1(G)$ is injective.

9. Use (7.1) to compute the fundamental groups of the classical groups and compare the result with that in the first chapter.

10. Classify the compact connected Lie groups of rank two. In other words, classify the Stiefel diagrams with integral lattices associated to the root systems of rank two. Verify the Stiefel diagrams (Figure 24) using §6.

11. Let Ω be the extended Weyl group of R, let $\Gamma \subset \Omega$ be the subgroup of translations, and let $W \subset \Omega$ be the Weyl group. Show that the following statements concerning $p \in LT$ are equivalent:

    (i) p is central.
    (ii) $\Omega_p \cong W$, where Ω_p is the isotropy group of p in Ω.
    (iii) $\Omega = \Omega_p \cdot \Gamma$.

8. The Structure of the Compact Groups

This section is not part of the mainstream of this book and may be taken as a sort of appendix. We will therefore feel free to use certain important facts which we have not properly discussed. Included in these are the correspondences between Lie algebras and simply connected Lie groups and between subalgebras of the Lie algebra and connected subgroups. Although the rest of the book is independent of this section, knowledge of the main result (8.1) can help clarify the general picture.

A semisimple compact connected Lie group G is the factor group of a simply connected group by a finite (central) subgroup (Remark (7.13)). The simply connected group is determined by the Lie algebra of G. Its center can be read off from the root system and hence from the Lie algebra as we have seen in the last section. This section concerns itself with the structure of an arbitrary compact connected Lie group.

(8.1) Theorem. *A compact connected Lie group G possesses a finite cover which is isomorphic to the direct product of a simply connected Lie group $\tilde{G}$ and a torus S. In particular, $\tilde{G}$ is also compact.*

PROOF. According to the theory of covering spaces, the finite covers of G correspond to the subgroups of $\pi_1(G)$ of finite index. Now $\pi_1(G) \cong I/\Gamma$, where I is the integral lattice of G and Γ, which is generated by the inverse roots, depends only on the Lie algebra.

Let $H \subset \mathrm{L}T$ be the subspace spanned by the inverse roots and let $H^\perp$ be its orthogonal complement with respect to a W-invariant inner product.

(i) **Lemma.** *$(I \cap H) \oplus (I \cap H^\perp)$ has finite index in I.*

PROOF OF (i). Since H is a W-submodule of $\mathrm{L}T$, so is $H^\perp$. In fact, $H^\perp$ is the fixed module of W since W leaves an element fixed precisely if it is orthogonal to every (inverse) root. Thus

$$p\colon \mathrm{L}T \to H^\perp, \qquad x \mapsto |W|^{-1} \sum_{w \in W} wx$$

is the orthogonal projection with kernel H. Since I is invariant under W, we have $|W|p(I) \subset I$. Thus, if $z \in I$,

$$|W|z = (|W|z - |W|pz) + |W|pz \in (I \cap H) \oplus (I \cap H^\perp).$$

Thus the index of $(I \cap H) \oplus (I \cap H^\perp)$ in I is at most $|W| \cdot \dim T$, which shows (i).

By (i), we can find a finite cover G_1 of G with integral lattice

$$\mathrm{I}(G_1) = (I \cap H) \oplus (I \cap H^\perp).$$

Thus $\mathrm{I}(G_1) = (\mathrm{I}(G_1) \cap H) \oplus (\mathrm{I}(G_1) \cap H^\perp)$. Replacing G by G_1, we may assume

(ii) $$I = (I \cap H) \oplus (I \cap H^\perp).$$

But $\Gamma \subset I \cap H$, and the rank of Γ is equal to the dimension of H. Hence Γ has finite index in $I \cap H$ and as above we may assume

(iii) $$\Gamma = I \cap H.$$

It is under these assumptions that we will show G is the product of a torus and a simply connected group.

Now, $H^\perp$ is the intersection of the kernels of the roots and hence the Lie algebra of the center Z of G. Let S be the connected component of the identity of Z. Then S is a torus and $\mathrm{L}S = H^\perp$.

Next, we find the complementary group as follows: We may suppose that the W-invariant metric on $\mathrm{L}T$ is induced by an inner product on $\mathrm{L}G$ which is invariant under the adjoint representation. Let K be the orthogonal complement of $H^\perp$ in $\mathrm{L}G$ with respect to this metric. Since S is central, $H^\perp$

is a (trivial) G-module. Therefore K is also a G-module for the adjoint representation and so K is a subalgebra (in fact an ideal) of LG. That is,

$$[X, Y] = \mathrm{ad}(X)Y = \frac{\partial}{\partial s}\bigg|_{s=0} \mathrm{Ad}(\exp(sX))Y \in K$$

for $Y \in K$, see I, (2.11).

We now quote the general fact that there is a Lie subgroup $\tilde{G}$ of G, corresponding to the subalgebra $K \subset \mathrm{L}G$, such that $\mathrm{L}\tilde{G} = K$. See, e.g., Lang [3], Ch. VI, 5, Th. 6, p. 148 or Chevalley [1], Ch. IV, §IV, Th. 1.

Since S is in the center of G, we have a homomorphism

(iv) $$\varphi\colon \tilde{G} \times S \to G, \qquad (g, s) \mapsto gs.$$

Furthermore, φ is surjective since $\mathrm{L}\varphi$ is surjective. We will show that φ is also injective. Let $x \in \tilde{G} \cap S$, where we think of $\tilde{G}$ as a (not necessarily closed) subset of G. Then x is in the center and hence in the maximal torus $\tilde{T}$ of $\tilde{G}$. Now $\mathrm{L}\tilde{T} \oplus H^{\perp} = \mathrm{L}\tilde{T} \oplus \mathrm{L}S$ is the Lie algebra of a torus in G. Since $H \oplus H^{\perp} = \mathrm{L}T$, it follows that $H = \mathrm{L}\tilde{T}$. Thus there are $h \in H$ and $v \in H^{\perp}$ with $\exp(h) = x = \exp(v)$. Since h and v both lie in LT, we have $\exp(h - v) = 1$ and so $h - v \in I$. Thus $h \in I$ by (ii) and so $x = \exp(h) = 1$.

Thus we have shown φ to be bijective and hence φ is an isomorphism of Lie groups (I, (2.22), Ex. 1). Finally, $\tilde{G}$ is simply connected since $\mathrm{I}(\tilde{G}) = I \cap H = \Gamma$ by (iii). □

Remark. A proof of the quoted theorem on integration of Lie subalgebras might go as follows: By a theorem of Frobenius, the Lie algebra K is integrable. And since K is left-invariant, the maximal integral submanifold $\tilde{G}$ of K containing the unit element is also left-invariant. But if $x \in \tilde{G}$, then both $\tilde{G}$ and $x\tilde{G}$ are maximal integral submanifolds containing x. We conclude that $x\tilde{G} = \tilde{G}$ and thus the inclusion $\tilde{G} \subset G$ is a subgroup of G. (Recall that by definition I, (3.9) a Lie subgroup is an injective homomorphism of Lie groups.)

We will now consider a short exact sequence

$$1 \to Z \to G \underset{p}{\to} H \to 1$$

with G connected and Z finite. The surjective (covering) map p induces an injection of representation rings $\mathrm{R}(H) \to \mathrm{R}(G)$. We think of these rings as rings of functions on G and H and, since distinct irreducible characters are linearly independent over $\mathbb{C}$, we obtain an embedding of $\mathrm{R}(G) \otimes \mathbb{C}$ into the ring $C(G)$ of continuous (or analytic) complex-valued functions on G. The same holds for H and p induces injections $C(H) \to C(G)$ and

$$p^*\colon \mathrm{R}(H) \otimes \mathbb{C} \to \mathrm{R}(G) \otimes \mathbb{C}.$$

Using this, we think of $\mathrm{R}(H) \otimes \mathbb{C}$ as a subring of $\mathrm{R}(G) \otimes \mathbb{C}$.

(8.2) Lemma. *Let χ be an irreducible character of G and suppose that z is in the center of G. Then for all $g \in G$:*

$$\chi(gz) = \chi(g) \cdot \chi(z)/\chi(1),$$

or, in perhaps more sensible notation,

$$\frac{\chi(gz)}{\chi(1)} = \frac{\chi(g)}{\chi(1)} \cdot \frac{\chi(z)}{\chi(1)}.$$

Proof. Let $g \mapsto D(g) \in U(n)$ be the representation corresponding to χ. Then $D(g) \cdot D(z) = D(z) \cdot D(g)$ and, by Schur's lemma, $D(z) = \zeta \cdot \mathrm{id}$ for some $\zeta \in \mathbb{C}$. Thus the claim says that

$$\mathrm{Tr}(\zeta \cdot D(g)) = \mathrm{Tr}(D(g)) \cdot \mathrm{Tr}(\zeta \cdot \mathrm{id})/\mathrm{Tr}(\mathrm{id}),$$

which is obvious. □

Now, the right translation $r_h: G \to G$, $g \mapsto gh$, induces a ring homomorphism

$$r_h^*: C(G) \to C(G), \quad f \mapsto f_h, \text{ with } f_h(g) = f(gh).$$

The lemma tells us:

(8.3) Corollary. *If z is in the center of G then r_z^* maps the subring $\mathrm{R}(G) \otimes \mathbb{C}$ into itself.*

Looking at the exact sequence above, for $z \in Z$ we obtain an automorphism r_z^* of $\mathrm{R}(G) \otimes \mathbb{C}$ which is the identity on $\mathrm{R}(H) \otimes \mathbb{C}$. We let

$$\mathrm{Aut}(\mathrm{R}G \otimes \mathbb{C}, \mathrm{R}H \otimes \mathbb{C})$$

denote the group of all automorphisms of the ring $\mathrm{R}G \otimes \mathbb{C}$ which leave $\mathrm{R}H \otimes \mathbb{C}$ pointwise fixed. We get a homomorphism of groups

$$\Theta: Z \to \mathrm{Aut}(\mathrm{R}G \otimes \mathbb{C}, \mathrm{R}H \otimes \mathbb{C}), \quad z \mapsto r_z^*. \tag{8.4}$$

(8.5) Proposition. *Let G be compact and connected and let $Z \subset G$ be a finite subgroup of the center, and let $H = G/Z$. Then Θ is an isomorphism.*

Proof. We will construct a homomorphism

$$\Theta': \mathrm{Aut}(\mathrm{R}G \otimes \mathbb{C}, \mathrm{R}H \otimes \mathbb{C}) \to \mathrm{Hom}(\mathrm{Hom}(Z, S^1), S^1).$$

The group on the right is canonically isomorphic to Z via the duality $z \mapsto (\varphi \mapsto \varphi(z))$.

In order to define Θ' we have to assign an element of S^1 to every pair (α, φ) consisting of an automorphism α of $\mathrm{R}G \otimes \mathbb{C}$ leaving $\mathrm{R}H \otimes \mathbb{C}$ pointwise fixed and a homomorphism $\varphi: Z \to S^1$. Since Z is finite, the assigned element must be a root of unity. This root of unity is obtained using the following

(8.6) Lemma. *Let χ be an irreducible character of G, let n be the order of Z, and let $\alpha \in \mathrm{Aut}(RG \otimes \mathbb{C}, RH \otimes \mathbb{C})$. Then there is an nth root of unity α_χ such that*

$$\alpha(\chi) = \alpha_\chi \cdot \chi.$$

PROOF. By (8.2)

$$\chi^n(gz) = \chi^n(g) \quad \text{for } z \in Z$$

since $(\chi(z)/\chi(1))^n = \chi(z^n)/\chi(1) = 1$. Thus χ^n as well as its associated representation is trivial on Z. Hence $\chi^n \in RH \otimes \mathbb{C}$ and $\alpha\chi^n = \chi^n$. For $g \in G$ with $\chi(g) \neq 0$ we define c_g by

$$(\alpha\chi)(g) = c_g \cdot \chi(g).$$

Then $c_g^n = 1$ and so c_g is constant on connected components of $\{g \in G \mid \chi(g) \neq 0\}$. But $\chi(1) \neq 0$ and hence the analytic functions $\alpha\chi$ and $c_1 \cdot \chi$ agree on a neighborhood of 1. It follows that $c_g = c_1 = \alpha\chi$ everywhere. □

Now we come to the construction of the homomorphism Θ' as promised. So suppose $\alpha \in \mathrm{Aut}(RG \otimes \mathbb{C}, RH \otimes \mathbb{C})$ and $\varphi: Z \to S^1$ are given. Then φ is an irreducible character of Z and using III, (4.5) we choose an irreducible character χ of G so that $\chi|Z$ contains the character φ. Notice that $\chi|Z$ is a multiple of φ since χ is irreducible and Z is in the center ((8.7), Ex. 1). We define

$$(\Theta'\alpha)(\varphi) = \alpha_\chi.$$

We now have to check the following points (i)–(v):

(i) The definition is independent of the choice of χ.

PROOF. Let χ' be another irreducible character such that $\chi'|Z$ is a multiple of φ. Then

$$(\chi(gz) - \chi'(gz))^n = (\chi(g)\chi(z)/\chi(1) - \chi'(g)\chi'(z)/\chi'(1))^n.$$

But $\chi|Z = \chi(1)\cdot\varphi$, $\chi'|Z = \chi'(1)\cdot\varphi$. Therefore $\chi(z)/\chi(1) = \chi'(z)/\chi'(1) = \varphi(z)$, which is an nth root of unity, and so $(\chi(gz) - \chi'(gz))^n = (\chi(g) - \chi'(g))^n$. Thus $(\chi - \chi')^n \in RH \otimes \mathbb{C}$. As above we conclude that

$$\alpha(\chi - \chi') = c \cdot (\chi - \chi')$$

for a constant c. On the other hand

$$\alpha(\chi - \chi') = \alpha_\chi \cdot \chi - \alpha_{\chi'} \cdot \chi'.$$

Since distinct irreducible characters are linearly independent, it follows that $\alpha_\chi = \alpha_{\chi'} = c$.

(ii) If $\alpha \in \mathrm{Aut}(RG \otimes \mathbb{C}, RH \otimes \mathbb{C})$, then $\Theta'\alpha\colon \mathrm{Hom}(Z, S^1) \to S^1$ is a homomorphism.

PROOF. Let $\varphi, \varphi' \in \mathrm{Hom}(Z, S^1)$. Choose irreducible characters χ and χ' such that $\chi|Z = m \cdot \varphi$ and $\chi'|Z = m' \cdot \varphi'$. Then $\chi \cdot \chi'|Z = m \cdot m' \cdot \varphi \cdot \varphi'$ and $\alpha(\chi \cdot \chi') = \alpha_\chi \cdot \alpha_{\chi'} \cdot \chi \cdot \chi'$. Although $\chi \cdot \chi'$ need not be irreducible, we clearly have $\alpha(\chi_1) = \alpha_\chi \cdot \alpha_{\chi'} \cdot \chi_1$ and $\varphi \cdot \varphi'$ contained in $\chi_1|Z$ for any irreducible summand χ_1 of $\chi \cdot \chi'$.

(iii) Θ' is a homomorphism.

PROOF. This is due to the formula $(\alpha \circ \beta)(\chi) = \alpha_\chi \cdot \beta_\chi \cdot \chi$.

(iv) Θ' is injective.

PROOF. Let $\alpha \in \ker(\Theta')$. If $\alpha \neq 1$ then $\alpha_\chi \neq 1$ for some irreducible character χ. Thus $\Theta'\alpha(\varphi) = \alpha_\chi \neq 1$ for any irreducible character φ of Z contained in χ.

(v) $\Theta' \circ \Theta\colon Z \to \mathrm{Hom}(\mathrm{Hom}(Z, S^1), S^1)$ is the canonical isomorphism.

PROOF. Let $z \in Z$ and $\alpha = \Theta(z)$. Then $\alpha_\chi = \chi(z)/\chi(1)$ by (8.2). Thus $(\Theta'\alpha)(\varphi) = \varphi(z)$ for $\varphi \in \mathrm{Hom}(Z, S^1)$. This finishes the proof of (8.5). □

(8.7) Exercises

1. Let V be an irreducible G-module and Z a central closed subgroup of G. Show that, as a Z-module, V is a multiple of an irreducible Z-module.

2. Let G be semisimple, T be the maximal torus of G, and V be an irreducible G-module. Show that if V as a T-module is a multiple of an irreducible T-module (isotypical), then V is trivial.

3. Let $\pi\colon G \to H$ be a surjective homomorphism of compact groups and χ be a character of G. Show that $\chi = \chi' \circ \pi$ for some character χ' of H if and only if χ factors through π as a map of sets.

4. Let G be semisimple. Show that G has no nontrivial abelian factor group. Consequently every representation $G \to \mathrm{U}(n)$ factors through $\mathrm{SU}(n)$. See also II, (5.13), Ex. 7.

5. Let G be a compact connected Lie group and $\langle\ ,\ \rangle$ an Ad-invariant inner product on the Lie algebra LG. Show that

$$\langle [X, Y], Z\rangle = \langle X, [Y, Z]\rangle,$$

and if H is an ideal in LG, so is the orthogonal complement $H^\perp$.

6. Show that a compact simply connected Lie group is a direct product of compact simply connected Lie groups which have no nontrivial proper connected normal subgroups. *Hint*: Exercise 5 and "integration of subalgebras" of the Lie algebra, i.e., every subalgebra of LG is the Lie algebra of a connected subgroup of G.

7. If we use the classification of root systems, the calculation of the center in (7.16), and Exercise 6, we may then completely enumerate all compact connected Lie groups. A typical conclusion would be: There are only finitely many isomorphism classes of compact Lie groups with a given dimension and number of components.

CHAPTER VI

Irreducible Characters and Weights

The central result of this chapter is the Weyl character formula. It establishes a bijection between the irreducible characters of a compact connected Lie group and the integral forms in a distinguished Weyl chamber. The character formula is stated and proved in the first section. In the second section we will introduce partial orderings on LT^* and analyze the structure of the character ring. The third section gives us some efficient formulas for computing the multiplicity of a weight in an irreducible representation. Finally, come the explicit calculations of representations and representation rings of classical groups.

1. The Weyl Character Formula

In this section we will work with a compact connected Lie group G for which the following have been chosen once and for all: The maximal torus T with Weyl group W, the W-invariant inner product on LT and LT^*, the set R of real roots, and the fundamental Weyl chamber K which we consider to be a subset of LT^*. This determines the basis S and the set of positive roots $R_+ = \{\alpha \in R \mid \langle \alpha, \beta \rangle > 0 \text{ for all } \beta \in K\}$. The lattice I^* of integral forms lies in LT^* and the roots are integral; they have integral values on the integral lattice.

In this chapter we have to deal with the global weights ϑ_α corresponding to real weights α. More precisely we look at functions

$$LT \to U(1) = S^1, \qquad H \mapsto \vartheta_\alpha \circ \exp(H) = e^{2\pi i \alpha(H)}.$$

We will use the notation

$$e(x) = e^{2\pi i x}.$$

Thus $\vartheta_\alpha \circ \exp(H) = e(\alpha H)$, i.e.,

$$\vartheta_\alpha \circ \exp = e(\alpha) = e \circ \alpha.$$

If $\eta = 2\pi i\alpha\colon LT \to i\mathbb{R} = LU(1)$ is the infinitesimal weight corresponding to α, then

$$e(\alpha) = e^\eta$$

and what appears here as convenient shorthand simultaneously reflects the transition from real to infinitesimal weights. In the literature computations are often carried out using the infinitesimal roots and in general formulas

$$e^\eta, \qquad \eta \in \mathrm{Hom}(LT, i\mathbb{R}), \qquad \eta I \subset 2\pi i\mathbb{Z}$$

appear instead of our

$$e(\alpha), \qquad \alpha \in LT^*, \qquad \alpha I \subset \mathbb{Z}.$$

A function $\varphi\colon LT \to \mathbb{C}$ or $\varphi\colon T \to \mathbb{C}$ is called ***symmetric*** if $\varphi \circ w = \varphi$ and ***alternating*** if $\varphi \circ w = \det(w) \cdot \varphi$ for all $w \in W$. Here $\det(w)$ is the determinant of the linear automorphism w of LT. Of course, $\det(w) = \pm 1$ since the Weyl group is generated by reflections.

The characters of representations are symmetric functions on the maximal torus. However, for the time being we are interested in alternating functions. If $\chi = \varphi/\psi$ is a quotient of alternating functions φ and ψ, then χ is symmetric. We shall see that every symmetric character of T is the quotient of an alternating character by a very special alternating function.

(1.1) Definition. Let $\lambda \in LT^*$ be a linear form. We define the ***alternating sum*** of λ to be the function $A(\lambda)\colon LT \to \mathbb{C}$,

$$A(\lambda)(H) = \sum_{w \in W} \det(w) \cdot e(\lambda w(H)).$$

(1.2) Lemma.

(i) *The function* $A(\lambda)$ *is alternating.*
(ii) $A(\lambda) \circ w = A(\lambda \circ w)$.
(iii) $A(\lambda) = 0$ *if and only if* $\lambda w = \lambda$ *for some* $w \in W$, $w \neq 1$ *if and only if* λ *is not contained in any Weyl chamber of* LT^*.

PROOF. (i) is trivial, in fact, W also operates on $C^\infty(LT)$ via $(w, \varphi) \mapsto \det(w) \cdot \varphi \circ w^{-1}$ and the alternating functions are those which are invariant under this operation. Thus if we let $p: C^\infty(LT) \to C^\infty(LT)^W$ be the canonical projection for this operation (see II, (4.1)), then $A(\lambda) = p(e \circ \lambda)$.

(ii) $$A(\lambda) \circ w = \sum_{v \in W} \det(v) \cdot e(\lambda v w)$$
$$= \sum_{v \in W} \det(w^{-1}vw) \cdot e(\lambda w(w^{-1}vw)) = A(\lambda w).$$

(iii) The second equivalence follows from the fact that W acts simply transitively on the set of Weyl chambers (see V, (2.12) or (4.1)). Thus if $\lambda w = \lambda$ for $w \in W$, $w \neq 1$, then λ is in a wall of LT^*. Hence $\lambda s_\alpha = \lambda$ for a reflection $s_\alpha \in W$, so $A(\lambda) = A(\lambda s_\alpha) = A(\lambda) \circ s_\alpha = -A(\lambda)$ and $A(\lambda) = 0$. Conversely, if $A(\lambda) = 0$, then the various λw, $w \in W$, cannot all be distinct by the following general

(1.3) Remark. Let V be a vector space and $\lambda_\nu: V \to \mathbb{C}$, $\nu = 1, \ldots, r$, distinct $\mathbb{R}$-linear maps. Then the functions $\exp \circ \lambda_\nu: V \to \mathbb{C}$, $\nu = 1, \ldots, r$, are linearly independent over $\mathbb{C}$.

Proof. The λ_ν are distinct on a suitable one-dimensional subspace, so we may reduce to the case $V = \mathbb{R}$, $\lambda_\nu \in \mathbb{C}$, and $\exp \circ \lambda_\nu$ is the function $t \mapsto e^{\lambda_\nu \cdot t}$. But these functions are eigenvectors associated to the eigenvalues λ_ν of the linear map $C^\infty(\mathbb{R}, \mathbb{C}) \to C^\infty(\mathbb{R}, \mathbb{C})$, $\varphi \mapsto (d/dt)\varphi$. □

The alternating function $\delta: LT \to \mathbb{C}$ is defined as follows:

(1.4) $$\delta(H) = \prod_{\alpha \in R_+} (e(\tfrac{1}{2}\alpha(H)) - e(-\tfrac{1}{2}\alpha(H))).$$

The index set for this product is the set of positive roots. Thus by V, (4.10) we have $\delta \circ s_\alpha = -\delta$ for $\alpha \in S$. And since these reflections generate W, we see that δ is alternating. Let $\varrho = \frac{1}{2}\sum_{\alpha \in R_+} \alpha$ be the half sum of the positive roots. Then

(1.5) $$\delta(H) = e\varrho(H) \cdot \prod_{\alpha \in R_+} (1 - e(-\alpha(H))).$$

Thus δ is nonzero off the Stiefel diagram $D = \bigcup_{\alpha \in R_+} \alpha^{-1}\mathbb{Z}$. That is to say, $\delta(H) \neq 0$ when $H \in LT_r$ is regular. Thus

$$c(\lambda) = A(\lambda)/\delta, \qquad \lambda \in LT^*,$$

is a continuous function on LT_r which has at most one continuous extension to all of LT. Such an extension must clearly be symmetric. We will show:

(1.6) Proposition. *Let $\gamma \in I^*$ be an integral form on* LT. *Then $c(\gamma + \varrho)$ has a continuous extension to* LT *which factors through the exponential map* $\exp: LT \to T$. *This defines a symmetric function in the character ring* R(T) *of the torus which we will also denote by* $c(\gamma + \varrho)$.

(1.7) The Weyl Character Formula. *Let* Irr$(G, T) \subset$ R(T) *denote the set of restrictions of irreducible characters of G to the maximal torus T.*

(i) *If γ is an integral form and $\langle \gamma, \alpha \rangle \geq 0$ for all positive roots α, that is $\gamma \in \bar{K} \cap I^*$, then* c$(\gamma + \varrho) \in$ Irr(G, T).

(ii) *The map $\gamma \mapsto$ c$(\gamma + \varrho)$ yields a bijection*

$$\bar{K} \cap I^* \to \mathrm{Irr}(G, T).$$

(iii) *We have*

$$\delta = \mathrm{A}(\varrho).$$

Thus by (ii) *the form* $0 \in \bar{K} \cap I^*$ *corresponds to the character* 1 *of the trivial one-dimensional representation of G.*

(iv) *If $\gamma \in \bar{K} \cap I^*$, the corresponding irreducible representation has dimension*

$$\prod_{\alpha \in R_+} \frac{\langle \alpha, \gamma + \varrho \rangle}{\langle \alpha, \varrho \rangle}.$$

The rest of this section is devoted to proving the proposition and the character formula. Recall that the restriction R$(G) \to$ R(T) is injective IV, (2.8). Thus if we know the position of the roots in the lattice I^* of integral forms then, with the help of the character formula, we can completely determine the irreducible characters.

First we pause to make a remark about the function δ which will make its appearance in this context more plausible.

(1.8) Lemma. *The function $\delta \cdot \bar{\delta}$ factors through* $\exp\colon LT \to T$; *in fact* $\delta \cdot \bar{\delta} = \eta \circ \exp$, *where* $\eta(t) = \det(\mathrm{Ad}_{G/T}(t^{-1}) - E_{G/T})$.

Proof. The eigenvalues of $\mathrm{Ad}_{G/T}(t^{-1})$, $t = \exp(H)$, are the values of the global roots $e(\pm\alpha H)$, $\alpha \in R_+$. Thus by (1.5)

$$\det(\mathrm{Ad}_{G/T} - E_{G/T}) \circ \exp = \prod_{\alpha \in R_+} (e(\alpha) - 1)(e(-\alpha) - 1) = \delta \cdot \bar{\delta}. \qquad \square$$

Consequently $\delta \cdot \bar{\delta} = \det(q) \circ \exp$ is the functional determinant appearing in the Weyl integral formula IV, (1.11), which transforms the integral of a class function over G into an integral over T.

We turn to proposition (1.6). We would like to divide A$(\gamma + \varrho)$ by δ. If $\vartheta_\alpha\colon T \to S^1$ is the global root associated to α,

$$\delta = e(\varrho) \cdot \prod_{\alpha \in R_+} (1 - \vartheta_{-\alpha}) = e(-\rho) \cdot \prod_{\alpha \in R_+} (\vartheta_\alpha - 1).$$

We start dividing by the individual terms of this product.

Recall II, (8.3)f that the character ring of the torus is

$$\mathrm{R}(T) \cong \mathbb{Z}[z_1, \ldots, z_k, z_1^{-1}, \ldots, z_k^{-1}], \qquad k = \mathrm{rank}(G).$$

(1.9) Lemma. *Let $\vartheta\colon T \to S^1$ be a homomorphism and suppose that the virtual character $g \in \mathrm{R}(T)$ vanishes on $U = \ker(\vartheta)$. Then there exists a virtual character $h \in \mathrm{R}(T)$ so that $g = (\vartheta - 1) \cdot h$ as a complex-valued function on T.*

PROOF. Choose coordinates on the torus so that $I = \mathbb{Z}^k$. The homomorphism ϑ induces a linear form $\Theta = \mathrm{L}\vartheta$ on the Lie algebra which is, in fact, integral:

$$\Theta\colon \mathbb{Z}^k \to \mathbb{Z}.$$

We may assume that $\Theta \neq 0$ since otherwise any h will do. Thus $\Theta(\mathbb{Z}^k) = n \cdot \mathbb{Z}$ for some $n > 0$ and the short exact sequence

$$0 \to \ker(\Theta) \to \mathbb{Z}^k \underset{\Theta}{\to} n\mathbb{Z} \to 0$$

splits. In other words, we have a decomposition $\mathbb{Z}^k \cong \mathbb{Z} \times \mathbb{Z}^{k-1}$ and the map $\mathbb{Z} \times \mathbb{Z}^{k-1} \to \mathbb{Z}$ induced by Θ maps $(\zeta_1, \ldots, \zeta_k)$ to $n \cdot \zeta_1$. Rephrasing, we have an isomorphism

$$T \cong S^1 \times \cdots \times S^1$$

so that ϑ is given by $\vartheta(z_1, \ldots, z_k) = z_1^n$, $n > 0$. After multiplying by a power of the unit $z_1 \cdot \ldots \cdot z_k \in \mathrm{R}(T)$ if necessary, we may assume that $g \in \mathbb{Z}[z_1, \ldots, z_k]$. Then we may write

$$g = (\vartheta - 1) \cdot h + \sum_{j=0}^{n-1} z_1^j h_j(z_2, \ldots, z_k),$$

where h and the h_j are polynomials and hence also in $\mathrm{R}(T)$. If $z = (z_1, \ldots, z_k)$ is a point with z_1 an nth root of unity, then z is in U and so $g = (\vartheta - 1) \cdot h = 0$ at z. The remainder term $\sum_{j=1}^{n-1} z_1^j h_j$ is therefore a polynomial in z_1 of degree $\leq n - 1$ with n distinct roots. Consequently the remainder term vanishes. □

(1.10) Lemma. *If $g \in \mathrm{R}(T)$ vanishes on all $U_\alpha = \ker(\vartheta_\alpha)$, $\alpha \in R$, then there is an $h \in \mathrm{R}(T)$ such that*

$$g = h \cdot \prod_{\alpha \in R_+} (\vartheta_\alpha - 1).$$

PROOF. Let $\alpha_1, \ldots, \alpha_m$ be the distinct positive roots. Suppose we already have an expression

$$g = h_r \cdot \prod_{j=1}^{r} (\vartheta_{\alpha_j} - 1), \qquad h_r \in \mathrm{R}(T), \quad r < m.$$

Since g vanishes on $U_{\alpha_{r+1}}$ and since the zeros $\bigcup_{j \leq r} U_{\alpha_j}$ of $\prod_{j=1}^{r} (\vartheta_{\alpha_j} - 1)$ in $U_{\alpha_{r+1}}$ have positive codimension, h_r also vanishes on $U_{\alpha_{r+1}}$. Thus the lemma follows by induction. □

The ring $\mathrm{R}(T)$ is factorial since it is the localization of the polynomial ring at the variables z_j. What we have just shown is that the elements $(1 - \vartheta_\alpha)$ have no multiple prime factors and are relatively prime for non-proportional $\alpha \in I^*$, see (1.20), Ex. 1.

Proof of (1.6). By (1.5) we have

(i) $$c(\gamma + \varrho) = \frac{e(\varrho)\cdot A(\gamma+\varrho)}{\prod_{\alpha\in R_+}(e(\alpha)-1)}.$$

We will show that the numerator $e(\varrho)\cdot A(\gamma+\varrho)$ factors through exp and defines a function in $R(T)$ which vanishes on all the U_α. Then the proposition follows from (1.10). Now

$$e(\varrho)\cdot A(\gamma+\varrho) = \sum_{w\in W} \det(w)e(\gamma\circ w)\cdot e(\varrho\circ w+\varrho)$$

and $\varrho w + \varrho = (\varrho w - \varrho) + 2\varrho$ is an integral form by V, (4.12)(iii). Thus the numerator of (i) factors through exp and defines an element on $R(T)$.

We have to show that this element vanishes on all the U_α. So let $X \in LT$ and $x = \exp(X) \in U_\alpha$. We show:

(ii) $$e(\varrho s_\alpha X)\cdot A(\gamma+\varrho)s_\alpha X = e(\varrho X)\cdot A(\gamma+\varrho)X,$$

(iii) $$A(\gamma+\varrho)s_\alpha X = -A(\gamma+\varrho)X,$$

(iv) $$e(\varrho s_\alpha X) = e(\varrho X).$$

These equations clearly imply the statement.

Proof of (ii). We have $x = \exp(X)\in U_\alpha$, so $s_\alpha x = x$, and since the function $e(\varrho)\cdot A(\gamma+\varrho)$ factors through exp, the equation follows.

Proof of (iii). $A(\gamma+\varrho)\circ s_\alpha = \det(s_\alpha)\cdot A(\gamma+\varrho) = -A(\gamma+\varrho)$.

Proof of (iv). $e(\varrho s_\alpha X) = e(\varrho - \varrho(\alpha^*)\alpha)X = e(\varrho X)\cdot e(-\varrho(\alpha^*)\alpha X)$ and by V, (4.12)(iv) $\varrho(\alpha^*) = \langle \varrho, \alpha^*\rangle = -r$ is an integer. Thus $e(-\varrho(\alpha^*)\alpha X) = e(\alpha X)^r = \vartheta_\alpha(x)^r = 1$ since $x \in U_\alpha$. □

Now we begin to tackle the main theorem (1.7). We start with a plausible remark.

(1.11) Lemma. *The abelian group of alternating complex-valued functions on* LT *of the form* $\sum_j n_j\cdot e(\lambda_j)$, $\lambda_j\in LT^*$, $n_j\in\mathbb{Z}$ *is free on the generators* $A(\gamma)$, $\gamma\in K$.

Proof. Let $g = \sum_j n_j\cdot e(\lambda_j)$ be alternating with $\lambda_j \in LT^*$ distinct. Then

$$|W|\cdot g = \sum_{w\in W}\det(w)\cdot g\circ w = \sum_j n_j\cdot A(\lambda_j)$$

and we obtain an equation

$$g = \sum_j r_j\cdot A(\lambda_j), \qquad r_j\in\mathbb{Q}.$$

Now $A(\lambda_j \circ w) = \det(w) \cdot A(\lambda_j)$ and $A(\lambda) = 0$ if λ is not in a Weyl chamber. Since W permutes the Weyl chambers transitively, we end up with an equation

$$g = \sum_j q_j \cdot A(\gamma_j), \qquad q_j \in \mathbb{Q}$$

with distinct $\gamma_j \in K$. If we substitute the definition of $A(\gamma_j)$ and g, both sides are linear combinations of functions $e(\gamma)$ with distinct $\gamma \in LT^*$. The left-hand side has integral coefficients and the right-hand side has rational coefficients $\pm q_j$. But since the functions $e(\gamma)$ are linearly independent, the q_j must also be integral. Hence the $A(\gamma)$ generate the group in question and they are linearly independent. □

Next, we recall the Weyl integral formula IV, (1.11). Applying it to a class function $\psi: G \to \mathbb{C}$ and using the formula (1.8) for $\det(\mathrm{Ad}_{G/T}(t^{-1}) - E_{G/T})$ it says:

(1.12) $$|W| \cdot \int_G \psi(g)\, dg = \int_T \psi(t) \cdot \eta(t)\, dt, \qquad \eta \circ \exp = \delta\bar{\delta}.$$

Since we obtain the integrands via the Lie algebra LT of the torus, we define the integral $\int_{LT} f$ of an arbitrary $f: LT \to \mathbb{C}$ which is a linear combination of functions

$$X \mapsto e(\alpha X), \qquad \alpha \in LT^*, \qquad X \in LT$$

as follows: We choose an isomorphism $T \cong \mathbb{R}^k/\mathbb{Z}^k$, $LT \cong \mathbb{R}^k$, $I \cong \mathbb{Z}^k$, so that the α may be read as linear forms and the f as functions on $\mathbb{R}^k$. We put

(1.13) $$\int_{LT} f = \lim_{N \to \infty} \frac{1}{(2N)^k} \int_{-N}^{N} \cdots \int_{-N}^{N} f(x_1, \ldots, x_k)\, dx_1 \ldots dx_k.$$

For the functions we consider, this integral is easy to compute. Indeed, for $a \in \mathbb{R}$

$$\lim_{N \to \infty} \frac{1}{2N} \int_{-N}^{N} e(at)\, dt = \begin{cases} 1 & \text{for } a = 0, \\ 0 & \text{for } a \neq 0. \end{cases}$$

Thus the integral on LT is

$$\int_{LT} e \circ \alpha = \begin{cases} 1 & \text{for } \alpha = 0, \\ 0 & \text{for } \alpha \neq 0. \end{cases}$$

In other words:

(1.14) Remark. The functions $e \circ \alpha$, $\alpha \in LT^*$ (also denoted by $e(\alpha)$) form an orthonormal system for the Hermitian product

$$\langle f, g \rangle = \int_{LT} f \cdot \bar{g}$$

of functions on LT.

Incidentally, this shows once again that the functions $e(\alpha)$ are linearly independent over $\mathbb{C}$. If $f \in \mathrm{R}(T)$ we clearly have

$$\int_{LT} f \circ \exp = \int_T f.$$

Having laid this groundwork we begin to prove parts of the character formula:

(1.15) Lemma. *Let* $\chi \in \mathrm{Irr}(G, T)$ and $\tilde{\chi} = \chi \circ \exp\colon LT \to \mathbb{C}$. *Then there is a linear form* $\gamma \in K$ *such that* $\tilde{\chi} \cdot \delta = \pm \mathrm{A}(\gamma)$.

PROOF. We have $\chi = \psi | T$ for some irreducible character ψ of G and by the Weyl integral formula

$$1 = \int_G \psi \cdot \bar{\psi}\, dg = |W|^{-1} \int_T \chi \cdot \bar{\chi} \cdot \eta\, dt = |W|^{-1} \int_{LT} \tilde{\chi} \cdot \bar{\tilde{\chi}} \cdot \delta \cdot \bar{\delta}.$$

Using the inner product (1.14), this simply says

$$\langle \tilde{\chi}\delta, \tilde{\chi}\delta \rangle = |W|.$$

On the other hand, $\tilde{\chi}\delta$ is an alternating function of the form considered in (1.11). Thus

$$\tilde{\chi}\delta = \sum_j n_j \cdot \mathrm{A}(\gamma_j), \qquad n_j \in \mathbb{Z}, \quad \gamma_j \in K.$$

Now, from the orthogonality relations (1.14) we immediately get

$$\langle \mathrm{A}(\gamma), \mathrm{A}(\lambda) \rangle = \begin{cases} 0 & \text{for } \gamma \neq \lambda, \\ |W| & \text{for } \gamma = \lambda, \text{ if } \gamma, \lambda \in K. \end{cases}$$

Hence, in our case, one of the coefficients n_j must be ± 1 and the others vanish. □

(1.16) Lemma. *Let* $\chi \in \mathrm{Irr}(G, T)$ and let $\gamma \in K$ *be the linear form (uniquely determined by* (1.11), (1.15)) *such that* $(\chi \circ \exp) \cdot \delta = \pm A(\gamma)$. *We set* $\gamma = \beta + \varrho$, *where* ϱ *is the half sum of the positive roots. Then* $\beta \in \bar{K} \cap I^*$.

PROOF. By assumption, the function $\mathrm{A}(\gamma)/\delta$ factors through exp. Thus for any $X \in LT$, $Y \in I$, we have

$$\mathrm{A}(\gamma)(X + Y) \cdot \delta(X) = \mathrm{A}(\gamma)(X) \cdot \delta(X + Y).$$

By the definition of $\mathrm{A}(\gamma)$ and (1.5), this means explicitly

$$\left\{\sum_w \det(w) e((\beta + \varrho)w(X + Y))\right\} \cdot e(\varrho X) \cdot \prod_{\alpha \in R_+} [1 - e(-\alpha X)]$$

$$= \left\{\sum_w \det(w) e((\beta + \varrho)wX)\right\} \cdot e(\varrho(X + Y)) \prod_{\alpha \in R_+} [1 - e(-\alpha(X + Y))].$$

Now, since $Y \in I$ we have $e(\varrho w Y) = e(\varrho Y)$ by V, (4.12)(iii), and $e(-\alpha Y) = 1$ for $\alpha \in R_+$. Hence if $X \in \mathrm{L}T_r$, we may divide both sides of the equation by the nonzero factor

$$e(\varrho(X + Y)) \cdot \prod_{\alpha \in R_+} (1 - e(-\alpha X)).$$

What we get is

$$\sum_w \det(w) \cdot e((\beta + \varrho)wX) \cdot (e(\beta w Y) - 1) = 0.$$

By continuity, this equation holds for every $X \in \mathrm{L}T$. Since $\beta + \varrho = \gamma \in K$, the functions $e((\beta + \varrho)w)$, $w \in W$, are linearly independent. Consequently $e(\beta w Y) = 1$ for $Y \in I$, and therefore $\beta \in I^*$. So we end up with $\beta \in \bar{K} \cap I^*$ by V, (4.14). □

We now show the converse and therewith—up to sign—the first two assertions of the character formula. If $\beta \in \bar{K} \cap I^*$, then $\mathrm{c}(\beta + \varrho) = \mathrm{A}(\beta + \varrho)/\delta$ is a symmetric virtual character on the torus by (1.6), and is nonzero, since $\beta + \varrho$ is in K and not on any wall, see (1.2)(iii).

(1.17) Lemma. *If* $\beta \in \bar{K} \cap I^*$, *then* $\mathrm{c}(\beta + \varrho)$ *or* $-\mathrm{c}(\beta + \varrho)$ *is in* $\mathrm{Irr}(G, T)$.

Proof. By IV, (2.7), there is a class function $f: G \to \mathbb{C}$ with $f \mid T = \mathrm{c}(\beta + \varrho)$, since $\mathrm{c}(\beta + \varrho)$ is symmetric. Now, the irreducible characters form a complete orthonormal system in the space of all class functions. So we compute the scalar product of f with an irreducible character χ of G. By the Weyl integral formula and (1.15) we get

$$|W| \cdot \langle f, \chi \rangle = |W| \cdot \int_G f \cdot \bar{\chi}\, dg = \int_T \mathrm{c}(\beta + \varrho) \cdot \bar{\chi} \cdot \eta\, dt$$

$$= \int_{\mathrm{L}T} \mathrm{c}(\beta + \varrho) \cdot \delta \cdot (\bar{\chi} \circ \exp) \cdot \bar{\delta} = \pm \int_{\mathrm{L}T} \mathrm{A}(\beta + \varrho) \cdot \bar{\mathrm{A}}(\gamma)$$

for some linear form $\gamma \in K$. But the last integral is $|W|$ if $\beta + \rho = \gamma$ and 0 otherwise. And it cannot always be zero, since f is nonzero and the irreducible characters form a complete orthonormal system. Thus there must be an irreducible character χ with $\chi \circ \exp | \mathrm{L}T = \pm \mathrm{A}(\gamma)/\delta$ and $\gamma = \beta + \varrho$. □

(1.18) Lemma. $\delta = \mathrm{A}(\rho)$.

Proof. If we expand the product $\delta = e(\varrho) \cdot \prod_{\alpha \in R_+} (1 - e(-\alpha))$, we get an expression

$$\delta = e(\varrho) + \sum_j a_j \cdot e(\varrho - \beta_j),$$

where each β_j is a nontrivial linear combination of positive roots with nonnegative integer coefficients. In particular, these β_j do not vanish. On the other hand, applying (1.15) to the character 1 of the trivial one-dimensional representation we get $\delta = \pm \mathrm{A}(\gamma + \varrho)$ for some integral form $\gamma \in \overline{K}$, and this is another expression of δ as a linear combination of functions $e(\lambda)$, $\lambda \in LT^*$. In these expressions, the only linear form λ appearing in some term $e(\lambda)$ and lying in K is ϱ in the first case and $\gamma + \varrho$ in the second. This is because $\varrho - \beta_j \in K \Rightarrow -\beta_j \in \overline{K} \Rightarrow -\langle \beta_j, \alpha \rangle \geq 0$ for all $\alpha \in R_+ \Rightarrow -\langle \beta_j, \beta_j \rangle \geq 0 \Rightarrow \beta_j = 0$, see V, (4.14). So we conclude $\varrho = \gamma + \varrho$ and $\gamma = 0$. Therefore $\delta = \mathrm{A}(\varrho)$, since the coefficient of $e(\varrho)$ is 1 in both cases. □

To finish, we prove the dimension formula (1.7)(iv) and at the same time we get rid of the ambiguity of sign.

(1.19) Lemma. *If $\gamma \in \overline{K} \cap I^*$, then at the point $1 \in T$ the function* $\mathrm{c}(\gamma + \varrho)$ *has the value*

$$\prod_{\alpha \in R_+} \frac{\langle \alpha, \gamma + \varrho \rangle}{\langle \alpha, \varrho \rangle} > 0.$$

The value of a character at 1 is the dimension of the corresponding representation. So the sign in (1.15) must be positive, since $\mathrm{c}(\gamma + \varrho)(1)$ is positive.

Proof of the Lemma. We identify LT with LT^* using the inner product. The point $1 \in T$ corresponds to $0 \in LT$. So we are going to calculate

$$\lim_{t \to 0} \mathrm{c}(\gamma + \varrho)(t\varrho).$$

We have $\delta(t\varrho) \neq 0$ for $t > 0$ since $\varrho \in K$. From $\delta = \mathrm{A}(\varrho)$ we get

$$\begin{aligned} \mathrm{A}(\gamma + \varrho)(t\varrho) &= \sum_w \det(w) \cdot e(t \langle \varrho w, \gamma + \varrho \rangle) = \mathrm{A}(\varrho)(t(\gamma + \varrho)) = \delta(t(\gamma + \varrho)) \\ &= \prod_{\alpha \in R_+} [\exp(\pi i t \langle \alpha, \gamma + \varrho \rangle) - \exp(-\pi i t \langle \alpha, \gamma + \varrho \rangle)] \\ &\equiv \prod_{\alpha \in R_+} 2\pi i t \langle \alpha, \gamma + \varrho \rangle \bmod t^{m+1}, \end{aligned}$$

where m is the cardinality of R_+. Now by l'Hospital's rule, the limit $\mathrm{c}(\gamma + \varrho)(0)$ is as stated in the lemma, and its value is positive since $\gamma + \varrho \in K$ and $\langle \alpha, \gamma + \varrho \rangle > 0$ for all $\alpha \in R_+$. □

Proof of the Character Formula (1.7). Nothing is missing: By (1.15), (1.16) we have a map

$$\mathrm{Irr}(G, T) \to \overline{K} \cap I^*, \qquad \chi \mapsto \gamma \quad \text{with} \quad \mathrm{c}(\gamma + \varrho) = \pm \chi.$$

This is well defined by (1.11). It is injective since γ determines $\mathrm{c}(\gamma + \varrho)$ and hence χ because only one of the possible signs gives a positive value at the unit element. It is surjective by (1.17), and the sign is positive by (1.19). The further formulas are in (1.18), (1.19). □

The calculations in this section would be somewhat simpler if the half sum of the positive roots were an integral form. This is in fact the case if the adjoint representation factors through Spin, see V, (6.8), Ex. 4. If the group is simply connected, then $I = \Gamma$ and ϱ is integral since it always takes integer values on Γ, see V, (4.12)(iv).

The main theorem (1.7) is due to Hermann Weyl [1]. For an algebraic treatment, see Freudenthal and de Vries [1], p. 48.

(1.20) Exercises

Let G be a compact connected Lie group.

1. Show that a localization of a factorial ring is factorial (i.e., it admits unique prime factorization). What are the prime elements? Let ϑ be an irreducible character in $R(T)$. Show that $\vartheta - 1$ has no multiple prime factors in $R(T)$. If ϑ_1 is another irreducible character in $R(T)$, then $\vartheta - 1$ and $\vartheta_1 - 1$ are relatively prime unless $L\vartheta$ and $L\vartheta_1$ are proportional.
2. Show that G possesses irreducible representations of arbitrarily high dimension if it is nonabelian.
3. Let G be semisimple. Show that G possesses only finitely many nonisomorphic representations of a given dimension.
4. Suppose that G is not semisimple. Show that G possesses infinitely many nonisomorphic irreducible representations of dimension 1.
5. Which irreducible root system contains a simple root which is also contained in the closed fundamental Weyl chamber?

2. The Dominant Weight and the Structure of the Representation Ring

We continue to use the notation $e(t) = e^{2\pi i t}$.

Let $j\colon T \to G$ be the inclusion of the maximal torus in a compact connected Lie group. It induces a monomorphism of the representation rings $j^*\colon R(G) \to R(T)$.

The Weyl group W acts on T and hence on $R(T)$. The image of j^* lies in the ring $R(T)^W$ of the symmetric characters. From the character formula we now obtain:

(2.1) Proposition. *The inclusion of the maximal torus induces an isomorphism*

$$j^*\colon R(G) \to R(T)^W.$$

Proof. Let $f \in R(T)$ be symmetric and $g = f \circ \exp\colon LT \to \mathbb{C}$. Then $g = \sum_s n_s \cdot e(\gamma_s)$, $\gamma_s \in I^*$, $n_s \in \mathbb{Z}$, and by (1.5)

$$g \cdot \delta = e(\rho) \cdot \sum_s m_s \cdot e(\lambda_s), \qquad m_s \in \mathbb{Z}, \quad \lambda_s \in I^*.$$

But since $g \cdot \delta$ is alternating, it may also be written in the form

$$g \cdot \delta = \sum_l r_l \cdot \mathrm{A}(\beta_l), \qquad r_l \in \mathbb{Z}, \quad \beta_l \in K,$$

see (1.11). Comparing coefficients, we see that every β_l must be of the form $\rho + \lambda_{s(l)}$ and hence the character formula (1.7) says that

$$f = \sum_l r_l \cdot c(\rho + \lambda_{s(l)})$$

is in the image of j^*. □

In other words: A class function $f: G \to \mathbb{C}$ is a virtual character if and only if $f \mid T$ is a virtual character of T. Of course, knowing the character ring is much less than knowing the characters themselves, let alone the representations. All the same, let us, once again, derive this proposition directly from the Weyl integral formula. This elucidates the basic ideas in the proof of the character formula:

Let $f \in \mathrm{R}(T)^W$. We need to show that $\int_G f \cdot \bar{\chi}\, dg$ is an integer for every irreducible character χ of G. The Weyl integral formula says

$$|W| \cdot \int_G f \cdot \bar{\chi}\, dg = \int_{LT} f \cdot \bar{\chi} \cdot \delta \cdot \bar{\delta} = \int_{LT} (f \cdot \delta) \cdot (f \cdot \delta)^-,$$

and the last integral is an integer multiple of the order $|W|$ of the Weyl group. Indeed, the factors of the integrand are alternating and for the basis $\mathrm{A}(\lambda)$, $\lambda \in LT^*$ of the alternating functions on LT we have the orthogonality relations

$$\int_{LT} \mathrm{A}(\lambda)\bar{\mathrm{A}}(\gamma) = 0 \text{ or } |W|.$$

The character formula yields a bijection between the integral forms in the closed fundamental Weyl chamber $\bar{K}$ and the irreducible characters. But how may we characterize the integral form $\gamma \in \bar{K}$ with respect to the corresponding representation? We will define certain partial orders on LT^*, and we will show that γ is the real weight of the corresponding representation which is maximal with respect to these orders. This will also yield information on the multiplication of characters.

In fact, there are two different reasonable orders for elements of LT^*.

(2.2) Definition. Let γ, $\lambda \in LT^*$, let K be the fundamental Weyl chamber, and let $\mathrm{Conv}(A)$ denote the convex closure of a subset $A \subset LT^*$. Then we define

(i) $\gamma \sqsubseteq \lambda$ if and only if $\gamma \in \mathrm{Conv}(W\lambda)$.
(ii) $\gamma \leq \lambda$ if and only if $\langle \gamma, \tau \rangle \leq \langle \lambda, \tau \rangle$ for every $\tau \in K$.

If $\gamma \leq \lambda$, $\gamma \neq \lambda$, we will say that γ is ***lower*** and λ is ***higher***.

The relation $\gamma \subseteq \lambda$ is equivalent to

$$\text{Conv}(W\gamma) \subset \text{Conv}(W\lambda),$$

and this defines a partial order on the space of W-orbits of linear forms. This order has the advantage that it is independent of the choice of K; it is determined by the root system itself (compare Adams [1]).

Nevertheless, the order $\leq$ is common in the literature. Although it is less invariant, it is quite useful in explicit calculations. It is reflexive: If $\langle \gamma, \tau \rangle = \langle \lambda, \tau \rangle$ for all $\tau \in K$, then $\gamma = \lambda$ since K is open. It is also compatible with algebraic operations:

$$\begin{aligned}
&\gamma \leq \lambda \quad \text{and} \quad \mu \leq \nu \Rightarrow \gamma + \mu \leq \lambda + \nu, \\
\textbf{(2.3)} \qquad &\gamma \leq \lambda \quad \text{and} \quad r \in \mathbb{R}_+ \Rightarrow r \cdot \gamma \leq r \cdot \lambda, \\
&\gamma \leq \lambda \Rightarrow -\lambda \leq -\gamma.
\end{aligned}$$

The following proposition explains how these two orders are related and gives a new description of the order $\leq$.

(2.4) Proposition. *Let S be the distinguished basis corresponding to the Weyl chamber K.*

(i) *If $\gamma \in \bar{K}$ and $w \in W$, then $w(\gamma) \leq \gamma$.*

(ii) *If $\gamma, \lambda \in \bar{K}$, then $\gamma \leq \lambda$ if and only if $\gamma \subseteq \lambda$.*

(iii) *We have $0 \leq \gamma$ if and only if $\gamma = \sum_{\alpha \in S} c_\alpha \cdot \alpha$ with nonnegative real c_α.*

PROOF. (i) Let $w(\gamma) \neq \gamma$ and $\tau \in K$. Then we have to show $\langle w(\gamma) - \gamma, \tau \rangle \leq 0$. Now, suppose $\langle w(\gamma) - \gamma, \tau \rangle > 0$ for some fixed $\tau \in K$. Choose $w \in W$ so that $\langle w(\gamma) - \gamma, \tau \rangle$ is maximal. Then $w(\gamma) \notin \bar{K}$ by V, (4.1)(iii) since $w(\gamma) \neq \gamma$. Hence by V, (4.5) there is a root $\alpha \in S$ such that $\langle w(\gamma), \alpha \rangle < 0$. However, if s_α is the reflection corresponding to α, we have

$$\langle s_\alpha w(\gamma), \tau \rangle = \langle w(\gamma) - \langle w(\gamma), \alpha^* \rangle \alpha, \tau \rangle > \langle w(\gamma), \tau \rangle,$$

contradicting the choice of w.

(ii) Suppose $\gamma, \lambda \in \bar{K}$ and $\gamma \subseteq \lambda$, that is $\gamma \in \text{Conv}(W\lambda)$. Explicitly, this means

$$\gamma = \sum_w c_w \cdot w(\lambda), \qquad \sum_w c_w = 1, \qquad c_w \geq 0.$$

Hence by (i) and (2.3) we get $\gamma \leq \sum_w c_w \cdot \lambda = \lambda$.

Conversely, suppose $\gamma \notin \text{Conv}(W\lambda)$. Then γ and the orbit $W\lambda$ lie on different sides of some affine hyperplane. In other words, there is some $\kappa \in LT^*$ such that $\langle \gamma, \kappa \rangle > \langle w(\lambda), \kappa \rangle$, and hence $\langle \gamma, \kappa \rangle > \langle \lambda, w(\kappa) \rangle$ for all $w \in W$. One of the $w(\kappa)$ lies in $\bar{K}$, and we may find some nearby $\tau \in K$ with $\langle \gamma, w^{-1}(\tau) \rangle > \langle \lambda, \tau \rangle$. But then by (i) we have $\langle \gamma, \tau \rangle \geq \langle w(\gamma), \tau \rangle = \langle \gamma, w^{-1}(\tau) \rangle > \langle \lambda, \tau \rangle$, and therefore $\gamma \nleq \lambda$.

(iii) Suppose $\gamma = \sum_{\alpha \in S} c_\alpha \cdot \alpha$ with $c_\alpha \geq 0$. Since for $\tau \in K$ we have $\langle \alpha, \tau \rangle > 0$ by definition of the basis S corresponding to K, we get $\langle \gamma, \tau \rangle \geq 0$ and $\gamma \geq 0$.

To prove the converse, we choose a set $B \subset LT^*$, disjoint from S, so that $S \cup B$ is a linear basis of LT^*. Let $\gamma = \sum_{\alpha \in S} c_\alpha \cdot \alpha + \sum_{\beta \in B} c_\beta \cdot \beta$. Now, suppose $c_\nu < 0$ for some fixed $\nu \in S$ or that $c_\nu \neq 0$ for some fixed $\nu \in B$. By the theory of linear equations, we may find an element $\tau \in LT^*$ so that $\langle \mu, \tau \rangle = 1$ for all $\mu \in S \cup B$, $\mu \neq \nu$, and so that $-c_\nu \cdot \langle \nu, \tau \rangle = x$ for some given $x > 0$. Then $\tau \in K$ since $\langle \alpha, \tau \rangle > 0$ for all $\alpha \in S$. But we may choose x so large that $\langle \gamma, \tau \rangle < 0$. Consequently $0 \not\leq \gamma$. □

From the description (iii)—or directly from the definition—one sees that

$$R_+ = \{\alpha \in R \mid \alpha \geq 0\}.$$

Due to (ii), the rules (2.3) are also valid for the relation $\subseteq$ if $\gamma, \lambda, \mu, \nu \in \bar{K}$. It also follows that the order on the W-orbits of forms induced by $\subseteq$ is reflexive. If $\gamma \subseteq \lambda$ and $\lambda \subseteq \gamma$, then there is a $w \in W$ with $w(\gamma) = \lambda$ since the orbits $W\gamma$ and $W\lambda$ have the same representatives in $\bar{K}$; see also (2.12), Ex. 1.

As usual we set

$$\gamma \leq \lambda \Leftrightarrow \lambda \geq \gamma, \qquad \gamma \subseteq \lambda \Leftrightarrow \lambda \supseteq \gamma,$$

$$\gamma < \lambda \Leftrightarrow \gamma \leq \lambda \text{ and } \gamma \neq \lambda, \qquad \gamma \subset \lambda \Leftrightarrow \gamma \subseteq \lambda \text{ and } W\gamma \neq W\lambda,$$

$$\gamma < \lambda \Leftrightarrow \lambda > \gamma, \qquad \gamma \subset \lambda \Leftrightarrow \lambda \supset \gamma.$$

Since the convex closure of $W\gamma$ is compact and I^* is discrete, there are only finitely many integral forms λ with $\lambda \subseteq \gamma$ for any given $\gamma \in LT^*$. This allows us to argue by induction with respect to the order $\subseteq$ (or $\leq$ in $\bar{K} \cap I^*$).

(2.5) Definition. The complex function

$$S(\gamma) = \sum_{\lambda \in W\gamma} e(\lambda), \qquad \gamma \in LT^*,$$

on LT is called the ***symmetric sum*** of γ.

Needless to say, $S(\gamma)$ is symmetric, and using V, (4.1)(iii) we see that the functions $S(\gamma)$, $\gamma \in \bar{K}$, form a basis of the abelian group of symmetric functions on LT of the form $\sum_j n_j e(\lambda_j)$, $\lambda_j \in LT^*$, $n_j \in \mathbb{Z}$.

(2.6) Proposition.

(i) *Let $\gamma \in \bar{K} \cap I^*$. Then*

$$c(\gamma + \varrho) = S(\gamma) + \sum_j n_j S(\gamma_j) \quad \text{with} \quad n_j \in \mathbb{Z}, \quad \gamma_j \in \bar{K} \cap I^*, \quad \gamma_j < \gamma.$$

(ii) *An irreducible representation possesses a highest real weight in $\bar{K}$. The corresponding weight space is one-dimensional.*

PROOF. Statement (ii) follows from (i) since the functions $c(\gamma + \varrho)$ are the restrictions of the irreducible characters to T. Thus the weights of the representation associated to $c(\gamma + \varrho)$ are the forms γ, γ_j together with their transforms under W.

For the proof of (i) we need the following

Lemma.

$$S(\gamma) \cdot \delta = A(\gamma + \varrho) + \sum_j m_j A(\gamma_j + \varrho), \qquad m_j \in \mathbb{Z}, \quad \gamma_j \in \bar{K} \cap I^*, \quad \gamma_j < \gamma.$$

PROOF.

$$S(\gamma) = \sum_{\lambda \in W\gamma} e(\lambda) \quad \text{and} \quad \delta = A(\varrho) = \sum_w \det(w) \cdot e(w(\varrho)).$$

Moreover, $S(\gamma) \cdot \delta$ is alternating and therefore of the form

$$S(\gamma) \cdot \delta = \sum_l \pm A(\tau_l), \qquad \tau_l \in K.$$

The (not necessarily distinct) τ_l may all be written as $\lambda + w(\varrho) \in K$, $\lambda \in W\gamma$, as can be seen by comparing coefficients. Now, $\gamma \in \bar{K}$ and $\varrho \in K$, so by (2.3), (2.4)(i).

$$\lambda + w(\varrho) \le \gamma + \varrho$$

and the form $\gamma + \varrho$ itself appears exactly once in the product $S(\gamma) \cdot \delta$, hence exactly once among the τ_l. Indeed, if $\lambda \neq \gamma$ or $w \neq 1$, then $\lambda < \gamma$ or $w(\varrho) < \varrho$ and $\lambda + w(\varrho) < \gamma + \varrho$. Hence $\tau_l < \gamma + \varrho$ for all $\tau_l \neq \gamma + \varrho$. Also, $\tau_l - \varrho = \lambda + (w(\varrho) - \varrho)$, $\lambda \in W\gamma$, is an integral form and $\tau_l - \varrho \in \bar{K}$ since $\tau_l \in K$; see V, (4.12), (4.14). This demonstrates the lemma.

Now the statement (i) of the proposition follows by induction since in $\bar{K} \cap I^*$ there are only finitely many forms lower than a given form:

$$c(\gamma + \varrho) = S(\gamma) - \sum_j m_j c(\gamma_j + \varrho), \qquad \gamma_j \in \bar{K} \cap I^*, \qquad \gamma_j < \gamma$$

by the lemma. And the induction hypothesis for forms lower than γ allows us to write this as

$$S(\gamma) + \sum_j n_j S(\lambda_j), \qquad \lambda_j \in \bar{K} \cap I^*, \quad \lambda_j < \gamma. \qquad \square$$

(2.7) Definition. The integral form $\gamma \in \bar{K} \cap I^*$ is called the ***dominant weight*** of the irreducible representation with the character χ_γ with $\chi_\gamma | T = c(\gamma + \varrho)$. If V and V' are irreducible representations with dominant weights γ and γ' then the irreducible representation with dominant weight $\gamma + \gamma'$ is called the ***Cartan composite*** of V and V' and is denoted by $V * V'$.

Thus an irreducible representation is uniquely determined by its dominant weight or, what amounts to the same thing, by its largest W-orbit of weights in the order $\subseteq$ above. On occasion we will index the irreducible characters by their dominant weights.

Any virtual character of G, that is any symmetric element of $R(T)$ may be recursively decomposed into its irreducible summands. This is the topic of the next section. Now we begin to analyze the multiplication of characters. The following lemma relates the tensor product and the Cartan composite of irreducible representations.

(2.8) Lemma. $\chi_\gamma \cdot \chi_\lambda = \chi_{\gamma+\lambda} + \sum_\mu n_\mu \cdot \chi_\mu$ *with coefficients* $n_\mu \in \mathbb{N}_0$ *and* $\mu < \gamma + \lambda$. *The indices are the dominant weights of the corresponding representations.*

PROOF. Let U and V be irreducible representations with dominant weights γ and λ. The global weights of $U \otimes V$ are the products of the global weights of U and V. Hence, the real weights are the sums $\alpha + \beta$ of a weight α of U and β of V. Among these, $\gamma + \lambda$ is the highest and has multiplicity 1 by (2.6)(ii). □

The lemma simply says that $U * V$ appears as the irreducible summand of the highest dominant weight in $U \otimes V$ and has multiplicity 1.

The set $\bar{K} \cap I^*$ is additively closed and forms an abelian monoid. If Λ is a finite set, then we use $\mathbb{N}_0^\Lambda$ to denote the additive abelian monoid of all maps $n\colon \Lambda \to \mathbb{N}_0$, that is, all Λ-tuples $n = (n(\lambda) \mid \lambda \in \Lambda)$ of nonnegative integers.

(2.9) Definition. A finite set of integral forms $\Lambda \subseteq \bar{K} \cap I^*$ is called a ***system of generators*** of $\bar{K} \cap I^*$ if the map

$$\varphi\colon \mathbb{N}_0^\Lambda \to \bar{K} \cap I^*, \qquad n \mapsto \sum_{\lambda \in \Lambda} n(\lambda) \cdot \lambda$$

is surjective. A system of generators is called ***free***, or a ***fundamental system of weights***, if φ is bijective.

The corresponding irreducible representations with dominant weights $\lambda \in \Lambda$ are then called ***fundamental representations***.

If $\bar{K} \cap I^*$ possesses a fundamental system (i.e., if $\bar{K} \cap I^* \cong \mathbb{N}_0^\Lambda$) then, clearly, it is uniquely determined and consists of the ***indecomposable*** elements of the monoid $\bar{K} \cap I^*$. These are the elements $\lambda \neq 0$ in $\bar{K} \cap I^*$ which cannot be written in the form

$$\lambda = \lambda_1 + \lambda_2, \qquad 0 \neq \lambda_j \in \bar{K} \cap I^*, \qquad j = 1, 2.$$

(2.10) Theorem. *Let Λ be a system of generators of $\bar{K} \cap I^*$ and let $\mathbb{Z}[X_\lambda \mid \lambda \in \Lambda]$ denote the ring of polynomials with indeterminates X_λ, $\lambda \in \Lambda$ with integer coefficients.*

(i) *The ring homomorphism*

$$\psi\colon \mathbb{Z}[X_\lambda \mid \lambda \in \Lambda] \to R(G), \qquad X_\lambda \mapsto \chi_\lambda,$$

is surjective.

(ii) *If Λ is free, then ψ is an isomorphism.*

PROOF. (i) If $\gamma \in \bar{K} \cap I^*$, then $\gamma = \sum_{\lambda \in \Lambda} n(\lambda) \cdot \lambda$ with $n(\lambda) \in \mathbb{N}_0$. Therefore by (2.8)

$$\chi_\gamma = \psi\left(\prod_{\lambda \in \Lambda} X_\lambda^{n(\lambda)}\right) - \sum_\mu m_\mu \chi_\mu \quad \text{with } m_\mu \in \mathbb{N}_0 \text{ and } \mu < \lambda.$$

Using the order on the forms in $\bar{K} \cap I^*$ and applying induction as before, we see that χ_γ is in the image of ψ.

(ii) If the system of generators is free, then the monomials in the polynomial ring correspond bijectively to their tuples of exponents in $\mathbb{N}_0^\Lambda$ and to the forms in $\bar{K} \cap I^*$ via φ. Using this, we order the monomials so that this correspondence is order-preserving. Suppose, then, that

$$\psi\left(\sum_{j=1}^{s} c_j \cdot R_j\right) = 0$$

for certain monomials R_j among which

$$R_s = \prod_{\lambda \in \Lambda} X_\lambda^{n(\lambda)} \quad \text{with} \quad \sum_{\lambda \in \Lambda} n(\lambda) \cdot \lambda = \gamma$$

is maximal. Then by (2.8)

$$\psi(R_s) = \chi_\gamma + \sum_\mu a_\mu \cdot \chi_\mu, \qquad a_\mu \in \mathbb{N}_0, \quad \mu < \gamma.$$

Thus the character χ_γ in $\psi(\sum_j c_j \cdot R_j)$ appears with the coefficient c_s. It follows that $c_s = 0$ and that ψ is injective. □

(2.11) Corollary. *Let G be simply connected with rank k. Then G possesses a fundamental system $\Lambda = \{\lambda_1, \ldots, \lambda_k\}$ with k elements. Consequently, there is a ring isomorphism*

$$\mathbb{Z}[X_1, \ldots, X_k] \cong \mathbb{Z}[X_\lambda \mid \lambda \in \Lambda] \to \mathrm{R}(G), \qquad X_\lambda \mapsto \chi_\lambda.$$

PROOF. By V, (7.1) we have $\pi_1(G) \cong I/\Gamma$. So in this case $I = \Gamma$. If S is the distinguished basis of the root system, then $S^* = \{\alpha^* \mid \alpha \in S\}$ is a basis of Γ and hence of I; see V, (4.13). Let $\Lambda = \{\lambda_1, \ldots, \lambda_k\}$ be the vector space basis of $\mathrm{L}T^*$ dual to S^*. Then Λ is a basis of the lattice of integral forms. The fundamental Weyl chamber corresponding to S is

$$K = \{\gamma \in \mathrm{L}T^* \mid \langle \gamma, \alpha^* \rangle > 0 \text{ for } \alpha^* \in S^*\}.$$

In other words, K (resp. $\bar{K}$) is the set of linear combinations $\sum_{\lambda \in \Lambda} x_\lambda \cdot \lambda$ with $x_\lambda > 0$ (resp. $x_\lambda \geq 0$). Thus Λ is a fundamental system of weights. □

For the general structure of the set of weights of an irreducible representation see Bourbaki [1], VIII, §7.2, Humphreys [1], 13 and 21.3. For the largest root, Bourbaki [1], VI, §1.8. For general results about the structure of the representation ring, also in connection with topology and K-theory, see

Atiyah [1], and Atiyah and Hirzebruch [1], Atiyah and Segal [1], Segal [1], Pittie [1], Steinberg [2], McLeod [1].

The determination of the characters does not settle the question of finding models of the irreducible representations themselves. This is the content of the Borel–Weil–Bott theorem; see Serre [1], Bott [1]. Roughly the result is:

For each choice of a system of positive roots the space G/T carries the structure of a homogeneous complex manifold; compare Bourbaki [1], Ch. IX, §4, Ex. 9. One can obtain the representation with highest weight $\gamma: T \to S^1$ as the space of holomorphic sections of the bundle $G \times_T \mathbb{C} \to G/T$, i.e., as holomorphically induced representation.

The Weyl character formula can also be interpreted as a computation of the character of an induced representation by an analytic Lefschetz fixed-point formula; see Atiyah and Bott [1].

(2.12) Exercises

1. Let W be a compact group, V a finite-dimensional real W-module, and $v \in V$. Show that Wv is the set of extremal points of $\mathrm{Conv}(Wv)$ in V. A point $p \in V$ is called an ***extremal point*** of a convex set $C \subset V$ if there is a hyperplane $H \subset V$ such that $H \cap C = \{p\}$.

2. Show that if G is semisimple, then 0 is the smallest integral form for the ordering $\subseteq$ in (2.2)(i). If G is not semisimple, then there are infinitely many minimal forms for this ordering. An integral form which is fixed under the Weyl group W is minimal. Is a minimal form necessarily fixed under W?

3. Let $\gamma, \lambda, \mu \in LT^*$. Show that if $\gamma \subseteq \lambda \subset \mu$ or $\gamma \subset \lambda \subseteq \mu$, then $\gamma \subset \mu$. If $\gamma \leq \lambda < \mu$ or $\gamma < \lambda \leq \mu$, then $\gamma < \mu$.

4. Let $\gamma, \lambda \in \bar{K}$. Show that
$$\mathrm{S}(\gamma) \cdot \mathrm{S}(\lambda) = \mathrm{S}(\gamma + \lambda) + \sum_{\mu} m_\mu \cdot \mathrm{S}(\mu) \quad \text{with } m_\mu \in \mathbb{N}_0 \text{ and } \mu < \gamma + \lambda.$$

5. Let G be a compact Lie group. The following argument shows that the representation ring $\mathrm{R}(G)$ is a finitely generated $\mathbb{Z}$-algebra and hence is Noetherian (Atiyah, compare Segal [1]). $\mathrm{R}(G)$ may be made into an $\mathrm{RU}(n)$-module by means of an embedding $G \to \mathrm{U}(n)$, and $\mathrm{RU}(n)$ is finitely generated over $\mathbb{Z}$. Thus it suffices to show that $\mathrm{R}(G)$ is finitely generated over $\mathrm{RU}(n)$. Choose one A_j from every conjugacy class of Cartan subgroups of G. Then we have
$$\mathrm{RU}(n) \to \mathrm{R}(G) \to \prod_{j=1}^{r} \mathrm{R}(A_j),$$
where the second map is injective. Thus we are reduced to showing that $\mathrm{R}(A_j)$ is a finitely generated $\mathrm{RU}(n)$-module. But the inclusion $A_j \to \mathrm{U}(n)$ factors through a maximal torus T of $\mathrm{U}(n)$, so we may accomplish our goal by showing that $\mathrm{R}(T)$ is finitely generated over $\mathrm{RU}(n)$ and $\mathrm{R}(A_j)$ is finitely generated over $\mathrm{R}(T)$.

3. The Multiplicities of the Weights of an Irreducible Representation

Let V be an irreducible representation with dominant weight $\gamma \in \bar{K} \cap I^*$. Then we know that this highest weight γ has multiplicity 1. We want to determine the multiplicities of the lower weights of V. To be specific, let

$$V\,|\,T = \bigoplus_{\lambda} V(\lambda), \qquad m(\lambda, \gamma) = \dim V(\lambda),$$

be the decomposition of V into its weight spaces. Then $m(\lambda, \gamma)$ is called the ***multiplicity*** of λ as a weight of V. We set $m(\lambda, \gamma) = 0$ if λ is not a weight of V. Of course, for a given $\gamma \in \bar{K} \cap I^*$, the multiplicity $m(\lambda, \gamma)$ only depends on the W-orbit of λ. Once one knows these numbers $m(\lambda, \gamma)$, one can decompose an arbitrary representation into its irreducible summands provided one is given its weights together with their multiplicities.

We continue to use the notation $e(\lambda) = e^{2\pi i\lambda}$. We will manipulate certain infinite series of the form

$$\varphi = \sum_{\lambda \in LT^*} m_\lambda \cdot e(\lambda), \qquad m_\lambda \in \mathbb{Z}.$$

Such a series should be interpreted strictly formally as an element of the additive group $\prod_{\lambda \in LT^*} \mathbb{Z} \cdot e(\lambda)$. The ***support*** of the series φ is the set

$$\{\lambda \in LT^* \,|\, m_\lambda \neq 0\}.$$

Let $I_+^* = \{\lambda \in I^* \,|\, \lambda \geq 0\}$ be the set of ***positive weights*** with respect to the ordering (2.2)(ii). Consider those series φ whose supports lie in a finite union of sets of the form $\mu - I_+^*$, $\mu \in LT^*$. With the Cauchy product as multiplication, these form a ring denoted by $\mathbb{Z}\langle I^* \rangle$. Thus, if the series φ above and $\psi = \sum n_\zeta \cdot e(\zeta)$ are elements of $\mathbb{Z}\langle I^* \rangle$, then

$$\varphi \cdot \psi = \sum_{\mu} \left(\sum_{\lambda + \zeta = \mu} m_\lambda \cdot n_\zeta \right) e(\mu).$$

One easily verifies that the interior sum is finite, so the product is well defined ((3.5), Ex. 1).

The restriction of a character to the maximal torus is a finite sum of functions $e(\lambda)$, and so each character defines an element of the ring $\mathbb{Z}\langle I^* \rangle$. The subsequent computations are to be read as equations in this ring.

A ***decomposition into positive roots*** of an integral form $\zeta \in I^*$ is a family of nonnegative integers $(n_\alpha \,|\, \alpha \in R_+)$ such that $\zeta = \sum_{\alpha \in R_+} n_\alpha \cdot \alpha$. Let $\mathfrak{p}(\zeta)$ denote the cardinality of the set of all decompositions of ζ into positive roots. Let $Q_+ = \{\zeta \in I^* \,|\, \mathfrak{p}(\zeta) > 0\}$ denote the set of linear combinations of simple roots with nonnegative integral coefficients and let $K \in \mathbb{Z}\langle I^* \rangle$ be the series

$$K = \sum_{\zeta \in Q_+} \mathfrak{p}(\zeta) \cdot e(-\zeta).$$

(3.1) Lemma. $K \cdot e(-\varrho) \cdot \delta = K \cdot \prod_{\alpha \in R_+} (1 - e(-\alpha)) = 1.$

PROOF. The first equation follows from (1.5). For the second, we have

$$K = \prod_{\alpha \in R_+} \left(\sum_{n=0}^{\infty} e(-n\alpha) \right)$$

and

$$(1 - e(-\alpha)) \cdot \sum_{n=0}^{\infty} e(-n\alpha) = 1. \qquad \square$$

(3.2) Theorem (Kostant). *Let $\gamma \in \bar{K} \cap I^*$ and let V be an irreducible representation with dominant weight γ. Then the multiplicity $m(\lambda, \gamma)$ of λ as a weight of V is*

$$m(\lambda, \gamma) = \sum_{w \in W} \det(w) \cdot p(w(\gamma + \varrho) - (\lambda + \varrho)).$$

PROOF. The character of V restricts to $c(\gamma + \varrho)$ on the maximal torus and, by the Weyl character formula and (3.1),

$$c(\gamma + \varrho) = K \cdot e(-\varrho) \cdot \delta \cdot c(\gamma + \varrho) = K \cdot e(-\varrho) \cdot A(\gamma + \varrho).$$

Hence if we substitute the definitions of $A(\gamma + \varrho)$ and K, we get

$$c(\gamma + \varrho) = \sum_{w \in W, \zeta \in Q_+} \det(w) \cdot p(\zeta) \cdot e(-\varrho + w(\gamma + \varrho) - \zeta).$$

Consequently

$$m(\lambda, \gamma) = \sum_{w, \zeta} \det(w) \cdot p(\zeta),$$

where w runs through the Weyl group and ζ runs through the elements of Q_+ satisfying

$$\lambda = -\varrho + w(\gamma + \varrho) - \zeta, \quad \text{or}$$
$$\zeta = w(\gamma + \varrho) - (\lambda + \varrho). \qquad \square$$

The theorem in particular says that the right side of the formula vanishes if λ is not a weight of V. This is not obvious from the expression itself and is still interesting even if $\gamma = 0$; see (3.5), Ex. 1.

(3.3) Proposition. *With the assumptions of* (3.2), *suppose $\lambda \subseteq \gamma$ and $\lambda \neq \gamma$. Then*

$$m(\lambda, \gamma) = -\sum_{w \neq 1} \det(w) \cdot m(\lambda + \varrho - w(\varrho), \gamma).$$

PROOF. By the Weyl character formula and the definition of $m(\zeta, \gamma)$, the character of V restricts on the maximal torus to

$$c(\gamma + \varrho) = A(\gamma + \varrho)/A(\varrho) = \sum_{\zeta} m(\zeta, \gamma) \cdot e(\zeta).$$

Multiplying the last equation by $A(\varrho)$, we obtain

(i) $$\sum_{w \in W} \sum_{\zeta} \det(w) \cdot m(\zeta, \gamma) \cdot e(\zeta + w(\varrho)) = \sum_{w \in W} \det(w) \cdot e(w(\gamma + \varrho)).$$

Now $\varrho \in K$ by V, (4.14). Therefore $\varrho \neq w(\varrho)$ and hence $w(\varrho) < \varrho$ if $w \neq 1$ by (2.4)(i). Consequently

(ii) $$\lambda + \varrho \notin W(\gamma + \varrho),$$

since $w(\lambda + \varrho) = w(\lambda) + w(\varrho) \leq \gamma + w(\varrho) < \gamma + \varrho$ for $w \neq 1$. By (ii), the coefficient of $e(\lambda + \varrho)$ vanishes on the right side of (i). Thus it must also vanish on the left, i.e.,

$$\sum_{w \in W} \sum_{\zeta \in M} \det(w) \cdot m(\zeta, \gamma) = 0, \qquad M = \{\zeta \mid \zeta + w(\varrho) = \lambda + \varrho\}.$$

In other words,

$$\sum_{w \in W} \det(w) \cdot m(\lambda + \varrho - w(\varrho), \gamma) = 0.$$ □

The formula in the proposition allows us recursively to compute the multiplicities $m(\lambda, \gamma)$, since the forms $\lambda + \varrho - w(\varrho)$ appearing on the right are integral and higher than λ, and we know the multiplicity $m(\gamma, \gamma) = 1$ of the highest possible weight.

The multiplicity formula (3.2) allows us to completely describe the decomposition of a product of irreducible characters into irreducible summands. Let χ_γ denote the character of the irreducible representation with dominant weight γ.

(3.4) Proposition (Steinberg). *If* $\gamma, \lambda \in \bar{K} \cap I^*$, *then*

$$\chi_\gamma \cdot \chi_\lambda = \sum_{\mu \in \bar{K} \cap I^*} m(\gamma, \lambda, \mu) \cdot \chi_\mu,$$

with

$$m(\gamma, \lambda, \mu) = \sum_{v, w \in W} \det(v \cdot w) \cdot p(v(\gamma + \varrho) + w(\lambda + \varrho) - (\mu + 2\varrho)).$$

PROOF. If we restrict the characters to the maximal torus, the proposition reads

$$\sum_{\mu} m(\gamma, \lambda, \mu) \cdot c(\mu + \varrho) = c(\gamma + \varrho) \cdot c(\lambda + \varrho)$$

with the $m(\gamma, \lambda, \mu)$ above. If we multiply by $A(\varrho)$ this becomes

$$\sum_{\mu} m(\gamma, \lambda, \mu) \cdot A(\mu + \varrho) = c(\gamma + \varrho) \cdot A(\lambda + \varrho).$$

If we write

$$c(\gamma + \varrho) = \sum_{\zeta \in I^*} m(\zeta, \gamma) \cdot e(\zeta),$$

this means that

$$\sum_{\mu} m(\gamma, \lambda, \mu) \cdot A(\mu + \varrho) = \left(\sum_{\zeta \in I^*} m(\zeta, \gamma) \cdot e(\zeta)\right) \cdot \sum_{w \in W} \det(w) \cdot e(w(\lambda + \varrho))$$

$$= \sum_{\eta \in I^*} \sum_{w \in W} \det(w) \cdot m(\eta + \varrho - w(\lambda + \varrho), \gamma) \cdot e(\eta + \varrho).$$

We once again use the fact that $\varrho - w(\varrho)$ is integral. The coefficient of $e(\mu + \varrho)$ on the left-hand side of this equation is $m(\gamma, \lambda, \mu)$. Comparing coefficients,

$$m(\gamma, \lambda, \mu) = \sum_{w \in W} \det(w) \cdot m(\mu + \varrho - w(\lambda + \varrho), \gamma).$$

The proposition now follows by substituting the multiplicities from (3.2). □

A special case of this theorem is the Clebsch–Gordan formula, see (3.5), Ex. 7.

The multiplicity formula, and in particular formula (3.2), contain all the information of the character formula. In fact, the Weyl character formula may be derived from (3.2). When it comes to practical explicit computations, the formulas (3.2) and (3.3) are considerably more efficient than the Weyl formula.

Original sources for this section are Kostant [1], Steinberg [1], Cartier [1]. There is another recursion formula due to Freudenthal [1]; see also Freudenthal and de Vries [1], p. 245, Jacobson [1], VIII.

(3.5) Exercises

1. Verify that the Cauchy product in $\mathbb{Z}\langle I^* \rangle$ is well defined and makes $\mathbb{Z}\langle I^* \rangle$ into a commutative ring with 1. Show that $\delta = A(\varrho)$ is a unit in this ring and that $\mathbb{Z}\langle I^* \rangle$ has no zero-divisors.

2. Derive the recursion formula (3.3) from (3.2). *Hint*: Set $\gamma = 0$ in (3.2). This yields a recursion formula for $\mathfrak{p}(\lambda)$ which may again be plugged into (3.2).

3. Let ϱ be the half sum of the positive roots. Show that $\mathfrak{p}(\varrho - w(\varrho)) > 0$ for all $w \in W$.

4. Let $\gamma \in \bar{K}$ and $\lambda \subset \gamma$. Show that $|\lambda + \varrho| < |\gamma + \varrho|$.

5. Using the formula in (3.2), show that

(i) $$m(\gamma, \gamma) = 1,$$

(ii) $$m(\lambda, \gamma) \neq 0 \Rightarrow \lambda \leq \gamma.$$

6. Use (3.4) to prove lemma (2.8), which says that if $\gamma, \lambda, \mu \in \bar{K} \cap I^*$, then

(i) $$m(\gamma, \lambda, \mu) \neq 0 \Rightarrow \mu \leq \gamma + \lambda,$$

(ii) $$m(\gamma, \lambda, \gamma + \lambda) = 1.$$

7. Derive the Clebsch–Gordan formula II, (5.5) from (3.4). You will find:

(i) SU(2) has exactly one positive root α.
(ii) I is generated by α^*, and I^* by $\alpha/2$.
(iii) $\quad p(k \cdot \alpha/2) = 1$ if k is nonnegative and even, whereas $p(k \cdot \alpha/2) = 0$ otherwise.

Let V_n be an irreducible representation with dominant weight $n \cdot \alpha/2$. Let $\gamma = n \cdot \alpha/2$, $\lambda = p \cdot \alpha/2$, $\mu = q \cdot \alpha/2$ with $n \geq p$.

(iv) $\quad m(\gamma, \lambda, \mu) = p((n + p - q)\alpha/2) - p((n - p - q - 2)\alpha/2).$
(v) $\quad V_n \otimes V_p \cong V_{n+p} \oplus V_{n+p-2} \oplus V_{n+p-4} \oplus \cdots \oplus V_{n-p}.$

4. Representations of Real or Quaternionic Type

An irreducible representation of a compact connected Lie group G is determined by its dominant weight. But how does one recognize the real, complex, or quaternionic representations from the dominant weights? We will now address this question.

(4.1) Proposition. *Let V be an irreducible representation with dominant weight γ. Then $V \cong \bar{V}$ if and only if $-\gamma = w(\gamma)$ for some w in the Weyl group W.*

Proof. The global weights of $\bar{V}$ are the conjugates of those of V. The real weights of $\bar{V}$ therefore are the negatives of the real weights of V. Since the convex closure of $W\gamma$ contains all weights of V by assumption, the weights of $\bar{V}$ lie in $-\mathrm{Conv}(W\gamma) = \mathrm{Conv}(W(-\gamma))$. Hence the weight $-w(\gamma)$, which lies in the closed fundamental Weyl chamber, is the dominant weight of $\bar{V}$. □

Now suppose that $\gamma_1, \ldots, \gamma_k$ form a fundamental system of weights, and let K be the fundamental Weyl chamber of G, see (2.9). If γ_j is indecomposable in $\bar{K} \cap I^*$, then $-\gamma_j$ is indecomposable in $-\bar{K} \cap I^*$, and if $w \in W$ transforms the chamber $-K$ into K, then we see that $-w(\gamma_j)$ is indecomposable in $\bar{K} \cap I^*$. Therefore $-w(\gamma_j) \in \{\gamma_1, \ldots, \gamma_k\}$. Since every W-orbit meets $\bar{K}$ in only one point, we conclude:

(4.2) Note. Let $\gamma_1, \ldots, \gamma_k$ be a fundamental system of weights of G. For every v with $1 \leq v \leq k$ there is a $w \in W$ and a $\bar{v}$, uniquely determined by v, such that $-\gamma_v = w(\gamma_{\bar{v}})$. The map $v \mapsto \bar{v}$ is an involution of the set $\{1, \ldots, k\}$.

Now, if $\gamma = \sum_{v=1}^k n_v \cdot \gamma_v$ is the dominant weight of a representation V, and $-wK = K$, then $-w(\gamma) = \sum_v n_v \cdot \gamma_{\bar{v}} = \sum_v n_{\bar{v}} \cdot \gamma_v$ is the dominant weight of $\bar{V}$. Thus (4.1) says:

(4.3) Proposition. *Let V be an irreducible representation with dominant weight $\gamma = \sum_v n_v \cdot \gamma_v$. Then $V \cong \bar{V}$ if and only if $n_v = n_{\bar{v}}$ for all v.* □

This is a condition on the ν with $\nu \neq \bar{\nu}$. The weights γ_ν corresponding to the fixed points $\nu = \bar{\nu}$ of the permutation $\nu \mapsto \bar{\nu}$ satisfy $-w(\gamma_\nu) = \gamma_\nu$. This means that the associated fundamental representations are self-conjugate, whereas the fundamental representations associated to the ν with $\nu \neq \bar{\nu}$ are of complex type.

If $V \cong \bar{V}$, we still have to determine whether V is of real or quaternionic type. These structures are given by a conjugate-linear structure map

$$\mathscr{J}: V \to V, \mathscr{J}^2 = \varepsilon \cdot \mathrm{id}, \varepsilon = \pm 1;$$

see II, §6. If V is irreducible and $V \cong \bar{V}$, then V possesses exactly one of these structures and the corresponding ε is called the ***index*** of V. If $\varepsilon = 1$, then V is real, and if $\varepsilon = -1$, then V is quaternionic.

The following lemmata (4.4), (4.5) show how to read off the index of an arbitrary irreducible representation from the dominant weight once we know the index of the fundamental representations. Recall that we introduced the Cartan composite $V * V'$ in (2.7).

(4.4) Lemma. *Let V and V' be irreducible self-conjugate representations with index ε, ε'. Then $V * V'$ is self-conjugate with index $\varepsilon \cdot \varepsilon'$.*

PROOF. If the group possesses a fundamental system of weights, then the self-conjugacy follows from (4.3). In general one may apply an argument analogous to the following ((4.7), Ex. 5):

Let γ and γ' be the dominant weights of V and V'. By (2.8), the term $V * V'$ appears with multiplicity 1 as the irreducible summand of highest weight in $V \otimes V'$, and this highest weight is $\gamma + \gamma'$. Now, if $\mathscr{J}$ and $\mathscr{J}'$ are the structure maps of V and V', then $\mathscr{J} \otimes \mathscr{J}'$ is a structure map for $V \otimes V'$ with

$$(\mathscr{J} \otimes \mathscr{J}')^2 = \varepsilon \cdot \varepsilon' \cdot \mathrm{id}.$$

Therefore we only have to convince ourselves that $\mathscr{J} \otimes \mathscr{J}'$ maps the summand $V * V'$ into itself. We decompose $V \otimes V'$ into irreducible summands and consider the composition.

$$f: V * V' \subset V \otimes V' \xrightarrow[\mathscr{J} \otimes \mathscr{J}']{} V \otimes V' \xrightarrow[\mathrm{pr}]{} U,$$

where pr is the projection onto an irreducible summand U. Assuming that $f \neq 0$, by Schur's lemma f defines an isomorphism $V * V' \to \bar{U}$. Therefore $\bar{U}$ has dominant weight $\gamma + \gamma'$ and U has dominant weight in the W-orbit of $-(\gamma + \gamma')$. It follows that $-(\gamma + \gamma') \subseteq \gamma + \gamma'$. In other words,

$$-(\gamma + \gamma') \in \mathrm{Conv}(W(\gamma + \gamma'))$$

and so $\gamma + \gamma' \in -\mathrm{Conv}(W(\gamma + \gamma')) = \mathrm{Conv}(W(-(\gamma + \gamma'))$. Hence

$$\gamma + \gamma' \subseteq -(\gamma + \gamma')$$

and we see that $\gamma + \gamma'$ and $-(\gamma + \gamma')$ define the same W-orbit. Thus $U = V * V'$. □

(4.5) Lemma. *Let V be an irreducible representation. Then $V * \bar{V}$ is self-conjugate of index* 1.

PROOF. $V * \bar{V}$ is self-conjugate by (4.3) and (4.7), Ex. 5. A structure map $\mathscr{J}: V \to \bar{V}$ is clearly also a structure map of $\bar{V}$ with the same index ε. Thus $V * \bar{V}$ has index $\varepsilon^2 = 1$ by (4.4). □

This allows us to read off the index of an irreducible representation from its dominant weight as follows:

(4.6) Proposition. *Let $Q \subset \{1, \ldots, k\}$ be the set of $\nu = \bar{\nu}$ such that the associated fundamental representation with dominant weight γ_ν is of quaternionic type, that is, has index -1. Let V be an irreducible representation with dominant weight $\gamma = \sum_{\nu=1}^{k} n_\nu \cdot \gamma_\nu$ with $n_\nu = n_{\bar{\nu}}$. Then V has index 1 or -1 according to whether $\sum_{\nu \in Q} n_\nu$ is even or odd.*

PROOF. (4.4), (4.5). □

To summarize, the index may be computed from the coefficients n_ν with $\nu = \bar{\nu}$ and is multiplicative

Knowledge of the irreducible characters themselves, along with their types, is more than just knowledge of the representation rings. Thus we may use what has been said to describe the real representation ring $R(G, \mathbb{R})$ and the abelian group $R(G, \mathbb{H})$ of virtual quaternionic representations.

Suppose that G admits a fundamental system of weights. We have embeddings II, (7.8)

$$R(G, \mathbb{R}) \to R(G, \mathbb{C}) \leftarrow R(G, \mathbb{H}),$$

induced by extension and restriction. We will view $R(G, \mathbb{C})$ as the ring of polynomials in the fundamental representations and describe $R(G, \mathbb{R})$ and $R(G, \mathbb{C})$ as a subring and a subgroup.

Additivity, $R(G, \mathbb{R})$ has a free system of generators, consisting of the following representations (see II, (6.10), Ex. 7 or 10):

$$\text{(4.7)}\qquad \begin{aligned} &\text{(i)} \quad V \in \operatorname{Irr}(G, \mathbb{C})_{\mathbb{R}}, \\ &\text{(ii)} \quad 2V \text{ for } V \in \operatorname{Irr}(G, \mathbb{C})_{\mathbb{H}}, \\ &\text{(iii)} \quad V + \bar{V} \text{ for } V \in \tfrac{1}{2} \operatorname{Irr}(G, \mathbb{C})_{\mathbb{C}}. \end{aligned}$$

Exchanging $\mathbb{R}$ and $\mathbb{H}$ correspondingly yields a free additive system of generators for $R(G, \mathbb{H})$.

Now, suppose we have been given a fundamental system of irreducible representations

$$X_1, \ldots, X_k$$

together with their types. In particular, we know the permutation $\nu \mapsto \bar{\nu}$ such that $X_{\bar{\nu}} \cong \bar{X}_\nu$. If

$$n = (n_1, \ldots, n_k), \qquad n_\nu \in \mathbb{N}_0,$$

we let $X^n \in \mathbb{Z}[X_1, \ldots, X_k]$ denote the monomial with exponent n and we define $\bar{n}$ by $\bar{n}_\nu = n_{\bar{\nu}}$. If we interpret a monomial X^n as a representation, it is clearly real, complex, or quaternionic precisely if its irreducible summand of highest weight is of the corresponding type. We may determine this using (4.3), (4.6).

(4.8) Proposition. *The real representation ring*

$$\mathrm{R}(G, \mathbb{R}) \subset \mathrm{R}(G, \mathbb{C}) = \mathbb{Z}[X_1, \ldots, X_k]$$

consists of the polynomials $f = \sum_n a_n X^n$, $n \in \mathbb{N}_0^k$, *whose coefficients satisfy*

(i) $a_n = a_{\bar{n}}$,

(ii) *if* $n = \bar{n}$ *and* $\sum_{\nu \in Q} n_\nu$ *is odd, then* $a_n \in 2\mathbb{Z}$.

Correspondingly, the abelian group $\mathrm{R}(G, \mathbb{H})$ *of the virtual quaternionic representations consists of those polynomials* f *whose coefficients* a_n *satisfy*:

(i) $a_n = a_{\bar{n}}$,

(ii) *if* $n = \bar{n}$ *and* $\sum_{\nu \in Q} n_\nu$ *is even, then* $a_n \in 2\mathbb{Z}$.

PROOF. Using (4.7) this follows exactly as (2.10). Condition (i) says that f is self-conjugate, while (ii) says that the monomial X^n is quaternionic (resp. real). □

The basic original source for this section is Malcev [1], §2. Compare also Iwahori [1].

(4.9) Exercises

1. Let G be compact. Consider the maps

$$\mathrm{R}(G) \xrightarrow[1+t]{} \mathrm{R}(G) \xrightarrow[1-t]{} \mathrm{R}(G),$$

where t is a conjugation. Let $H(G) = \ker(1 - t)/\mathrm{im}(1 + t)$. Show that $H(G)$ is a $\mathbb{Z}/2$-algebra and that the self-conjugate irreducible representations form a $\mathbb{Z}/2$-basis of $H(G)$.

2. Let $\gamma_1, \ldots, \gamma_k$ be a fundamental system of weights for G. Show that the algebra $H(G)$ in Exercise 1 is a polynomial algebra in the equivalence classes of irreducible characters with dominant weights

$$\gamma_\nu, \nu = \bar{\nu} \quad \text{and} \quad \gamma_\nu + \gamma_{\bar{\nu}}, \qquad \nu \neq \bar{\nu}.$$

3. Consider a product $G \times H$ of compact Lie groups. Let V and V' be irreducible representations of G and H. Show that the irreducible representation $V \otimes V'$ of $G \times H$ is self-conjugate if and only if V and V' are self-conjugate. In this case $V \otimes V'$ is real if and only if V and V' are of the same type; see II, (4.15).

4. Let G be compact and connected and let U be a representation of G which has a highest weight γ with multiplicity 1. Let V be an irreducible representation of G with the same dominant weight γ. Show that if U is self-conjugate, real, or quaternionic, the corresponding holds for V.

5. Let V and V' be self-conjugate irreducible representations of a compact connected Lie group. Show that $V * V'$ is self-conjugate. Let V be an arbitrary irreducible representation. Show that $V * \bar{V}$ is self-conjugate. Note that G need not possess a fundamental system.

6. Show that if $R(G)$ is generated by real representations as a ring, then every representation of G is real. *Hint*: II, (6.10), Ex. 10.

5. Representations of the Classical Groups

We continue to use the notation $e(t) = e^{2\pi i t}$, and we work with the real weights II, (9.7).

(5.1) *The special unitary group* $\mathrm{SU}(n + 1)$.

It has type A_n. The root system is specified in V, (6.2), (6.3). Since the group is simply connected, the representation ring is a polynomial ring

(i) $$\mathrm{RSU}(n+1) \cong \mathbb{Z}[\Lambda^1, \ldots, \Lambda^n]$$

in the fundamental representations; see (2.11) and (iv) below. Recall that we have described LT as $\{\vartheta \in \mathbb{R}^{n+1} \mid \vartheta_1 + \cdots + \vartheta_{n+1} = 0\}$, and I as

$$\{\vartheta \in \mathbb{Z}^{n+1} \mid \vartheta_1 + \cdots + \vartheta_{n+1} = 0\}.$$

Thus

$$LT^* = \mathbb{R}^{n+1}/\mathbb{R}\cdot(1, \ldots, 1),$$
$$I^* = \mathbb{Z}^{n+1}/\mathbb{Z}\cdot(1, \ldots, 1).$$

Now, the projection $(\vartheta_1, \ldots, \vartheta_{n+1}) \mapsto (\vartheta_1, \ldots, \vartheta_n)$ induces isomorphisms $LT \xrightarrow{\cong} \mathbb{R}^n$ and $I \xrightarrow{\cong} \mathbb{Z}^n$. Dually, we have

$$\begin{aligned} &\mathbb{R}^n \xrightarrow{\subset} \mathbb{R}^{n+1} \to LT^*, \\ &\mathbb{Z}^n \xrightarrow{\subset} \mathbb{Z}^{n+1} \to I^*. \end{aligned} \qquad (\zeta_1, \ldots, \zeta_n) \mapsto (\zeta_1, \ldots, \zeta_n, 0),$$

We will later use these maps as coordinates of LT^* and I^* and will thus denote elements by n-tuples. However, for the time being it seems more convenient to work with (cosets of) elements in $\mathbb{R}^{n+1} = L\Delta(n)^*$, since the description of the roots and the action of the Weyl group is simpler for $\mathrm{U}(n + 1)$. The (integral) forms on LT then uniquely correspond to the (integral) elements $\zeta \in \mathbb{R}^{n+1} = L\Delta(n)^*$ with $\zeta_{n+1} = 0$. If $\vartheta \in LT$, then $\zeta(\vartheta) = \langle \zeta, \vartheta \rangle$ is the standard inner product on $\mathbb{R}^{n+1}$. Using these conventions, we have

(ii) $$\bar{K} \cap I^* = \{\zeta \in \mathbb{Z}^{n+1} \mid \zeta_1 \geq \zeta_2 \geq \cdots \geq \zeta_n \geq \zeta_{n+1} = 0\}.$$

Let e_μ denote the μth standard unit vector in $\mathbb{R}^{n+1}$. Then the weights

(iii) $$\lambda_\nu = e_1 + \cdots + e_\nu, \qquad 1 \le \nu \le n,$$

form a fundamental system.

Let Λ^ν denote a fundamental representation for λ_ν. We compute its dimension using the Weyl formula (1.7) (iv):

$$\dim \Lambda^\nu = \prod_{\alpha \in R_+} \frac{\langle \lambda_\nu + \varrho, \alpha \rangle}{\langle \varrho, \alpha \rangle} = \prod_{1 \le i < j \le n+1} \frac{\langle \lambda_\nu + \varrho, e_i - e_j \rangle}{j - i}$$
$$= \prod_{i \le \nu < j} \frac{j - i + 1}{j - i} = \prod_{i \le \nu} (n + 2 - i) / \prod_{i \le \nu} (\nu + 1 - i) = \binom{n+1}{\nu}.$$

Let $V = \mathbb{C}^{n+1}$ with the standard SU$(n+1)$-action. Then V has the weights $e_1, \ldots, e_n, e_{n+1} = -(e_1 + \cdots + e_n)$. The νth standard basis vector v_ν of V is a weight vector for e_ν. Now, the exterior power $\Lambda^\nu V$ has the same dimension as Λ^ν and the same highest weight λ_ν with weight vector $v_1 \wedge \cdots \wedge v_\nu$. Thus we have (see also (5.6), Ex. 1):

(iv) The exterior power $\Lambda^\nu V$ is a fundamental representation for λ_ν.

The exterior product defines the dual pairing

$$\wedge : \Lambda^\nu V \otimes \Lambda^{n+1-\nu} V \to \Lambda^{n+1} V \cong \mathbb{C}$$

since the determinant is 1 on SU$(n+1)$. Hence $\Lambda^{n+1-\nu} \cong \overline{\Lambda}^\nu$ or, in the notation of (4.2),

(v) $$\bar{\nu} = n + 1 - \nu.$$

If $n+1$ is odd, then all fundamental representations are of complex type. If $n+1$ is even and $\mu = (n+1)/2$, then the exterior product

$$\wedge : \Lambda^\mu \otimes \Lambda^\mu \to \mathbb{C}$$

is a regular bilinear form which is symmetric if μ is even and skew-symmetric otherwise. Thus by II, (6.4):

(vi) Λ^μ is real if $\mu = (n+1)/2$ is even and quaternionic if μ is odd.

The alternating sums may be computed as follows: Let $\zeta \in \mathrm{L}T^*$ be given by an element of $\mathbb{R}^{n+1}$ as explained above. Suppose $\vartheta \in \mathrm{L}T$. Then

$$\mathrm{A}(\zeta)(\vartheta) = \sum_{w \in W} \det(w) \cdot e(\zeta w(\vartheta)) = \sum_{\sigma \in S(n+1)} \mathrm{sign}(\sigma) \cdot e\Big(\sum_j \zeta_j \vartheta_{\sigma(j)}\Big)$$
$$= \sum_\sigma \mathrm{sign}(\sigma) \prod_j e(\zeta_j \vartheta_{\sigma(j)}).$$

That is

(vii) $$A(\zeta)(\vartheta) = \det(e(\zeta_j \vartheta_k)).$$

In the case of SU(2) we recover some of our previous results II, §5. The irreducible characters are labeled by the integers. We have $\varrho = \vartheta_1$, and if $\gamma \in \mathbb{Z}$, then

$$c(\gamma + \varrho) = A(\gamma + \varrho)/A(\varrho) = \frac{e((\gamma + 1)\vartheta_1) - e((\gamma + 1)\vartheta_2)}{e(\vartheta_1) - e(\vartheta_2)}$$
$$= \frac{\sin(2\pi(\gamma + 1)\vartheta_1)}{\sin(2\pi\vartheta_1)}.$$

(5.2) *The unitary group* $U(n + 1)$, $n \geq 0$.

It has type A_n, but it is not semisimple. We identify S^1 with the center $\{zE_{n+1} \mid z \in S^1\}$ of $U(n + 1)$ via $z \mapsto zE_{n+1}$. We have the short exact sequence

(i) $$1 \to C_{n+1} \to SU(n + 1) \times S^1 \underset{p}{\to} U(n + 1) \to 1.$$

Here C_{n+1} is the group of $(n + 1)$st roots of unity. It is embedded into $SU(n + 1) \times S^1$ via $z \mapsto (zE_{n+1}, \bar{z})$, and p maps (A, z) to $z \cdot A$.

Since p is surjective, a representation V of $U(n + 1)$ is irreducible, self-conjugate, real, or quaternionic, if and only if the same holds for p^*V. Now, the irreducible representations of $SU(n + 1) \times S^1$ have the form $V' \otimes V''$, where V' and V'' are irreducible representations of $SU(n + 1)$ and S^1; see II, (4.15). If V is an irreducible representation of $U(n + 1)$ and

$$p^*V \cong V' \otimes V'',$$

then

(ii) $$V' \cong V \mid SU(n + 1), \qquad \dim(V') \cdot V'' \cong V \mid S^1.$$

The tensor product $V' \otimes V''$ is self-conjugate if and only if both factors are. But only the trivial irreducible representation of S^1 is self-conjugate. So we have:

(iii) An irreducible representation V of $U(n + 1)$ is self-conjugate if and only if $V \mid SU(n + 1)$ is self-conjugate and $V \mid S^1$ is trivial.

Now we look at the dominant weights. Choosing coordinates as in V, (6.1), (6.2), the projection p in (i) induces the map of Lie algebras

$$Lp: ((\vartheta_1, \ldots, \vartheta_{n+1}), t) \mapsto (\vartheta_1 - t, \ldots, \vartheta_{n+1} - t),$$

where $\sum_\nu \vartheta_\nu = 0$. The dual map is

$$Lp^*: (\gamma_1, \ldots, \gamma_{n+1}) \mapsto \left((\gamma_1 - \gamma_{n+1}, \ldots, \gamma_n - \gamma_{n+1}), - \sum_\nu \gamma_\nu\right).$$

Hence, if $\gamma = (\gamma_1, \ldots, \gamma_{n+1})$ is the dominant weight of V:

(iv) $\quad V \mid S^1$ is trivial if and only if $\gamma_1 + \cdots + \gamma_{n+1} = 0$.

Now suppose that the dominant weight

$$\zeta = (\zeta_1, \ldots, \zeta_n) = (\gamma_1 - \gamma_{n+1}, \ldots, \gamma_n - \gamma_{n+1})$$

of $V \mid \mathrm{SU}(n+1)$ is given. Then there is at most one corresponding

$$\gamma = (\gamma_1, \ldots, \gamma_{n+1})$$

with $\sum_\nu \gamma_\nu = 0$, since $\zeta_1 + \cdots + \zeta_n = \gamma_1 + \cdots + \gamma_{n+1} - (n+1)\gamma_{n+1}$. More precisely, we see:

(v) Let V' be an irreducible representation of $\mathrm{SU}(n+1)$ with dominant weight $(\zeta_1, \ldots, \zeta_n)$. Then there is a (unique) irreducible representation V of $\mathrm{U}(n+1)$ such that $V \mid S^1$ is trivial and $V \mid \mathrm{SU}(n+1) \cong V'$ if and only if $\zeta_1 + \cdots + \zeta_n$ is divisible by $n+1$.

Now consider self-conjugate irreducible representations of $\mathrm{SU}(n+1)$. Then condition (v) automatically holds for the weights $\lambda_\nu + \lambda_{n+1-\nu}$ in (5.1)(iii), (v). If $n + 1 = 2\mu$ is even, then there is also the fundamental weight λ_μ of a self-conjugate representation, and the sum of the components of λ_μ is μ. Hence we have:

(vi) Let $\zeta = (\zeta_1, \ldots, \zeta_n)$ be the dominant weight of a self-conjugate irreducible representation of $\mathrm{SU}(n+1)$. If $n+1$ is odd, then condition (v) always holds for ζ. If $n+1$ is even, then condition (v) holds if and only if

$$\zeta = 2n_\mu \cdot \lambda_\mu + \sum_{\nu<\mu} n_\nu \cdot (\lambda_\nu + \lambda_{n+1-\nu}), \qquad \mu = (n+1)/2.$$

Therefore by (4.6), (4.9), Ex. 3:

(vii) All irreducible self-conjugate representations of $\mathrm{U}(n+1)$ are real.

(viii) Let $V = \mathbb{C}^{n+1}$ with the standard action of $\mathrm{U}(n+1)$, and let $\Lambda^\nu = \Lambda^\nu V$. The representation

$$\Lambda^\nu \otimes \Lambda^{n+1-\nu} \otimes \Lambda^{n+1}$$

has trivial restriction to S^1, and the restriction to $\mathrm{SU}(n+1)$ has highest weight $\lambda_\nu + \lambda_{n+1-\nu}$ of multiplicity 1.

Note, however, that this representation is not irreducible; see (5.6), Ex. 4.

Recall that we have computed the representation ring of U($n + 1$) in IV, (3.13):

(ix) $\mathrm{RU}(n+1) = \mathbb{Z}[\Lambda^1, \ldots, \Lambda^{n+1}, (\Lambda^{n+1})^{-1}], \qquad (\Lambda^{n+1})^{-1} = \bar{\Lambda}^{n+1}.$

Finally, the alternating sums have the same description as those of SU($n + 1$)

(x) $$A(\gamma)(\vartheta) = \det(e(\gamma_j \cdot \vartheta_k)).$$

(5.3) *The symplectic group* Sp(n), $n \geq 2$.

It has type C_n ($= B_n$ for $n = 2$); see V, (6.6).
We have

(i) $$\bar{K} \cap I^* = \{(\zeta_1, \ldots, \zeta_n) | \zeta_1 \geq \zeta_2 \geq \cdots \geq \zeta_n \geq 0, \zeta_\nu \in \mathbb{Z}\}.$$

Let e_μ be the μth standard unit vector of $\mathbb{R}^n$. Then the weights

(ii) $$\lambda_\nu = e_1 + \cdots + e_\nu = (1, \ldots, 1, 0, \ldots, 0), \qquad 1 \leq \nu \leq n,$$

constitute a fundamental system of weights.

(iii) The fundamental representations are self-conjugate, $\bar{\nu} = \nu$, since $-\lambda_\nu = w\lambda_\nu$ for some w in the Weyl group G(n); see (4.3) and IV, (3.8).

Let $V = \mathbb{H}^n \cong \mathbb{C}^{2n}$ as a complex vector space, where $\mathbb{C}$ acts by right multiplication. Then Sp(n) acts on V in the obvious way, and V is quaternionic. Let $v_1, \ldots, v_n$ be the standard basis of the $\mathbb{H}$-module V. Then v_μ and $v_{-\mu} = j \cdot v_\mu$ (where $j \in \mathbb{H}$ is the basic quaternion) are weight vectors of the Sp(n)-module V corresponding to the weights e_μ and $-e_\mu$. Let $\Lambda^\nu = \Lambda^\nu V$ denote the complex νth exterior power of V.

(iv) The Sp(n)-module Λ^ν, $1 \leq \nu \leq n$, has highest weight λ_ν with multiplicity 1.

Note, however, that for $\nu > 1$ these representations are ***not irreducible***. Let P^ν denote the fundamental representation with dominant weight λ_ν. Then Λ^ν and P^ν only differ by summands with lower weights than λ_ν. This enables us to replace the fundamental representations P^ν by the Λ^ν and vice versa in some general situations. For instance, by (2.10) the P^ν are algebraically independent generators of the representation ring. Hence the same holds for the Λ^ν:

(v) $$\mathrm{RSp}(n) = \mathbb{Z}[\Lambda^1, \ldots, \Lambda^n].$$

Let $\mathcal{J}: V \to V$ be the stucture map of the quaternionic structure of V. Then $\Lambda^\nu \mathcal{J}$ is a structure map for $\Lambda^\nu V$ with $(\Lambda^\nu \mathcal{J})^2 = (-1)^\nu \cdot \mathrm{id}$. Hence Λ^ν is real if ν is even and quaternionic if ν is odd. By (4.9), Ex. 4:

(vi) The fundamental representation P^ν is real if ν is even and quaternionic if ν is odd.

Before we proceed to find the representations P^{ν}, we compute the alternating sums as follows: Let $\gamma = (\gamma_1, \ldots, \gamma_n)$. We have the general identity

$$\sum_{\varepsilon_\nu = \pm 1} \varepsilon_1 \cdot \ldots \cdot \varepsilon_n \cdot e(\gamma_1\varepsilon_1\vartheta_1 + \cdots + \gamma_n\varepsilon_n\vartheta_n) = \prod_{\nu=1}^{n} (e(\gamma_\nu\vartheta_\nu) - e(-\gamma_\nu\vartheta_\nu))$$

$$= (2i)^n \prod_{\nu=1}^{n} \sin(2\pi\gamma_\nu\vartheta_\nu).$$

Also, by IV, (3.8) the Weyl group is

$$W = \mathrm{G}(n) = (\mathbb{Z}/2) \wr \mathrm{S}(n) = \{(\varepsilon_1, \ldots, \varepsilon_n, \sigma) \mid \varepsilon_\nu = \pm 1, \sigma \in \mathrm{S}(n)\}.$$

Using this, we obtain for $\vartheta = (\vartheta_1, \ldots, \vartheta_n) \in \mathrm{L}T^n$:

$$\mathrm{A}(\gamma)(\vartheta) = \sum_{w} \det(w) \cdot e(\gamma w(\vartheta)) = (2i)^n \sum_{\sigma \in S(n)} \operatorname{sign}(\sigma) \prod_{\nu=1}^{n} \sin(2\pi\gamma_\nu\vartheta_{\sigma(\nu)}).$$

Thus,

(vii) $$\mathrm{A}(\gamma)(\vartheta) = (2i)^n \cdot \det(\sin(2\pi\gamma_\nu\vartheta_\mu)).$$

We proceed to consider the fundamental representations P^{ν}, $\nu \geq 2$.

(viii) $$\dim \mathrm{P}^{\nu} = \binom{2n}{\nu} - \binom{2n}{\nu - 2} \quad \text{for } \nu \geq 2.$$

Proof. We use the Weyl formula (1.7)(iv). From the data on the root system in V, (6.6) we obtain

$$\dim \mathrm{P}^{\nu} = \prod_{j<k} \frac{\langle \lambda_\nu + \varrho, \vartheta_j + \vartheta_k \rangle \cdot \langle \lambda_\nu + \varrho, \vartheta_j - \vartheta_k \rangle}{\langle \varrho, \vartheta_j + \vartheta_k \rangle \cdot \langle \varrho, \vartheta_j - \vartheta_k \rangle} \cdot \prod_{k=1}^{n} \frac{\langle \lambda_\nu + \varrho, \vartheta_k \rangle}{\langle \varrho, \vartheta_k \rangle}$$

$$= \prod_{j=1}^{\nu} \prod_{k=\nu+1}^{n} \frac{(2n - (j + k) + 3)(k + 1 - j)}{(2n - (j + k) + 2)(k - j)}$$

$$\cdot \prod_{j=1}^{\nu-1} \prod_{k=j+1}^{\nu} \frac{2n - (j + k) + 4}{2n - (j + k) + 2} \cdot \prod_{k=1}^{\nu} \frac{n - k + 2}{n - k + 1}.$$

The interior products are

$$\prod_{k=\nu+1}^{n} \frac{2n - (j + k) + 3}{2n - (j + k) + 2} = \frac{2n + 2 - (j + \nu)}{n + 2 - j},$$

$$\prod_{k=\nu+1}^{n} \frac{k + 1 - j}{k - j} = \frac{n + 1 - j}{\nu + 1 - j},$$

$$\prod_{k=j+1}^{\nu} \frac{2n - (j + k) + 4}{2n - (j + k) + 2} = \frac{(2n + 3 - 2j)(2n + 2 - 2j)}{(2n + 2 - (j + \nu))(2n + 3 - (j + \nu))},$$

$$\prod_{k=1}^{\nu} \frac{n - k + 2}{n - k + 1} = \frac{n + 1}{n + 1 - \nu}.$$

Substituting this above and simplifying a bit, we end up with

$$\begin{aligned}
\dim \mathsf{P}^\nu &= 2(n+1-\nu)\cdot\prod_{j=1}^{\nu-1}\frac{(2n+3-2j)(2n+2-2j)}{(\nu+1-j)(2n+3-(j+\nu))}\\
&= \frac{2(n+1-\nu)}{\nu}\cdot\prod_{j=1}^{\nu-1}\frac{(2n+3-2j)(2n+2-2j)}{j(2n+3-(j+\nu))}\\
&= \frac{2(n+1-\nu)}{\nu}\prod_{j=1}^{\nu-1}\frac{2n+2-j}{j}\\
&= \frac{2(n+1-\nu)(2n+1)}{\nu(\nu-1)}\cdot\prod_{j=1}^{\nu-2}\frac{2n+1-j}{j}\\
&= \left(\frac{(2n+1-\nu)(2n+2-\nu)}{\nu(\nu-1)}-1\right)\cdot\prod_{j=1}^{\nu-2}\frac{2n+1-j}{j}\\
&= \binom{2n}{\nu}-\binom{2n}{\nu-2}.
\end{aligned}$$

(ix) The construction of the fundamental representations P^ν of $\mathrm{Sp}(n)$ for $n, \nu \geq 2$ (compare Bourbaki [1], Ch. VIII, §13, IV).

As above, let $v_1, \ldots, v_n$ be the standard basis of the $\mathbb{H}$-module $V = \mathbb{H}^n$, and let $v_1, \ldots, v_n, v_{-1}, \ldots, v_{-n}$ with $v_{-\mu} = j \cdot v_\mu$ be the corresponding complex orthonormal basis of $V \cong \mathbb{C}^{2n}$. Let $v^*_{-n}, \ldots, v^*_n$ be the dual basis of V^*.

The group $\mathrm{Sp}(n)$ leaves the alternating two-form

$$\Gamma^* = \sum_{\mu=1}^{n} v^*_\mu \wedge v^*_{-\mu}$$

invariant, since this form corresponds to the matrix $-J$; see I, (1.12). On the other hand, the Hermitian inner product gives us a conjugate linear $\mathrm{Sp}(n)$-equivariant isomorphism $V \cong V^*$ which maps v_μ to v^*_μ. Thus from Γ^* we obtain the element

$$\Gamma = \sum_{\mu=1}^{n} v_\mu \wedge v_{-\mu} \in \Lambda^2 V,$$

which is invariant under the action of $\mathrm{Sp}(n)$ on $\Lambda^2 V$. We define endomorphisms X^+ and X^- of $\Lambda(V)$ by

$$X^-(u) = \Gamma \wedge u, \qquad X^+(u) = \Gamma^* \llcorner u.$$

The bilinear product $\mathrm{Alt}^r(V) \otimes \Lambda^s(V) \to \Lambda^{s-r}(V)$ appearing in the definition of X^+ is determined by the following properties:

If $\alpha, \beta \in \mathrm{Alt}(V) \cong \Lambda(V^*)$ and $u \in \Lambda(V)$, then

$$(\alpha \wedge \beta) \llcorner u = \alpha \llcorner (\beta \llcorner u),$$

and if $\alpha \in \mathrm{Alt}^1(V) = V^*$, $x_\mu \in V$ for $\mu = 1, \ldots, r$, then

$$\alpha \llcorner (x_1 \wedge \cdots \wedge x_r) = \sum_\mu (-1)^{\mu-1} \alpha(x_\mu) \cdot x_1 \wedge \cdots \wedge \hat{x}_\mu \wedge \cdots \wedge x_r.$$

In order to compute the action of X^+ and X^-, we specify a basis of $\Lambda(V)$ as follows: For each triple (A, B, C) of disjoint subsets of $\{1, \ldots, n\}$, we set $v_{A,B,C} = v_{a_1} \wedge \cdots \wedge v_{a_k} \wedge v_{-b_1} \wedge \cdots \wedge v_{-b_l} \wedge v_{c_1} \wedge v_{-c_1} \wedge \cdots \wedge v_{c_m} \wedge v_{-c_m}$, where $a_1 < a_2 < \cdots < a_k$ are the elements of A, and analogously for B and C. The set of all these $v_{A,B,C}$ is, in fact, a basis of $\Lambda(V)$ and one may check:

$$X^+(v_{A,B,C}) = -\sum_{\mu \in C} v_{A,B,C \setminus \{\mu\}},$$

$$X^-(v_{A,B,C}) = \sum_{\mu \notin A \cup B \cup C} v_{A,B,C \cup \{\mu\}}.$$

Let H be the endomorphism of $\Lambda(V)$ which is multiplication by $n - v$ on $\Lambda^v V$. Then, using this explicit description of X^+ and X^-, the reader may easily verify:

$$[H, X^+] = 2X^+, \qquad [H, X^-] = -2X^-, \qquad [X^+, X^-] = -H.$$

In other words: the elements X^+, X^-, H generate a Lie subalgebra of $\mathrm{End}(\Lambda(V))$ which is isomorphic to $\mathrm{sl}(2, \mathbb{C})$. The action of this subalgebra makes $\Lambda(V)$ into an $\mathrm{sl}(2, \mathbb{C})$-module, and $\Lambda^v V$ is the subspace of the elements of weight $n - v$ of this module. However, we know the structure of finite-dimensional $\mathrm{sl}(2, \mathbb{C})$-modules; see II, (10.4), (10.5), ((10.18), Ex. 5). Let $E_v = (\Lambda^v V) \cap \ker X^+$ be the subspace of primitive elements in $\Lambda^v V$. Then we have:

For $v < n$ the restriction of X^- to $\Lambda^v V$ is injective, and for $v \leq n$ one has a direct sum decomposition

$$\Lambda^v V = E_v \oplus X^- E_{v-2} \oplus (X^-)^2 E_{v-4} \oplus \cdots = E_v \oplus X^- \Lambda^{v-2} V.$$

In particular, $\dim E_v = \binom{2n}{v} - \binom{2n}{v-2} = \dim \mathsf{P}^v$.

By construction, the space E_v is invariant under the $\mathrm{Sp}(n)$-action on $\Lambda^v V$. Moreover, it contains the weight vector $v_{A,\varnothing,\varnothing}$, $A = \{1, 2, \ldots, v\}$, corresponding to the highest weight λ_v of $\Lambda^v V$, since $X^+ v_{A,\varnothing,\varnothing} = 0$.

So we conclude $E_v \cong \mathsf{P}^v$ as an $\mathrm{Sp}(n)$-module.

(5.4) *The special orthogonal group* $\mathrm{SO}(2n + 1)$.

It has type B_n for $n \geq 2$; see V, (6.5). The maximal torus and the Weyl group are the same as for $\mathrm{Sp}(n)$; see II, (3.7), (3.8). This suffices to conclude that the representation rings are isomorphic. We will see this more explicitly. As for $\mathrm{Sp}(n)$, we have

(i) $$\bar{K} \cap I^* = \{(\zeta_1, \ldots, \zeta_n) \mid \zeta_1 \geq \cdots \geq \zeta_n \geq 0, \zeta_v \in \mathbb{Z}\},$$

and the forms

(ii) $\lambda_\nu = e_1 + \cdots + e_\nu = (1, 1, \ldots, 1, 0, \ldots, 0), \qquad 1 \le \nu \le n,$

constitute a fundamental system of weights. If Λ^ν is a fundamental representation with dominant weight λ_ν, then

(iii) $$\mathrm{RSO}(2n+1) = \mathbb{Z}[\Lambda^1, \ldots, \Lambda^n].$$

Let $V = \mathbb{C} \otimes \mathbb{R}^{2n+1} = \mathbb{C}^{2n+1}$ with the standard action of $\mathrm{SO}(2n+1)$. We will show:

(iv) $$\Lambda^\nu \cong \Lambda^\nu(V).$$

The $\mathrm{SO}(2n+1)$-module V has the weights 0 and $\pm e_\nu$, $1 \le \nu \le n$. Let $u_1, v_1, \ldots, u_n, v_n, u_{n+1}$ be the standard basis of $\mathbb{R}^{2n+1}$. Then $u_\nu \mp iv_\nu$ is a weight vector for the weight $\pm e_\nu$ and u_{n+1} has weight 0. Hence the vector $(u_1 - iv_1) \wedge \cdots \wedge (u_\nu - iv_\nu)$ in $\Lambda^\nu V$ has weight λ_ν. So we need to show that $\Lambda^\nu V$ is irreducible.

(v) If $\nu < m/2$, then $\mathbb{C} \otimes \Lambda^\nu \mathbb{R}^m$ is an irreducible $\mathrm{SO}(m)$-module.

Proof. Let $w_1, \ldots, w_m$ be the standard basis of $\mathbb{R}^m$. Then an element of $\mathbb{C} \otimes \Lambda^\nu \mathbb{R}^m$ may be written in the form $\sum_S a_S \cdot w_S$, where S runs through the subsets of $\{1, \ldots, m\}$ with ν elements, $a_S \in \mathbb{C}$, and $w_S = w_{\mu_1} \wedge \cdots \wedge w_{\mu_\nu}$, $\mu_1 < \cdots < \mu_\nu$, $S = \{\mu_1, \ldots, \mu_\nu\}$. Now let U be an $\mathrm{SO}(m)$-submodule of $\mathbb{C} \otimes \Lambda^\nu \mathbb{R}^m$. Let $x = \sum_S a_S \cdot w_S, x \neq 0$, be in U. Using induction on the number of coefficients $a_S \neq 0$ of x, we will show $U = \mathbb{C} \otimes \Lambda^\nu \mathbb{R}^m$. Suppose this number is 1 and $a_T \neq 0$. Then w_T is in U. But, up to sign, all the w_S are obtained from w_T by permuting the basis $w_1, \ldots, w_m$ of $\mathbb{R}^m$ and, in the case of odd permutations, substituting $-w_1$ for w_1. Hence all w_S are in U and $U = \mathbb{C} \otimes \Lambda^\nu \mathbb{R}^m$. Now, suppose that x has at least two nonvanishing coefficients a_S and a_T. Choose indices $j \in S, j \notin T$ and $k \notin S \cup T$. This is possible, since $S \neq T$ and $2\nu < m$. Then consider the transformation $A \in \mathrm{SO}(m)$ with $Aw_j = -w_j$, $Aw_k = -w_k$, $Aw_l = w_l$ for $l \neq j, k$. This transforms w_S to $-w_S$, fixes w_T, and all the other w_R are either fixed or transformed to their negatives. Hence $x + Ax \in U$ has fewer nonvanishing coefficients than x, but still at least one. □

Thus we have proved (iv) and (v) and, by construction:

(vi) Λ^ν is self-conjugate and real for $1 \le \nu \le n$.

The alternating sums have the same description as in the case of $\mathrm{Sp}(n)$. Hence the same result: Let $\vartheta = (\vartheta_1, \ldots, \vartheta_n) \in LT^n$ and $\gamma = (\gamma_1, \ldots, \gamma_n) \in (LT^n)^*$, then

(vii) $$A(\gamma)(\vartheta) = (2i)^n \cdot \det(\sin(2\pi\gamma_j \vartheta_k)).$$

(5.5) *The special orthogonal group* SO(2*n*), $n \geq 2$.

It has type D_n; see V, (6.4). We postpone the computation of the representation ring until the next section (6.6). We have

(i) $$\bar{K} \cap I^* = \{(\zeta_1, \ldots, \zeta_n) | \zeta_1 \geq \cdots \geq \zeta_{n-1} \geq |\zeta_n|, \zeta_\nu \in \mathbb{Z}\}.$$

A system of generators of this semigroup is

(ii) $$\lambda_\nu = e_1 + \cdots + e_\nu, \quad 1 \leq \nu \leq n-1,$$
$$\lambda_n^\pm = e_1 + \cdots + e_{n-1} \pm e_n,$$

where e_ν is the νth standard unit vector in $\mathbb{R}^n$. Let

$$\Lambda^\nu \text{ for } 1 \leq \nu \leq n-1, \quad \text{and} \quad \Lambda_+^n, \Lambda_-^n$$

denote the corresponding irreducible representations with dominant weights $\lambda_\nu, \lambda_n^+, \lambda_n^-$. By (2.10),

(iii) $\quad RSO(2n)$ is generated by $\Lambda^1, \ldots, \Lambda^{n-1}, \Lambda_+^n, \Lambda_-^n$.

We give an explicit construction of these representations:

Let $V = \mathbb{C} \otimes \mathbb{R}^{2n}$ and let $u_1, v_1, \ldots, u_n, v_n$ be the standard basis of $\mathbb{R}^{2n}$.

(iv) $\quad \Lambda^\nu \cong \Lambda^\nu V$ is the νth exterior power for $1 \leq \nu \leq n-1$.

In fact, $\Lambda^\nu V$ is irreducible for $\nu \leq n-1$ by (5.4)(v), and

$$(u_1 - iv_1) \wedge \cdots \wedge (u_\nu - iv_\nu)$$

is a weight vector for the dominant weight λ_ν.

The exterior power $\Lambda^n V$, however, is reducible. It has two maximal weights λ_n^+, λ_n^-, both with multiplicity one. Corresponding weight vectors are $(u_1 - iv_1) \wedge \cdots \wedge (u_{n-1} - iv_{n-1}) \wedge (u_n \mp iv_n)$. We will show:

(v) The SO(2n)-module $\Lambda^n V$ is the sum of two irreducible submodules $\Lambda_+^n V$ and $\Lambda_-^n V$ with dominant weights λ_n^+ and λ_n^-.

These summands are the eigenspaces of the $*$-operator. We briefly recall what that is.

In general, let E be an m-dimensional oriented Euclidean vector space. Then E has a canonical volume form ω determined by the metric and orientation. In other words, we have a canonical isomorphism

$$\mathbb{R} \to \Lambda^m E, t \mapsto t \cdot \omega.$$

Thus the exterior product gives us a dual pairing

$$\wedge : \Lambda^\nu E \otimes \Lambda^{m-\nu} E \to \Lambda^m E = \mathbb{R}.$$

Rewriting this, we have an isomorphism

$$d: \Lambda^\nu E \to (\Lambda^{m-\nu} E)^* \cong \Lambda^{m-\nu} E^*, \qquad dx(y) = x \wedge y.$$

On the other hand, the Euclidean inner product on E induces the isomorphism

$$\kappa: E \to E^*, \qquad x \mapsto \langle x, \ \rangle.$$

Combining these, we obtain the isomorphism

$$* = (\Lambda^{m-\nu}\kappa)^{-1} \circ d: \Lambda^\nu E \to \Lambda^{m-\nu} E.$$

If $e_1, \ldots, e_m$ is a positively oriented orthonormal basis of E, then

$$*(e_1 \wedge \cdots \wedge e_\nu) = e_{\nu+1} \wedge \cdots \wedge e_m.$$

From this one easily gets the more general formula

$$*(e_{\sigma(1)} \wedge \cdots \wedge e_{\sigma(\nu)}) = \operatorname{sign}(\sigma) \cdot e_{\sigma(\nu+1)} \wedge \cdots \wedge e_{\sigma(m)}$$

for any permutation $\sigma \in S(m)$, and

$$* \circ * = (-1)^{\nu(m-\nu)} \cdot \mathrm{id}.$$

Now, in our case, $m = 2n$, $\nu = n$, $E = \mathbb{R}^{2n}$, and we choose the orientation so that $u_1 \wedge \cdots \wedge u_n \wedge v_1 \wedge \cdots \wedge v_n$ is positive. Hence

$$*(u_1 \wedge \cdots \wedge u_n) = v_1 \wedge \cdots \wedge v_n.$$

Tensoring with $\mathbb{C}$, we obtain the automorphism

(vi) $$\tau = (-i)^n \cdot *: \Lambda^n V \to \Lambda^n V, \qquad \tau \circ \tau = \mathrm{id}.$$

From this, we have the canonical decomposition

(vii) $$\Lambda^n V = \Lambda^n_+ V \oplus \Lambda^n_- V$$

into the eigenspaces of τ corresponding to the eigenvalues 1 and -1. These are SO(2n)-modules by construction, and we show:

(viii) $$\Lambda^n_+ \cong \Lambda^n_+ V, \qquad \Lambda^n_- \cong \Lambda^n_- V.$$

PROOF. Modifying the proof of (5.4)(v), one easily shows that $\Lambda^n V$ is irreducible as an O(2n)-module. Moreover, transformation by an element $T \in O(2n)$ of determinant -1 interchanges the summands of (vii). In fact,

$$T: \Lambda^n_+ V \xrightarrow{\cong} \Lambda^n_- V,$$

since $T \circ \tau = -\tau \circ T$. Now let $U \subset \Lambda^n_+(V)$ be an SO(2n)-submodule. Then $U \oplus TU$ is an O(2n)-module, and so it is either 0 or equal to $\Lambda^n V$. Thus U is either 0 or equal to $\Lambda^n_+ V$. This shows that the SO(2n)-modules $\Lambda^n_\pm V$ are irreducible. We still have to check that

$$x = (u_1 - iv_1) \wedge \cdots \wedge (u_n - iv_n) \in \Lambda^n_+ V.$$

But this weight vector must lie in either $\Lambda^n_+ V$ or $\Lambda^n_- V$. In other words, $\tau(x) = x$ or $\tau(x) = -x$. Expanding

$$x = u_1 \wedge \cdots \wedge u_n + \cdots + (-i)^n v_1 \wedge \cdots \wedge v_n,$$

we find that

$$\tau(x) = (-i)^n * (x) = (-i)^n v_1 \wedge \cdots \wedge v_n + \cdots,$$

and hence $\tau(x) = x$. □

(ix) If n is even, then all irreducible representations of SO(2n) are self-conjugate and real.

Indeed, in this case τ is real, and the real SO(2n)-module $\Lambda^n\mathbb{R}^{2n}$ splits into the eigenspaces of τ. The decomposition of $\Lambda^n V$ arises from this by tensoring with $\mathbb{C}$. The $\Lambda^\nu V = \mathbb{C} \otimes \Lambda^\nu \mathbb{R}^{2n}$, $\nu < n$, are, of course, also real.

(x) If n is odd, then an irreducible SO(2n)-module is self-conjugate if and only if the last component of its dominant weight vanishes. Also, all self-conjugate SO(2n)-modules are real.

Indeed, let $(\gamma_1, \ldots, \gamma_n)$ be the dominant weight of the irreducible representation U. Then $\bar{U}$ has dominant weight $(\gamma_1, \ldots, \gamma_{n-1}, -\gamma_n)$, since this lies in $\bar{K} \cap I^*$ and in the orbit of $-(\gamma_1, \ldots, \gamma_n)$ under the action of the Weyl group $SG(n)$. Thus $U \cong \bar{U}$ if and only if $\gamma_n = 0$. But in this case U is an iterated Cartan composite of the real representations $\Lambda^1, \ldots, \Lambda^{n-1}$, and hence is real.

As in the case of SO(2n + 1), one may describe the alternating sums by suitable determinants. Let $E = \{(\varepsilon_1, \ldots, \varepsilon_n) \mid \varepsilon_j = \pm 1, \varepsilon_1 \cdot \ldots \cdot \varepsilon_n = 1\}$. Then

$$\begin{aligned} 2\sum_E e(\varepsilon_1\gamma_1\vartheta_1 + \cdots + \varepsilon_n\gamma_n\vartheta_n) &= \prod_{j=1}^n (e(\gamma_j\vartheta_j) + e(-\gamma_j\vartheta_j)) + \prod_{j=1}^n (e(\gamma_j\vartheta_j) - e(-\gamma_j\vartheta_j)) \\ &= 2^n \prod_{j=1}^n \cos(2\pi\gamma_j\vartheta_j) + (2i)^n \prod_{j=1}^n \sin(2\pi\gamma_j\vartheta_j). \end{aligned}$$

Now recall the structure of the Weyl group

$$SG(n) = E \cdot S(n) \subset C_2 \wr S(n), \quad \text{with} \quad C_2 = \{1, -1\}.$$

So let $\vartheta = (\vartheta_1, \ldots, \vartheta_n) \in LT^n$, and $\gamma = (\gamma_1, \ldots, \gamma_n) \in (LT^n)^*$. Then

$$\begin{aligned} 2A(\gamma)(\vartheta) &= 2 \sum_{w \in SG(n)} \det(w) e(\gamma w \vartheta) \\ &= \sum_{\sigma \in S(n)} \operatorname{sign}(\sigma) \cdot 2 \sum_E e(\varepsilon_1\gamma_1\vartheta_{\sigma(1)} + \cdots + \varepsilon_n\gamma_n\vartheta_{\sigma(n)}). \end{aligned}$$

Substituting the above, we get

(xi) $$2A(\gamma)(\vartheta) = 2^n \det(\cos(2\pi\gamma_j\vartheta_k)) + (2i)^n \det(\sin(2\pi\gamma_j\vartheta_k)).$$

There are methods for the explicit construction of the irreducible representations for the classical groups which are based on the fact that such representations are contained in iterated tensor products $V \otimes \cdots \otimes \bar{V} \otimes \cdots$ of the standard representation V. Such tensor representations are decomposed using the representation theory of the symmetric group. This method has been developed by H. Weyl. For a detailed elementary exposition of Weyl's method see Boerner [1]. See also Naimark and Stern [1].

For computations for classical groups from the Lie algebra point of view, see Bourbaki [1], VIII, §13. This is also relevant for our V, §6.

(5.6) Exercises

1. Show that the representation $\Lambda^\nu V$ of $SU(n)$ has the weights $w\lambda_\nu$, each with multiplicity 1, where w runs through the Weyl group $S(n)$. This shows once more that $\Lambda^\nu V$ is irreducible.

2. Show that $HSU(n)$ in (4.9), Ex. 1 is a polynomial algebra. The $SU(n)$-modules $\Lambda^\nu \otimes \Lambda^{n-\nu}$, $\nu < n - \nu$, together with $\Lambda^{n/2}$ if n is even, represent algebraically independent generators.

3. Similarly, show that $HU(n)$ is polynomial. The $U(n)$-modules $\Lambda^\nu \otimes \Lambda^{n-\nu} \otimes \bar{\Lambda}^n$, $\nu \le n/2$, represent algebraically independent generators.

4. Show that the representations $\Lambda^\nu \otimes \Lambda^{n-\nu} \otimes \bar{\Lambda}^n$ of $U(n)$ contain a nonzero trivial summand.

5. Show in two ways that every (irreducible) representation of $SU(n)$ may be extended to all of $U(n)$:

   (i) By considering the dominant weight.
   (ii) Directly, by inspecting the action of the center C_n of $SU(n)$.

6. Show that $HSp(n)$ and $HSO(2n+1)$ in (4.9), Ex. 1 are polynomial algebras. In each case $\Lambda^1, \ldots, \Lambda^n$ represent algebraically independent generators.

7. Check the computation of $\dim P^\nu$ in (5.3)(viii).

8. Show that the $Sp(n)$-module Λ^ν has the weights $\lambda_{\nu-2\mu}$ with multiplicities $\binom{n-\nu+2\mu}{\mu}$.

9. Characterize those irreducible $SU(n)$-representations which factor through the projective group $SU(n)/C_n$ by a condition on their dominant weights.

10. Show that the fundamental representation Λ^ν of $SO(2n+1)$ has the weights $\lambda_{\nu-2k}$ and $\lambda_{\nu-2k-1}$, each with multiplicity $\binom{n-\nu+2k}{k}$.

11. Compute the dimension of the fundamental $SO(2n)$-representations Λ^ν, Λ^n_+, Λ^n_- using the Weyl formula (1.7)(iv). This gives a new proof of irreducibility. Similarly for $SO(2n+1)$.

12. Let n be odd. Show directly that the SO(2n)-representations Λ^n_+ and Λ^n_- are conjugate.

13. For n even, show $\mathrm{HSO}(2n) \cong \mathrm{RSO}(2n) \otimes_{\mathbb{Z}} (\mathbb{Z}/2)$. For n odd, show $\mathrm{HSO}(2n) = \mathbb{Z}/2[\Lambda^1, \ldots, \Lambda^{n-1}]$.

14. Show that the SO(2n)-representation Λ^ν has the weights $\lambda_{\nu-2k}$ with multiplicities $\binom{n-\nu+2k}{k}$. Also, Λ^n_+ has the weights λ_{n-4k} with multiplicities $\binom{4k}{2k}$, and Λ^n_- has the weights λ_{n-2k-2} with multiplicities $\binom{2k+2}{k+1}$.

15. Show that the inclusion $j_{2n}\colon \mathrm{SO}(2n) \to \mathrm{SO}(2n+1)$ induces on the representation rings the injective homomorphism

$$j_{2n}^*\colon \mathrm{RSO}(2n+1) \to \mathrm{RSO}(2n), \qquad \Lambda^\nu \mapsto \Lambda^\nu + \Lambda^{\nu-1},$$

for $1 \le \nu \le n$, with $\Lambda^0 = 1$ and $\Lambda^n = \Lambda^n_+ + \Lambda^n_-$ in RSO(2n). Similarly, the inclusion $j_{2n-1}\colon \mathrm{SO}(2n-1) \to \mathrm{SO}(2n)$ induces

$$j_{2n-1}^*\colon \mathrm{RSO}(2n) \to \mathrm{RSO}(2n-1), \qquad \Lambda^\nu \mapsto \Lambda^\nu + \Lambda^{\nu-1} \quad \text{for } \nu < n,$$

and $\Lambda^n_\pm \mapsto \Lambda^{n-1}$.

6. Representations of the Spinor Groups

We start by describing the spin representations in terms of their weights. This leads to the computation of the representation ring in an easy way. Later we construct the so-called half-spin representations as modules over Clifford algebras as in Atiyah, Bott and Shapiro [1] and determine their real, complex, or quaternionic type. The computations of the representation rings, in particular theorems (6.2) and (6.6), are taken from mimeographed notes of J. Milnor.

We recall the following conventions. Let $m = 2n$ or $m = 2n + 1$. The maximal torus $\tilde{T}(n)$ of Spin(m) with Lie algebra $L\tilde{T}(n)$, as well as the maximal torus $T(n)$ of SO(m) with Lie algebra $LT(n)$, are described in coordinates as in IV, §3. We have $LT(n) \cong \mathbb{R}^n$ with integral lattice $I \cong \mathbb{Z}^n$ for SO(m). The canonical projection $\rho\colon \mathrm{Spin}(m) \to \mathrm{SO}(m)$ induces a double cover $\tilde{T}(n) \to T(n)$ through which $L\tilde{T}(n)$ is identified with $LT(n)$. The invariant inner product on $LT(n)$ corresponds to the standard Euclidean inner product on $\mathbb{R}^n$. Thus we have fixed isomorphisms

$$L\tilde{T}(n)^* \cong L\tilde{T}(n) \cong LT(n)^* \cong LT(n) \cong \mathbb{R}^n.$$

A representation of SO(m) induces one of Spin(m) by means of ρ. The induced representation has the same real weights (see II, (9.7)) as elements of $LT(n)^* \cong \mathbb{R}^n$. The integral lattice of Spin(m) is a subgroup of index 2 in that of SO(m). The lattice of integral forms of SO(m) is thus a subgroup of index 2 in the lattice of integral forms of Spin(m). With our choice of coordinates, the lattice of integral forms of Spin(m) is given by

$$\mathbb{Z}^n + \varepsilon \cdot (1, \ldots, 1), \qquad \varepsilon = 0 \text{ or } \varepsilon = \tfrac{1}{2};$$

see V, (6.7).

From this, together with the description of the root system of SO(m) in V, (6.4), (6.7), we get

(6.1) Proposition. *The group* Spin($2n$) *has as fundamental system of weights*

$$\lambda_\nu = e_1 + \cdots + e_\nu, \qquad 1 \le \nu \le n-2,$$
$$\partial_n^\pm = \tfrac{1}{2}(e_1 + \cdots + e_{n-1} \pm e_n),$$

where e_ν is the νth standard unit vector of $\mathbb{R}^n$. The group Spin($2n+1$) *has as fundamental system of weights*

$$\lambda_\nu, 1 \le \nu \le n-1,$$
$$\partial_n = \tfrac{1}{2}(e_1 + \cdots + e_n).$$

These fundamental weights are dominant for fundamental representations

$$\Lambda^1, \Lambda^2, \ldots, \Lambda^{n-2}, \Delta_+^n, \Delta_-^n \quad \text{of Spin}(2n), \quad \text{and}$$
$$\Lambda^1, \Lambda^2, \ldots, \Lambda^{n-1}, \Delta^n \quad \text{of Spin}(2n+1).$$

The representations Λ^ν arise from the representations of SO(m) with identical notation via the projection ρ: Spin(m) $\to$ SO(m). In other words, Λ^ν comes from the exterior powers $\Lambda^\nu = \Lambda^\nu V$, where V is $\mathbb{C}^{2n}$ or $\mathbb{C}^{2n+1}$. The representations Δ, Δ_+, and Δ_- are the ***half-spin representations*** of the spinor groups mentioned above.

Now, in order to compute the representation rings, we will only need to know the dominant weights.

We will often omit the index n from $\Delta, \partial, \ldots$, so $\Delta_+ = \Delta_+^n$, and so on. We will also denote the irreducible representation of Spin($2n+1$) with dominant weight 2∂ by Λ^n, and similarly, the irreducible representations of Spin($2n$) with dominant weights $2\partial^+$, $2\partial^-$, and $\partial^+ + \partial^-$ are denoted by Λ_+^n, Λ_-^n, and Λ^{n-1}. These representations are induced by those of SO(m) with the same names.

(6.2) Theorem. *The representation ring of* Spin($2n+1$) *is the polynomial ring*

$$\text{RSpin}(2n+1) = \mathbb{Z}[\Lambda^1, \ldots, \Lambda^{n-1}, \Delta].$$

Moreover

(i) $$\Delta \cdot \Delta = 1 + \Lambda^1 + \cdots + \Lambda^{n-1} + \Lambda^n.$$

The representation ring of Spin(2*n*) *is the polynomial ring*

$$\mathrm{RSpin}(2n) = \mathbb{Z}[\Lambda^1, \ldots, \Lambda^{n-2}, \Delta_+, \Delta_-].$$

Moreover

$$\text{(ii)} \qquad \Delta_+ \cdot \Delta_+ = \Lambda^n_+ + \Lambda^{n-2} + \Lambda^{n-4} + \cdots,$$
$$\text{(iii)} \qquad \Delta_+ \cdot \Delta_- = \Lambda^{n-1} + \Lambda^{n-3} + \Lambda^{n-5} + \cdots,$$
$$\text{(iv)} \qquad \Delta_- \cdot \Delta_- = \Lambda^n_- + \Lambda^{n-2} + \Lambda^{n-4} + \cdots.$$

The sums end in $\Lambda^4 + \Lambda^2 + 1$ or $\Lambda^3 + \Lambda$. Note that (i)–(iv) describe Λ^n, $\Lambda^n_\pm$, and Λ^{n-1} as polynomials.

PROOF. See (2.11). We only have to check the relations (i)–(iv), and we do this by comparing the weights of the representations on both sides of the equation. We pause for a lemma:

(6.3) Lemma. *The weights of the half-spin representations are conjugate to the dominant weights under the action of the Weyl group. Thus they all have multiplicity* 1 *and* $\dim \Delta = 2^n$, *while* $\dim \Delta_+ = \dim \Delta_- = 2^{n-1}$.

PROOF. Except for the trivial weight 0, the convex closure of the orbit of ∂ under the action of the Weyl group G(*n*) contains no additional weights.

If 0 were a weight of Δ, then $\partial = 0 + \partial$ would be a weight of $\Delta \otimes \Delta$. But the central element $-1 \in \mathrm{Spin}(2n+1)$ operates on an irreducible representation via multiplication by either $+1$ or -1 due to Schur's lemma. At any rate, it acts trivially on $\Delta \otimes \Delta$, so $\Delta \otimes \Delta$ comes from a representation of SO(2*n* + 1). Therefore it only has integral weights, a contradiction. The same argument may be applied to Δ_+ and Δ_-, from which the lemma follows. □

Back to the proof of (i)–(iv). We know the weights on the left-hand side of the equation and must only show that the weights λ_ν, λ_ν^+, and λ_ν^- appear equally often on both sides, since all the other weights which appear are conjugate to those under the operation of the Weyl group.

We choose notation as in (5.4), so the Spin(2*n* + 1)-module $V = \mathbb{C} \otimes \mathbb{R}^{2n+1}$ has as standard basis $u_1, v_1, \ldots, u_n, v_n, u_{n+1}$. The vectors $u_j \pm iv_j$ generate the weight spaces of V associated to the weights $\mp e_j$, and u_{n+1} generates the weight space associated to the weight 0 of V. Accordingly, the weight spaces of $\Lambda^\nu = \Lambda^\nu V$ are generated by those weight vectors which are ν-fold exterior products with factors $u_j \pm iv_j$ and u_{n+1}. For Spin(2*n*) the notation is the same—only the u_{n+1} is gone—as in (5.5).

(i) The weight λ_ν arises in $\Delta \cdot \Delta$ as the sum

$$\tfrac{1}{2}(1, \ldots, 1, \varepsilon_{\nu+1}, \ldots, \varepsilon_n) + \tfrac{1}{2}(1, \ldots, 1, -\varepsilon_{\nu+1}, \ldots, -\varepsilon_n), \qquad \varepsilon_j = \pm 1,$$

of two weights of Δ in $2^{n-\nu}$ different ways. On the righthand side of (i), the weight λ_ν appears exactly half as often as in the representation

$$1 + \wedge^1 V + \cdots + \wedge^{2n+1} V,$$

since $\wedge^j V \cong \wedge^{2n+1-j} V$.

In this representation, λ_ν appears as a product

$$(u_1 - iv_1) \wedge \cdots \wedge (u_\nu - iv_\nu) \wedge w,$$

where w is itself an exterior product with factors $(u_j + iv_j) \wedge (u_j - iv_j)$, $\nu < j \le n$, and u_{n+1}. There are clearly $2 \cdot 2^{n-\nu}$ such products w, from which we see that λ_ν also has multiplicity $2^{n-\nu}$ on the right-hand side of (i).

(ii) On both sides, only the weights conjugate to λ_{n-2j} and λ_n^+ appear, the latter exactly once. The weight λ_ν with even $n - \nu$ arises on the left as a sum

$$\tfrac{1}{2}(1, \ldots, 1, \varepsilon_{\nu+1}, \ldots, \varepsilon_n) + \tfrac{1}{2}(1, \ldots, 1, -\varepsilon_{\nu+1}, \ldots, -\varepsilon_n), \qquad \varepsilon_j = \pm 1,$$

in $2^{n-\nu-1}$ ways. Now, for $\nu < n$, the weight λ_ν occurs the same number of times in $\wedge_+^n$ and $\wedge_-^n$, since changing the orientation of $\mathbb{R}^n$ switches these summands of $\wedge^n V$. Consequently λ_ν occurs on the right-hand side of (ii) exactly half as often as in

$$\cdots + \wedge^{n+2} V + (\wedge_+^n V + \wedge_-^n V) + \wedge^{n-2} V + \cdots + \wedge^\nu V.$$

Here the λ_ν come from weight vectors

$$(u_1 - iv_1) \wedge \cdots \wedge (u_\nu - iv_\nu) \wedge w$$

with w an exterior product of the factors $(u_j + iv_j) \wedge (u_j - iv_j)$, $\nu < j \le n$. Since there are $2^{n-\nu}$ such products w, the multiplicity of λ_ν on the right-hand side of (ii) is also $2^{n-\nu-1}$.

(iii) and (iv) may be proved analogously. $\square$

We will use the information about the spinor groups to describe the representation ring $RSO(2n)$. The inclusion $j: SO(2n) \to SO(2n + 1)$ is the identity on the maximal torus $T(n)$ and therefore induces an injective ring homomorphism $j^*: RSO(2n + 1) \to RSO(2n)$. This makes $RSO(2n)$ into an $RSO(2n + 1)$-module. From the natural isomorphism $\wedge^\nu(V \oplus \mathbb{C}) \cong \wedge^\nu(V) \oplus \wedge^{\nu-1}(V)$ we see that $j^* \wedge^\nu = \wedge^\nu + \wedge^{\nu-1}$, $1 \le \nu \le n$, with $\wedge^0 = 1$ and $\wedge^n = \wedge_+^n + \wedge_-^n$ in $RSO(2n)$. Thus $j^* RSO(2n + 1)$ is the subring of $RSO(2n)$ generated by $\wedge^1, \ldots, \wedge^n$, and these elements are algebraically independent in $RSO(2n)$:

(6.4) $j^*: RSO(2n + 1) \xrightarrow{\cong} \mathbb{Z}[\wedge^1, \ldots, \wedge^n] \subset RSO(2n)$

$$\wedge^\nu \mapsto \wedge^\nu + \wedge^{\nu-1}.$$

The character ring of the maximal torus $T(n)$ of $SO(2n)$ and $SO(2n+1)$ has an algebraic description as the ring of finite Laurent series in n indeterminates $z_1, \ldots, z_n$. The ring $RSO(2n+1)$ sits inside this as the ring of those elements left invariant under all permutations and inversions of the variables z_ν, while $RSO(2n)$ consists of those elements left fixed by arbitrary permutations and by inversions of an even number of the z_ν. In this description j^* is the inclusion. Observe, however, that $\Lambda^\nu \in RSO(2n+1)$ is different from $\Lambda^\nu \in RSO(2n)$ by (6.4).

Inverting an odd number of the z_ν defines a ring automorphism

(6.5) $$\varphi\colon RSO(2n) \to RSO(2n), \qquad \varphi^2 = \mathrm{id},$$

and $j^*RSO(2n+1)$ is the fixed ring of φ. The automorphism φ is induced via conjugation by an $A \in SO(2n+1)$ where, for example,

$$A = \left[\begin{array}{cc|cccc} & 1 & & & & \\ 1 & & & & & \\ \hline & & 1 & & & \\ & & & \ddots & & \\ & & & & & 1 \end{array}\right] \quad \text{(blank places are zero).}$$

(6.6) Theorem.

(i) *The representation ring of* $SO(2n)$ *is*

$$RSO(2n) \cong \mathbb{Z}[\Lambda^1, \ldots, \Lambda^{n-1}, \Lambda^n_+, \Lambda^n_-]/R$$

with the relation

$$R = (\Lambda^n_+ + \Lambda^{n-2} + \Lambda^{n-4} + \cdots)(\Lambda^n_- + \Lambda^{n-2} + \Lambda^{n-4} + \cdots) - (\Lambda^{n-1} + \Lambda^{n-3} + \cdots)^2.$$

The sums in parentheses end as in (6.2), *and under the isomorphism the indeterminates* Λ^ν, $\Lambda^n_\pm$ *correspond to the representations with the same names.*

(ii) $RSO(2n)$ *is a free* $RSO(2n+1)$*-module generated by* 1 *and* Λ^n_+.

Proof. If we use (6.2)(ii–iv) to expand $(\Delta_+)^2(\Delta_-)^2 - (\Delta_+ \cdot \Delta_-)^2$, we obtain the relation R. Thus R is in the kernel of the homomorphism

$$\mathbb{Z}[\Lambda^1, \ldots, \Lambda^{n-1}, \Lambda^n_+, \Lambda^n_-] \to RSO(2n)$$

sending Λ^ν, $\Lambda^n_\pm$ to the corresponding representations. We already know (5.5)(iii) that this homomorphism is surjective. Using the relation $\Lambda^n = \Lambda^n_+ + \Lambda^n_-$, we may write

$$\begin{aligned} \mathbb{Z}[\Lambda^1, \ldots, \Lambda^{n-1}, \Lambda^n_+, \Lambda^n_-]/R &= \mathbb{Z}[\Lambda^1, \ldots, \Lambda^{n-1}, \Lambda^n, \Lambda^n_+]/F \\ &= RSO(2n+1)[\Lambda^n_+]/F, \end{aligned}$$

where F is a certain monic quadratic polynomial in the variable Λ_+^n with coefficients in $RSO(2n+1)$. The reader has the opportunity to compute F from R in (6.20), Ex. 1. We thus have a surjective ring homomorphism $RSO(2n+1)[\Lambda_+^n]/F \to RSO(2n)$ which allows us to express every $g \in RSO(2n)$ in the form

$$g = a + b\Lambda_+^n, \qquad a, b \in RSO(2n+1).$$

Both statements (i) and (ii) are now equivalent to the statement that a and b are uniquely determined by g. But if $0 = a + b\Lambda_+^n$ with $a, b \in RSO(2n+1)$, then a and b are fixed by the automorphism φ of (6.5). And since $\varphi\Lambda_+^n = \Lambda_-^n \neq \Lambda_+^n$, we conclude that $a = b = 0$. □

(6.7) Remark. The irreducible representations Λ^ν of the spinor group are real. The representation Δ^n of Spin$(2n+1)$ is self-conjugate. If n is even, then the half-spin representations Δ_+^n, Δ_-^n of Spin$(2n)$ are self-conjugate. For n odd, Δ_+^n is conjugate to Δ_-^n.

Indeed, the Λ^ν come from the representations $\mathbb{C} \otimes \Lambda^\nu \mathbb{R}^m$ of $SO(m)$, and the statements just made about the half-spin representations follow from (4.1) by looking at the dominant weights.

To read off the exact types of the half-spin representations, we need to explicitly construct them. They are obtained as irreducible modules over certain Clifford algebras, as we will now explain. First we recall a bit from I, §8:

The real Clifford algebra $C_m = C_m^0 \oplus C_m^1$ associated to the quadratic form $Q(x) = -|x|^2$ is generated by monomials

$$x_1 x_2 \ldots x_{2k+s} \in C_m^s, \qquad s \in \{0, 1\}, \quad 0 \le 2k + s \le m, \quad x_\nu \in S^{m-1}.$$

These monomials are units in the algebra and form the group Pin(m). Moreover, Spin$(m) = $ Pin$(m) \cap C_m^0$ consists of the monomials of the form above with $s = 0$. The product in the Clifford algebra satisfies the relation

$$xy + yx = -2\langle x, y\rangle \quad \text{for } x, y \in \mathbb{R}^m.$$

Thus $x^2 = -1$ for $x \in S^{m-1}$ and $x, y \in \mathbb{R}^m$ anticommute precisely if they are orthogonal. If $e_1, \ldots, e_m$ is any orthonormal basis for $\mathbb{R}^m$, then the 2^m monomials (including 1)

$$e_{\nu_1} e_{\nu_2} \ldots e_{\nu_k}, \qquad 1 \le \nu_1 < \cdots < \nu_k \le m$$

form a vector space basis for C_m. There is the canonical anti-automorphism

$$t\colon C_m \to C_m, \qquad (x_1 \ldots x_k)^t = x_k \cdot x_{k-1} \ldots x_1 \quad \text{for } x_\nu \in \mathbb{R}^m$$

and this yields the projection ((6.20), Ex. 2)

$$\rho\colon \mathrm{Pin}(m) \to \mathrm{O}(m), \qquad \rho(a)v = ava^t.$$

If $x \in S^{m-1}$ and $v = rx + y \in \mathbb{R}^m$ with $r \in \mathbb{R}$ and $\langle x, y\rangle = 0$, then $\rho(x)v = x(rx + y)x = rx^3 + xyx = -rx + y$. Thus $\rho(x)$ is the reflection of $\mathbb{R}^m$ in the hyperplane orthogonal to x.

The even part C_m^0 is a subalgebra of C_m, and there is an isomorphism of algebras

(6.8) $$C_{m-1} \xrightarrow{\cong} C_m^0, \qquad a^0 + a^1 \mapsto a^0 + a^1 e_m \quad \text{for } a^\nu \in C_{m-1}^\nu;$$

see I, (6.6). We are particularly interested in complex modules. We obtain the complex Clifford algebra $\mathbb{C} \otimes C_m$ from C_m by extending coefficients. Of course, the basis and relations remain unchanged.

We now come to spin representations. The group Spin(m) acts on the real vector space C_m^0 by left multiplication, and this representation cannot come via ρ^* from a representation of SO(m) since the element $-1 \in \mathrm{Spin}(m)$ does not act trivially. More generally, if V is any left module over the algebra C_m^0 or its complexification $\mathbb{C} \otimes C_m^0$, then Spin(m) acts on V by left multiplication and this yields a representation. We will obtain the still missing half-spin representations through such C_m^0-modules. Therefore we are interested in the structure of these real algebras and hence by (6.8) in the structure of the Clifford algebra C_m.

We know from I, (6.3) that

$$C_0 = \mathbb{R}, \qquad C_1 = \mathbb{C}, \qquad C_2 = \mathbb{H}.$$

Let $K(n)$ denote the algebra of $(n \times n)$-matrices over K, where K is $\mathbb{R}$, $\mathbb{C}$, or $\mathbb{H}$. We then have the following standard isomorphisms of real algebras:

(6.9)
$$\begin{array}{ll} \text{(i)} & \mathbb{R}(n) \otimes_{\mathbb{R}} K \cong K(n),\\ \text{(ii)} & \mathbb{R}(n) \otimes_{\mathbb{R}} \mathbb{R}(m) \cong \mathbb{R}(nm),\\ \text{(iii)} & \mathbb{C} \otimes_{\mathbb{R}} \mathbb{C} \cong \mathbb{C} \times \mathbb{C},\\ \text{(iv)} & \mathbb{H} \otimes_{\mathbb{R}} \mathbb{C} \cong \mathbb{C}(2),\\ \text{(v)} & \mathbb{H} \otimes_{\mathbb{R}} \mathbb{H} \cong \mathbb{R}(4). \end{array}$$

These isomorphisms are defined as follows:

(i) $(a_{ij}) \otimes k \mapsto (ka_{ij})$.
(ii) The map is induced by the isomorphism $\mathbb{R}^n \otimes \mathbb{R}^m \cong \mathbb{R}^{nm}$.
(iii) $z \otimes w \mapsto (zw, \bar{z}w)$.
(iv) View $\mathbb{H}$ as a subalgebra of $\mathbb{C}(2)$. So, if $A = (a_{ij}) \in \mathbb{H}$ is a complex (2×2)-matrix, then $(a_{ij}) \otimes z \mapsto (z \cdot a_{ij})$.
(v) $A \otimes B$ is sent to the real-linear map $\mathbb{H} \to \mathbb{H}$, $X \mapsto AX{}^*B$, and $\mathbb{H} \cong \mathbb{R}^4$. Note that the adjoint complex matrix ${}^*B$ is conjugate to B in $\mathbb{H}$.

It is easy to check that the maps given are actually isomorphisms ((6.20), Ex. 5).

The Clifford algebras C_n may now be computed as follows:

Let $C'_n = C(-Q_n)$ be the Clifford algebra associated to the quadratic form

$$-Q_n\colon \mathbb{R}^n \to \mathbb{R}, \qquad x \mapsto |x|^2,$$

giving the square of the Euclidean absolute value. Then C'_n is generated by basis elements e'_ν, $\nu = 1, \ldots, n$, with the relations

$$(e'_\nu)^2 = 1, \qquad e'_\nu e'_\mu + e'_\mu e'_\nu = 0 \quad \text{for } \nu \neq \mu.$$

(6.10) Proposition. *There are isomorphisms*

$$C_n \otimes C'_2 \cong C'_{n+2},$$
$$C'_n \otimes C_2 \cong C_{n+2}.$$

PROOF. Let V' denote the linear subspace of C'_{n+2} spanned by the e'_ν. Consider the linear map

$$\psi\colon V' \to C_n \otimes C'_2,$$
$$\psi(e'_\nu) = \begin{cases} e_{\nu-2} \otimes e'_1 e'_2 & \text{for } 2 < \nu \leq n+2, \\ 1 \otimes e'_\nu & \text{for } \nu = 1, 2. \end{cases}$$

This map satisfies the hypotheses for the universal property I, (6.1) of the Clifford algebra $C'_n = C(-Q_n)$, as may easily be verified. Therefore ψ extends to a homomorphism of algebras

$$\psi\colon C'_{n+2} \to C_n \otimes C'_2.$$

It is clear that ψ is surjective since the $\psi(e'_\nu)$, $\nu = 1, \ldots, n+2$, generate $C_n \otimes C'_2$. Since both algebras have the same dimension 2^{n+2}, we conclude that ψ is an isomorphism. Exchanging primed for unprimed symbols in the proof will give the second isomorphism. □

Now, we have

$$\text{(6.11)} \qquad \begin{matrix} C_0 = \mathbb{R}, & C_1 = \mathbb{C}, & C_2 = \mathbb{H}, \\ C'_0 = \mathbb{R}, & C'_1 = \mathbb{R} \times \mathbb{R}, & C'_2 = \mathbb{R}(2). \end{matrix}$$

The last two isomorphisms in the second row are given by

$$e'_1 \mapsto (1, -1) \quad \text{and} \quad e'_1 \mapsto \begin{pmatrix} 0 & 1 \\ 1 & 0 \end{pmatrix}, \qquad e'_2 \mapsto \begin{pmatrix} 1 & 0 \\ 0 & -1 \end{pmatrix}.$$

Starting with these data we may use (6.9) and the inductive procedure in (6.10) to compute the following table of Clifford algebras:

(6.12)

n	C_n	C'_n	$\mathbb{C} \otimes C_n \cong \mathbb{C} \otimes C'_n$
0	$\mathbb{R}$	$\mathbb{R}$	$\mathbb{C}$
1	$\mathbb{C}$	$\mathbb{R} \times \mathbb{R}$	$\mathbb{C} \times \mathbb{C}$
2	$\mathbb{H}$	$\mathbb{R}(2)$	$\mathbb{C}(2)$
3	$\mathbb{H} \times \mathbb{H}$	$\mathbb{C}(2)$	$\mathbb{C}(2) \times \mathbb{C}(2)$
4	$\mathbb{H}(2)$	$\mathbb{H}(2)$	$\mathbb{C}(4)$
5	$\mathbb{C}(4)$	$\mathbb{H}(2) \times \mathbb{H}(2)$	$\mathbb{C}(4) \times \mathbb{C}(4)$
6	$\mathbb{R}(8)$	$\mathbb{H}(4)$	$\mathbb{C}(8)$
7	$\mathbb{R}(8) \times \mathbb{R}(8)$	$\mathbb{C}(8)$	$\mathbb{C}(8) \times \mathbb{C}(8)$
8	$\mathbb{R}(16)$	$\mathbb{R}(16)$	$\mathbb{C}(16)$

...

$$C_{n+8} \cong C_n \otimes \mathbb{R}(16), \qquad C'_{n+8} \cong C'_n \otimes \mathbb{R}(16),$$

$$\mathbb{C} \otimes C_{n+2} \cong (\mathbb{C} \otimes C_n) \otimes_{\mathbb{C}} \mathbb{C}(2).$$

For the following, the reader should always keep this important table at hand.

The periodicity follows directly from (6.10):

$$C_{n+8} \cong C_n \otimes C'_2 \otimes C_2 \otimes C'_2 \otimes C_2 \cong C_n \otimes \mathbb{R}(16).$$

In the same way, $C'_{n+8} \cong C'_n \otimes \mathbb{R}(16)$. In the third column are the complex Clifford algebras associated to the unique equivalence class of regular complex quadratic forms. Here the analogue of (6.10) reads

$$\mathbb{C} \otimes C_{n+2} \cong (\mathbb{C} \otimes C_n) \otimes_{\mathbb{C}} (\mathbb{C} \otimes C_2) \cong (\mathbb{C} \otimes C_n) \otimes_{\mathbb{C}} \mathbb{C}(2).$$

The sequence of real Clifford algebras is periodic with period 8 in that raising the index n by 8 fixes the field $\mathbb{R}$, $\mathbb{C}$, or $\mathbb{H}$ and multiplies the "dimension" by 16. Thus $C_{8k+1} \cong \mathbb{C}(16^k)$, $C_{8k+3} \cong \mathbb{H}(16^k) \times H(16^k)$, and so on. Similarly, the sequence of complex Clifford algebras is periodic with period 2.

This also answers the question of which modules exist over the algebras C_n and $\mathbb{C} \otimes C_n$. Namely, one makes the following purely formal and general remark: Let R and S be rings or, in our case, real algebras with unit (associative but not necessarily commutative) and let $R(n)$ be the ring of $(n \times n)$-matrices with coefficients in R. Let $\mathrm{Mod}(R)$ denote the category of (left) R-modules.

(6.13) Proposition. *There are equivalences of categories:*

(i) $$\mathrm{Mod}(R) \times \mathrm{Mod}(S) \cong \mathrm{Mod}(R \times S),$$

(ii) $$\mathrm{Mod}(R) \cong \mathrm{Mod}(R(n)).$$

Proof. We do not wish to get lost in formal details but here are the main points.

(i) If U is an R-module and V an S-module, then $U \times V$ is an $(R \times S)$-module in an obvious way. Conversely, every $(R \times S)$-module M is of this form with $U = (1, 0) \cdot M$ and $V = (0, 1) \cdot M$. The equivalence (i) is therefore defined on the objects by $(U, V) \Rightarrow U \times V$ and on morphisms by $(\varphi, \psi) \mapsto \varphi \times \psi$. The inverse functor is easy to write down.

(ii) If V is an R-module, then $V^n = V \oplus \cdots \oplus V$ (n summands) is an $R(n)$-module in the obvious way. And if $f: V \to U$ is an R-module homomorphism, then $f^n: V^n \to U^n, (v_1, \ldots, v_n) \mapsto (fv_1, \ldots, fv_n)$ is an $R(n)$-module homomorphism. Properly understood, the claim is that these are all the $R(n)$-modules and $R(n)$-module homomorphisms. In other words, the equivalence (ii) is defined by the functor $V \mapsto V^n, f \mapsto f^n$. The inverse functor is obtained as follows: Let $P_\nu \in R(n)$ be the matrix associated to the projection onto the νth component, so the νth column of P_ν is the νth standard unit vector and all other entries are zero. Then

$$P_\nu^2 = P_\nu, \qquad P_\nu P_\mu = 0 \quad \text{for } \nu \neq \mu, \quad \text{and} \quad \sum_{\nu=1}^{n} P_\nu = 1 \in R(n).$$

Also, there are invertible permutation matrices $A_\nu \in R(n)$ (so A_ν has coefficients in the center of R) such that

$$A_\nu P_1 = P_\nu A_\nu.$$

From this it follows that if M is an $R(n)$-module, then $M = \bigoplus_{\nu=1}^{n} P_\nu M$, and $A_\nu: P_1 M \xrightarrow{\cong} P_\nu M$ as R-modules. Thus armed, one verifies that

$$(f: M \to M') \mapsto (P_1 f \,|\, P_1 M: P_1 M \to P_1 M')$$

defines a functor $\mathrm{Mod}(R(n)) \Rightarrow \mathrm{Mod}(R)$ which, up to natural transformation, is the inverse we want. □

In our case we start with a field or skew field, and every module is therefore a direct sum of ***irreducible*** (simple) modules, i.e., modules which contain no nontrivial proper submodules; see II, §2. And, up to isomorphism, there is only one irreducible module over $\mathbb{R}$, $\mathbb{C}$, or $\mathbb{H}$. This, together with proposition (6.13) and a look at table (6.12), shows:

(6.14) Proposition. *Every module over the Clifford algebra* C_n *or* $\mathbb{C} \otimes C_n$ *is the direct sum of irreducible modules (is semisimple). Up to isomorphism there is precisely one irreducible module over* C_n *for* $n = 0, 1, 2, 4, 5, 6 \bmod 8$, *and there are precisely two for* $n = 3, 7 \bmod 8$. *Up to isomorphism there is exactly one irreducible module* D^n *over* $\mathbb{C} \otimes C_{2n}$ *and* $\dim_{\mathbb{C}} D^n = 2^n$. *Up to isomorphism there are exactly two irreducible modules* D^n_+ *and* D^n_- *over* $\mathbb{C} \otimes C_{2n-1}$ *and*

$$\dim_{\mathbb{C}} D^n_+ = \dim_{\mathbb{C}} D^n_- = 2^{n-1}. \qquad \square$$

As a $\mathbb{C}(2^n)$-module, the irreducible module D^n is the complex vector space $\mathbb{C}^{2^n}$ and is determined up to isomorphism by its dimension. The modules D^n_+ and D^n_- may be distinguished as follows: In $\mathbb{C} \otimes C_{2n-1}$ we have the element

(6.15) $\tau = i^n \cdot e_1 e_2 \ldots e_{2n-1}$ with $\tau^2 = 1$ and $\tau e_\nu = e_\nu \tau$ for $1 \le \nu \le 2n - 1$.

A module V over $\mathbb{C} \otimes C_{2n-1}$ decomposes into the eigenspaces V_+ and V_- of τ associated to the eigenvalues $+1$ and -1. Since τ is central, these are submodules of V. Neither summand is trivial in $\mathbb{C} \otimes C_{2n-1}$ considered as a left module over itself, since

$$\tau(\tau + 1) = \tau + 1, \qquad \tau(\tau - 1) = -(\tau - 1).$$

We fix the notation so that τ acts on D^n_+ via multiplication by $+1$ and on D^n_- via multiplication by -1.

If n is even, and hence $2n - 1 \equiv 3, 7 \bmod 8$, then τ is already defined in C_{2n-1}. By the same argument, any C_{2n-1}-module V decomposes into $V_+ \oplus V_-$. As in the case of $\mathbb{C} \otimes C_{2n-1}$, this allows us to distinguish the two irreducible C_{2n-1}-modules (also see (6.20), Ex. 7).

For $m \equiv 2, 3, 4 \bmod 8$, every C_m-module has a symplectic structure, that is, the structure of an $\mathbb{H}$-vector space, on which C_m acts by $\mathbb{H}$-linear maps. This is evident from table (6.12). For $m \equiv 1, 5 \bmod 8$ we see that every C_m-module has a complex structure. We have an inclusion of real algebras

$$r: C_m = \mathbb{R} \otimes C_m \to \mathbb{C} \otimes C_m \tag{6.16}$$

induced by the inclusion $\mathbb{R} \subset \mathbb{C}$. This leads to an inclusion of the corresponding algebras which are explicitly specified in table (6.12). As such, r is given in the obvious way. Namely: for $m \equiv 2, 3, 4 \bmod 8$, it is induced by the inclusion $\mathbb{H} \subset \mathbb{C}(2)$. For $m \equiv 0, 6, 7 \bmod 8$, it is induced by the inclusion $\mathbb{R} \subset \mathbb{C}$. And for $m \equiv 1, 5 \bmod 8$, it is induced by the inclusion $\mathbb{C} = \mathbb{R} \otimes \mathbb{C} \subset \mathbb{C} \otimes \mathbb{C} \cong \mathbb{C} \times \mathbb{C}$; see the definition of the isomorphisms (6.9). The inclusion r makes every $(\mathbb{C} \otimes C_m)$-module V into a C_m-module r^*V by pulling back the structure, and r^*V possesses a complex structure.

(6.17) Remark. If $m \equiv 2, 3, 4 \bmod 8$ and if D (resp. D_+, D_-) are the irreducible $(\mathbb{C} \otimes C_m)$-modules, then r^*D (resp. r^*D_+, r^*D_-) are the irreducible modules over C_m. The complex structure comes from restricting the symplectic structure to $\mathbb{C}$. If $m \equiv 0, 6, 7 \bmod 8$, then r^*D (resp. r^*D_+, r^*D_-) come from the real irreducible modules over C_m by tensoring with $\mathbb{C}$. Finally, if $m \equiv 1, 5 \bmod 8$ and if V is the irreducible module over C_m with its complex structure, we have $r^*D_+ \cong V$ and $r^*D_- \cong \bar{V}$, or vice versa. Observe that V and $\bar{V}$ are isomorphic as real C_m-modules.

PROOF. Consider table (6.12). For even m it suffices to check dimensions. For $m \equiv 3, 7 \bmod 8$, the two irreducible C_m-modules may be distinguished by means of the action of the central element $\tau \in C_m$. For $m \equiv 1, 5 \bmod 8$ the dimension is correct, and the definition of the isomorphism $\mathbb{C} \otimes_{\mathbb{R}} \mathbb{C} \to \mathbb{C} \times \mathbb{C}$, $z \otimes w \mapsto (zw, \bar{z}w)$, shows that r^*D_+ and r^*D_- have conjugate complex structures. □

(6.18) Proposition. *As a* Spin$(2n + 1)$*-module,* D^n *is the half-spin representation* Δ^n. *As* Spin$(2n)$*-modules,* D^n_+ *and* D^n_- *are the half-spin representations* Δ^n_+ *and* Δ^n_-.

PROOF. The Spin-modules D^n and $D^n_\pm$ certainly have the correct dimensions (6.13), (6.14), so we only have to show that the appropriate dominant weights ∂_n, ∂_n^+, or ∂_n^- appear among their weights. If this is so, all the conjugates under the Weyl group also appear. By reasons of dimension, these must then be all the weights and each must have multiplicity one.

Thus, for the group Spin$(2n + 1)$, all we really need to show is the existence of any $(\mathbb{C} \otimes C^0_{2n+1})$-module (and hence $(\mathbb{C} \otimes C_{2n})$-module) which, as a Spin$(2n + 1)$-module, has the weight ∂ with positive multiplicity.

Now consider $\mathbb{C} \otimes C^0_{2n+1}$ as a left-module over itself. We use the standard basis for this module consisting of 2^{2n} products of unit vectors of even degree. An element of the maximal torus $\tilde{T}(n)$ of Spin$(2n)$ or Spin$(2n + 1)$ is given by

$$\xi = (\cos(\vartheta_1/2) - e_1 e_2 \sin(\vartheta_1/2)) \cdot \ldots \cdot (\cos(\vartheta_n/2) - e_{2n-1} e_{2n} \sin(\vartheta_n/2))$$

according to IV, (3.9). The projection ρ: Spin$(2n) \to$ SO$(2n)$ maps ξ to the element in the maximal torus $T(n)$ of SO$(2n)$ and SO$(2n + 1)$ whose components are

$$\begin{bmatrix} \cos \vartheta_\nu & -\sin \vartheta_\nu \\ \sin \vartheta_\nu & \cos \vartheta_\nu \end{bmatrix}.$$

Using the coordinates as in V, (6.7), the element ξ is described in the Lie algebra $LT(n) = L\tilde{T}(n)$ by

$$t = 1/2\pi \cdot (\vartheta_1, \ldots, \vartheta_n) \bmod \tilde{I}^n$$

with $\tilde{I}^n = \{(\zeta_1, \ldots, \zeta_n) \in \mathbb{Z}^n \mid \sum_\nu \zeta_\nu \in 2\mathbb{Z}\}$.

Under our choice of basis of $\mathbb{C} \otimes C^0_{2n+1}$, the action of ξ is given by a $(2^{2n} \times 2^{2n})$-matrix whose diagonal entries are obviously all equal to $\cos(\vartheta_1/2) \cdot \ldots \cdot \cos(\vartheta_n/2)$. Therefore, as a representation of the maximal torus $\tilde{T}(n)$ of Spin$(2n+1)$, the module $\mathbb{C} \otimes C^0_{2n+1}$ has character

$$\begin{aligned} &2^{2n} \cdot \cos(\vartheta_1/2) \cdot \ldots \cdot \cos(\vartheta_n/2) \\ &\quad = 2^n(e^{i\vartheta_1/2} + e^{-i\vartheta_1/2}) \cdot \ldots \cdot (e^{i\vartheta_n/2} + e^{-i\vartheta_n/2}) \\ &\quad = 2^n \exp(i\vartheta_1/2 + \ldots + i\vartheta_n/2) + \ldots = 2^n \exp(2\pi i\, \partial_n(t)) + \ldots . \end{aligned}$$

Thus ∂_n is a weight of $\mathbb{C} \otimes C^0_{2n+1}$ with multiplicity 2^n. Naturally, $\mathbb{C} \otimes C^0_{2n+1} \cong 2^n \Delta^n$ as a Spin$(2n+1)$-module.

The story for Spin$(2n)$ is similar: The Spin$(2n+1)$-module D^n yields the Spin$(2n)$-module $j^* D^n$ through the inclusion $j\colon \mathrm{Spin}(2n) \to \mathrm{Spin}(2n+1)$. Moreover, $j^* D^n \cong D^n_+ \oplus D^n_-$. Indeed, $2^n D^n = \mathbb{C} \otimes C^0_{2n+1}$ contains $\mathbb{C} \otimes C^0_{2n}$ as a $(\mathbb{C} \otimes C^0_{2n})$-module and this, in turn, contains both summands D^n_+ and D^n_- because $\mathbb{C} \otimes C_{2n-1}$ contains both irreducible modules as a module over itself, as has been pointed out above. Thus $j^* D^n$ must contain D^n_+ and D^n_- as summands. Counting dimensions, these occur exactly once.

Now, the calculation above shows that $j^* D^n$ has the weights

$$\tfrac{1}{2}(\varepsilon_1, \ldots, \varepsilon_n), \qquad \varepsilon_\nu = \pm 1.$$

It follows that

$$D^n_+ \oplus D^n_- = \Delta^n_+ \oplus \Delta^n_-,$$

and the question is only which is which. Consider the element $\xi \in \tilde{T}(n)$ with $\vartheta_1 = \vartheta_2 = \cdots = \vartheta_n = \pi$, which is associated to $t = \frac{1}{2}(1, \ldots, 1) \in L\tilde{T}(n)$. Then

$$\xi = (-1)^n e_1 e_2 \ldots e_{2n-1} e_{2n} = i^n \cdot \tau \cdot e_{2n},$$

with τ as in (6.15). Under the isomorphism (6.8), the element ξ corresponds to the element $i^n \tau \in \mathbb{C} \otimes C_{2n-1}$ and therefore, by definition, ξ operates on D^n_+ via multiplication by i^n. On the other hand, ξ operates on the weight space of Δ^n_+ associated to the weight $\partial^+_n = \frac{1}{2}(1, \ldots, 1)$ via multiplication by $\exp(2\pi i\, \partial t) = \exp(2\pi i n/4) = \exp(\pi i/2)^n = i^n$. □

We can now read off the types of the half-spin representations from (6.17). Note the shift of dimensions caused by $\mathrm{Spin}(m) \subset C^0_m \cong C_{m-1}$.

(6.19) Proposition. *For* $m \equiv \nu \bmod 8$, *the irreducible half-spin representations* Δ *or* Δ_+, Δ_- *of* Spin(m) *have the following type*:

ν	0	1	2	3	4	5	6	7
Type	$\mathbb{R}$	$\mathbb{R}$	$\mathbb{C}$	$\mathbb{H}$	$\mathbb{H}$	$\mathbb{H}$	$\mathbb{C}$	$\mathbb{R}$.

Proof. Compare (6.17) and the first column of (6.12) with ν shifted by 1. □

Even though we consider the fundamental representations of Spin(n) to be either SO(n) representations or as coming from modules over $C_{n-1} \cong C_n^0 \supset \mathrm{Spin}(n)$, not every irreducible representation of Spin(n) arises in this fashion. For example, the weight $3\,\partial$ of Spin($2n+1$) belongs to a representation which cannot come from a Clifford module (since then the representation would be a multiple of Δ) or from SO($2n+1$) (since $3\,\partial$ is integral only in $L\tilde{T}(n)$ and not in $LT(n)$).

(6.20) Exercises

1. Compute the polynomials F in the proof of (6.6).
2. Show that $ava^t = \alpha(a)va^{-1}$ for $a \in \mathrm{Pin}(m)$, $v \in \mathbb{R}^m$.
3. What happens to the fundamental representations under the map $R\,\mathrm{Spin}(m+1) \to R\,\mathrm{Spin}(m)$ induced by the inclusion of groups for $m = 2n$ or $2n+1$?
4. Describe $R\mathrm{Spin}(m)$ as a module over $R\mathrm{SO}(m)$ analogously to (6.6).
5. Show that the maps given in (6.9) are isomorphisms of real algebras.
6. Let K be a field and let $K(n)$ be the K-algebra of $(n \times n)$-matrices with coefficients in K. Give a decomposition of the $K(n)$-module $K(n)$ into irreducible modules.
7. Show that the decomposition of a $(\mathbb{C} \otimes C_{2n-1})$-module V into the eigenspaces V_+ and V_- of τ corresponds to the decomposition (6.13)(i) of V into two $\mathbb{C}(2^{n-1})$-modules via the isomorphism $\varphi\colon \mathbb{C} \otimes C_{2n-1} \cong \mathbb{C}(2^{n-1}) \times \mathbb{C}(2^{n-1})$. Where can τ be sent by φ?

For the following Exercises 8–11 compare Löffler in Karoubi *et al.* [1].

8. Let A be a $\mathbb{Z}/2$-graded algebra; see I, §6. Grade $A^n = \bigoplus_{\nu=1}^n A$ by
$$(A^n)^0 = A^0 \oplus A^1 \oplus A^0 \oplus \cdots \quad (n \text{ summands}),$$
$$(A^n)^1 = A^1 \oplus A^0 \oplus A^1 \oplus \cdots \quad (n \text{ summands}).$$
Let $A(n)$ be the algebra of $(n \times n)$-matrices with coefficients in A with the chessboard grading: $(a_{ij}) \in A(n)^k$ with $k = 0, 1$ if and only if $a_{ij} \in A^l$ for $l \equiv i + j + k \bmod 2$, $i, j = 1, \ldots, n$. Show that A^n is a $\mathbb{Z}/2$-graded module over $A(n)$; in other words,
$$(A(n))^\mu \cdot (A^n)^\nu \subset (A^n)^{\mu+\nu} \quad \text{for} \quad \mu, \nu, \mu + \nu \in \mathbb{Z}/2.$$
9. Let $C_{p,q}$ be the real Clifford algebra associated to the quadratic form
$$\mathbb{R}^{p+q} = \mathbb{R}^p \times \mathbb{R}^q \to \mathbb{R}, \qquad (x, y) \mapsto |x|^2 - |y|^2.$$
Show that there is an isomorphism of graded algebras $C_{1,1} \cong \mathbb{R}(2)$. For this $\mathbb{R}$ is graded such that only 0 is odd, and $\mathbb{R}(2)$ is then graded as in Exercise 8.
10. Let A be a $\mathbb{Z}/2$-graded algebra and $\mathbb{R}(n)$ be graded as in Exercise 8. Show that there is an isomorphism of graded algebras $\varphi\colon \mathbb{R}(n) \hat{\otimes} A \cong \mathbb{R}(n) \otimes A \cong A(n)$. *Hint*: Grade $\mathbb{R}^n$ as in Exercise 8. Show that $\mathbb{R}(n) \hat{\otimes} A$ acts on $\mathbb{R}^n \otimes A$ by
$$(B \hat{\otimes} \alpha)(v \otimes a) = (-1)^{\mu\nu} Bv \otimes \alpha a$$
for $\alpha \in A^\mu$ and $v \in (\mathbb{R}^n)^\nu$. Then set $\varphi(B \hat{\otimes} \alpha) = B_\alpha \otimes \alpha$ with $B_\alpha v = (-1)^{\mu\nu} Bv$.

11. Compute all the algebras $C_{p,q}$ with the help of Exercises 9 and 10. In particular, you should find that $C_{1,3} \cong \mathbb{H}(2)$. This is the Clifford algebra for Minkowski space. Complexification yields an embedding $C_{1,3} \to \mathbb{C} \otimes C_{1,3} \cong \mathbb{C} \otimes C_4 \cong \mathbb{C}(4)$. The basis elements e_0, e_1, e_2, e_3 are sent to four complex (4×4)-matrices $\gamma_0, \gamma_1, \gamma_2, \gamma_3$ with $\gamma_1^2 = \gamma_2^2 = \gamma_3^2 = -E$, $\gamma_0^2 = E$ and $\gamma_\nu\gamma_\mu = -\gamma_\mu\gamma_\nu$ for $\mu \neq \nu$. Such a quadruple is called a system of ***Dirac matrices***. Write down an explicit system of Dirac matrices. *Hint*: Consider the elements $1 \otimes e_1, i \otimes e_2, j \otimes e_2, k \otimes e_2$ in $\mathbb{H}(2) \cong \mathbb{H} \otimes C_2'$.

12. Show that $\mathrm{Spin}(4) \cong \mathrm{Spin}(3) \times \mathrm{Spin}(3)$ as follows: Let A^* denote the group of units in any given algebra A. There is an injective homomorphism $\mathrm{Spin}(4) \subset (C_4^0)^* \cong C_3^* \cong \mathbb{H}^* \times \mathbb{H}^*$. Show that the image of Spin(4) is $\mathrm{Sp}(1) \times \mathrm{Sp}(1) \cong \mathrm{Spin}(3) \times \mathrm{Spin}(3)$, where Sp(1) is the group of elements of norm 1 in $\mathbb{H}^*$. The inverse isomorphism may be described as follows: $\mathrm{Sp}(1) \times \mathrm{Sp}(1)$ acts on $\mathbb{R}^4 \cong \mathbb{H}$ by $(a, b)h = ahb^{-1}$. Show that this defines a surjective homomorphism $\mathrm{Sp}(1) \times \mathrm{Sp}(1) \to \mathrm{SO}(4)$ which, by the theory of coverings, may be lifted to an isomorphism $\mathrm{Sp}(1) \times \mathrm{Sp}(1) \to \mathrm{Spin}(4)$. *Hint*: Show that the action of $\mathrm{Sp}(1) \times \mathrm{Sp}(1)$ on $\mathbb{H}$ is norm-preserving and transitive, and that all rotations about the $\mathbb{R}$-axis are in the image of $\mathrm{Sp}(1) \times \mathrm{Sp}(1) \to \mathrm{SO}(4)$; see I, (6.18).

13. Show that $\mathrm{Spin}(6) \cong \mathrm{SU}(4)$ as follows: There is an injective homomorphism $\mathrm{Spin}(6) \subset (C_6^0)^* \cong C_5^* \cong \mathbb{C}(4)^* = GL(4, \mathbb{C})$. The image of Spin(6) is compact and has the right dimension. The same argument yields the isomorphism $\mathrm{Spin}(3) \cong \mathrm{Sp}(1)$ which we explicitly described in I, (6.18). Also see V, (8.7), Ex. 4.

7. Representations of the Orthogonal Groups

Since the group $O(2n + 1)$ is isomorphic to the direct product $SO(2n + 1) \times \mathbb{Z}/2$, we may determine its irreducible representations from those of $SO(2n + 1)$ and $\mathbb{Z}/2$; see II, (4.14). In particular, there is an isomorphism

(7.1) $$RO(2n + 1) \cong RSO(2n + 1) \otimes R(\mathbb{Z}/2)$$

of representation rings; see (7.7). Furthermore $R(\mathbb{Z}/2) \cong \mathbb{Z}[\omega]/(\omega^2 - 1)$, and, in the isomorphism (7.1), the element $1 \otimes \omega$ corresponds to the determinant representation Λ^{2n+1}, i.e., the $(2n + 1)$st exterior power of the standard representation of $O(2n + 1)$ on $\mathbb{C}^{2n+1} = \mathbb{C} \otimes \mathbb{R}^{2n+1}$. Thus using (5.4)(iii) we may write

(7.2) $$RO(2n + 1) \cong \mathbb{Z}[\Lambda^1, \ldots, \Lambda^n, \Lambda^{2n+1}]/I,$$

where I is the ideal generated by $(\Lambda^{2n+1})^2 - 1$. Since the representations Λ^ν are real, all representations of $O(2n + 1)$ are real ((4.9), Ex. 6).

We will now investigate the representations of $O(2n)$. Since some irreducible representations of $O(2n)$ are induced from $SO(2n)$-representations, we first make a general remark on induced representations.

Inducing representations of G from those of a subgroup H is an important means of constructing irreducible representations; particularly if H has finite index in G. We will look more closely at the case where H is a normal subgroup of G with prime index p, and we will compare the irreducible representations of H with those of G. First some notation:

Let U be an H-module and $g \in G$. Then the H-module U_g (U ***twisted*** by g) is defined to be U as an abelian group, with the H-action

$$H \times U \to U, \qquad (h, u) \mapsto ghg^{-1}u.$$

The isomorphism type of U_g depends only on the coset $x = gH \in G/H$, so it is simply denoted by U_x with $x = gH$. With this notation, $(U_x)_y = U_{xy}$ and $U_1 = U$. In other words: The factor group G/H acts on the set of isomorphism types of (say, finite-dimensional) H-modules from the right by $(U, x) \mapsto U_x$. The isotropy group of the isomorphism type of U is either 0 or G/H. Hence the twisted isomorphism types U_x, $x \in G/H$, are either all distinct or all equal. Note that, in the following, U_x may denote an H-module or an isomorphism type of H-modules.

Let V be a complex G-module. Let x be a fixed generator of $G/H \cong \mathbb{Z}/p$, and let $\Omega(k)$ be the representation

$$G/H \times \mathbb{C} \to \mathbb{C}, \qquad (x, z) \mapsto \exp(2\pi ik/p) \cdot z.$$

This depends only on $k \in \mathbb{Z}/p$. We also consider $\Omega(k)$ to be a G-module via the projection $G \to G/H$. Thus $\Omega(k) \otimes \Omega(l) \cong \Omega(k + l)$ and $\Omega(0) = \mathbb{C}$. Hence the G-modules $V \otimes \Omega(k)$, $k \in \mathbb{Z}/p$, are either all isomorphic, or V gives rise to p mutually nonisomorphic G-modules

$$V \cong V \otimes \Omega(0), \qquad V \otimes \Omega(1), \ldots, V \otimes \Omega(p - 1).$$

Note that if U is an irreducible H-module, then so is U_x for all $x \in G/H$, and if V is an irreducible G-module, then so is $V \otimes \Omega(k)$ for all $k \in \mathbb{Z}/p$.

We use these considerations to partition the irreducible complex H-modules and G-modules into two types:

	type I	type II
H-modules U	all U_x isomorphic	all U_x distinct
G-modules V	all $V \otimes \Omega(k)$ distinct	all $V \otimes \Omega(k)$ isomorphic

The reason for this arrangement will be apparent from the next theorem. Let $\text{res}_H V$ denote the restriction of a G-module V to H, and let $\text{ind}_H U$ be the G-module induced from the H-module U.

(7.3) Theorem.

(i) *If $V \in \operatorname{Irr}(G, \mathbb{C})$ is of type* I, *then* $\operatorname{res}_H V = U$ *is irreducible of type* I *and* $\operatorname{ind}_H U \cong \bigoplus_{k \in \mathbb{Z}/p} V \otimes \Omega(k)$. *If $V \in \operatorname{Irr}(G, \mathbb{C})$ is of type* II, *then* $\operatorname{res}_H V \cong \bigoplus_{x \in G/H} U_x$ *with U_x irreducible of type* II *and* $\operatorname{ind}_H U_x \cong V$ *for all* $x \in G/H$.

(ii) *If $U \in \operatorname{Irr}(H, \mathbb{C})$ is of type* I, *then* $\operatorname{ind}_H U \cong \bigoplus_{k \in \mathbb{Z}/p} V \otimes \Omega(k)$ *with V irreducible of type* I *and* $\operatorname{res}_H(V \otimes \Omega(k)) \cong U$ *for all* $k \in \mathbb{Z}/p$. *If $U \in \operatorname{Irr}(H, \mathbb{C})$ is of type* II, *then* $\operatorname{ind}_H U = V$ *is irreducible of type* II *and* $\operatorname{res}_H V = \bigoplus_{x \in G/H} U_x$.

(iii) *Let* $V, V' \in \operatorname{Irr}(G, \mathbb{C})$ *and suppose* $V \not\cong V' \otimes \Omega(k)$ *for all* $k \in \mathbb{Z}/p$. *Then* $\operatorname{res}_H V$ *and* $\operatorname{res}_H V'$ *have no irreducible summand in common.*

(iv) *Let* $U, U' \in \operatorname{Irr}(H, \mathbb{C})$ *and suppose* $U \not\cong U'_x$ *for all* $x \in G/H$. *Then* $\operatorname{ind}_H U$ *and* $\operatorname{ind}_H U'$ *have no irreducible summand in common.*

The proof of the theorem is based on the following:

(7.4) Lemma. *Let U be a complex H-module and V a complex G-module. Then*

(i)
$$\operatorname{res}_H \operatorname{ind}_H U \cong \bigoplus_{x \in G/H} U_x,$$

(ii)
$$\operatorname{ind}_H \operatorname{res}_H V \cong \bigoplus_{k \in \mathbb{Z}/p} V \otimes \Omega(k).$$

Proof. Let H act from the left on G by $(h, g) \mapsto gh^{-1}$ for $h \in H$ and $g \in G$. Then one may take $\operatorname{ind}_H U$ as the vector space $C^0_H(G, U)$ of all continuous H-maps from G to U. The G-action is given by $(gf)(x) = f(g^{-1}x)$; see III, §6. Now, an H-map is determined by its values on a system of representatives of G modulo H. Since H is normal in G, the left operation of H on G/H is trivial, and the H-space of all H-maps $gH \to U$ is isomorphic to the H-module U_d, $d = g^{-1}$, via $f \mapsto f(g)$. This shows (i).

Next, consider the group ring $\mathbb{C}[G/H]$ as a G-module and a G/H-module. Then, in our case, $G/H \cong \mathbb{Z}/p$, and hence $\mathbb{C}[G/H] \cong \bigoplus_k \Omega(k)$. This comes from III, (1.6), applied to the group $\mathbb{Z}/p$; compare III, (1.8), Ex. 12. Thus we must show that $\operatorname{ind}_H \operatorname{res}_H V$ is isomorphic to $\mathbb{C}[G/H] \otimes V$. So let $f\colon G \to V$ be an H-map, that is $h \cdot f(gh) = f(g)$ or $gh \cdot f(gh) = g \cdot f(g)$ for all $g \in G$ and $h \in H$. Then $g \cdot f(g)$ depends only on the coset gH, and

$$\varphi(g) = \sum_{gH \in G/H} gH \otimes g \cdot f(g)$$

is a well-defined element of $\mathbb{C}[G/H] \otimes V$. In this way we obtain a G-isomorphism

$$\varphi\colon C^0_H(G, V) \to \mathbb{C}[G/H] \otimes V.$$

This proves (ii). □

Proof of Theorem (7.3). We will use the notation

$$\langle A, B\rangle_H = \dim \operatorname{Hom}_H(A, B)$$

for H-modules A and B. This coincides with the inner product $\langle \chi_A, \chi_B\rangle$ of the corresponding characters of H-modules. Note that, if U is an H-module and V is a G-module, then

$$\langle U, \operatorname{res}_H V\rangle_H = \langle \operatorname{ind}_H U, V\rangle_G$$

by Frobenius reciprocity III, (6.2).

Now let $V \in \operatorname{Irr}(G, \mathbb{C})$. We will first show that one of the following two cases holds:

First Case: $\operatorname{res}_H V = U$ is irreducible of type I.
Second Case: $\operatorname{res}_H V \cong \bigoplus_{x \in G/H} U_x$, and all U_x are distinct (type II).

Indeed, if $\operatorname{res}_H V = U$ is irreducible, then U is of type I, since $U_x \cong \operatorname{res}_H V_x \cong \operatorname{res}_H V$. Conversely, suppose $\operatorname{res}_H V$ is reducible. Thus

$$\operatorname{res}_H V = U_1 \oplus \cdots \oplus U_j, j > 1,$$

and the U_i are irreducible H-modules. Then $\bigoplus_k V \otimes \Omega(k) \cong \bigoplus_{i=1}^j \operatorname{ind}_H U_i$ by (7.4)(ii). Each $V \otimes \Omega(k)$ is an irreducible G-module. Hence there is a k such that $V \otimes \Omega(k)$ is contained in $\operatorname{ind}_H U_1$. But then the H-module $\bigoplus_{i=1}^j U_i = \operatorname{res}_H V \cong \operatorname{res}_H(V \otimes \Omega(k))$ is contained in $\operatorname{res}_H \operatorname{ind}_H U_1 = \bigoplus_x (U_1)_x$. This shows that all the U_i, $i = 1, \ldots, j$ are twisted modules of $U = U_1$. Also, all U_x, $x \in G/H$, must appear among the U_i, since $\operatorname{res}_H V \cong \operatorname{res}_H V_x$ for all x. Now $\operatorname{ind}_H \operatorname{res}_H V$ splits into p irreducible summands by (7.4)(ii). Thus the only alternative to the second case above would be $\operatorname{res}_H V = p \cdot U$ with U of type I. However in that case we would have

$$\begin{aligned} p^2 = \langle pU, pU\rangle_H &= \langle \operatorname{res}_H V, \operatorname{res}_H V\rangle_H = \langle V, \operatorname{ind}_H \operatorname{res}_H V\rangle_G \\ &= \sum_k \langle V, V \otimes \Omega(k)\rangle_G \le p, \end{aligned}$$

which is absurd.

We proceed to show (i). In the first case above we have

$$1 = \langle \operatorname{res}_H V, \operatorname{res}_H V\rangle_H = \sum_k \langle V, V \otimes \Omega(k)\rangle_G.$$

Hence the $V \otimes \Omega(k)$ must all be distinct and V is of type I. Note that in this case, if $U = \operatorname{res}_H V$, then $\operatorname{ind}_H U = \bigoplus_k V \otimes \Omega(k)$ splits into p distinct summands. In the second case we have

$$p = \langle \operatorname{res}_H V, \operatorname{res}_H V\rangle_H = \sum_k \langle V, V \otimes \Omega(k)\rangle_G.$$

Hence all $V \otimes \Omega(k)$ are isomorphic and V is of type II. Note that in this case $\operatorname{ind}_H U_x = V$ for all $x \in G/H$.

Proof of (ii). Given $U \in \operatorname{Irr}(H, \mathbb{C})$, we choose $V \in \operatorname{Irr}(G, \mathbb{C})$ so that U is contained in $\operatorname{res}_H V$; see III, (4.5). Then (ii) easily follows from (i).

Proof of (iii). We have

$$\langle \operatorname{res}_H V, \operatorname{res}_H V' \rangle_H = \langle V, \operatorname{ind}_H \operatorname{res}_H V' \rangle_G = \sum_k \langle V, V' \otimes \Omega(k) \rangle_G = 0.$$

The proof of (iv) is similar. □

Suppose, now, that the irreducible representations of the subgroup H are known. Then, according to the theorem, we may itemize all irreducible representations of G as follows:

Each irreducible H-representation U of type I gives rise to p different irreducible G-representations $V \otimes \Omega(0), \ldots, V \otimes \Omega(p-1)$, all of which have restriction U.

Each "twisting class" $\{U_x \mid x \in G/H\}$ of irreducible H-representations of type II gives rise to one further irreducible G-representation $\operatorname{ind}_H U$.

(7.5) We apply this to the inclusion $H = \mathrm{SO}(2n) \subset \mathrm{O}(2n) = G$. According to (5.5), the irreducible representations of $\mathrm{SO}(2n)$, $n \geq 2$, may be labeled by their dominant weights in

$$\bar{K} \cap I^* = \{(\zeta_1, \ldots, \zeta_n) \mid \zeta_1 \geq \zeta_2 \geq \cdots \geq \zeta_{n-1} \geq |\zeta_n|, \zeta_\nu \in \mathbb{Z}\}.$$

This description is also valid for $n = 1$, where we need just one integer $\zeta = \zeta_n$ to label the irreducible representations of $\mathrm{SO}(2) \cong S^1$. Now, suppose the representation U of $\mathrm{SO}(2n)$ has dominant weight $(\zeta_1, \ldots, \zeta_n)$ and g is a generator of $\mathrm{O}(2n)/\mathrm{SO}(2n)$. Then the twisted module U_g has dominant weight $(\zeta_1, \ldots, \zeta_{n-1}, -\zeta_n)$. Hence the representations of type I are those with $\zeta_n = 0$. Each of these is a restriction of two nonisomorphic irreducible $\mathrm{O}(2n)$-representations V and $V \otimes \Omega(1)$. On the other hand, if $\zeta_n \neq 0$, then the irreducible representations with dominant weights $(\zeta_1, \ldots, \zeta_n)$ and $(\zeta_1, \ldots, \zeta_{n-1}, -\zeta_n)$ yield the same irreducible induced representation of $\mathrm{O}(2n)$.

Finally, we determine the representation ring of $\mathrm{O}(2n)$. Let Λ^i denote the ith exterior power of the standard representation of $\mathrm{O}(2n)$ on $\mathbb{C}^{2n} = \mathbb{C} \otimes \mathbb{R}^{2n}$. The exterior product defines a pairing

$$\wedge : \Lambda^i \otimes \Lambda^{2n-i} \to \Lambda^{2n}.$$

The adjoint of this is an isomorphism

$$d : \Lambda^i \to \operatorname{Hom}(\Lambda^{2n-i}, \Lambda^{2n}) \cong (\Lambda^{2n-i})^* \otimes \Lambda^{2n},$$

and the Euclidean inner product on $\mathbb{R}^{2n}$ yields an isomorphism $\Lambda^{2n-i} \cong (\Lambda^{2n-i})^*$. Combining these, we have an isomorphism

$$\Lambda^i \cong \Lambda^{2n-i} \otimes \Lambda^{2n}. \tag{7.6}$$

Compare the definition of the $*$-operator in (5.5)(v).

Accordingly, we have a well-defined homomorphism

$$\mu\colon S = \mathbb{Z}[\Lambda^1, \ldots, \Lambda^n, \Lambda^{2n}]/J \to \mathrm{RO}(2n),$$

where J is the ideal generated by $(\Lambda^{2n})^2 - 1$ and $\Lambda^n\Lambda^{2n} - \Lambda^n$.

(7.7) Theorem. *The homomorphism μ is an isomorphism.*

(7.8) Corollary. *All representations of* O(2n) *are of real type, since the exterior powers are real* ((4.9), Ex. 6).

The proof of (7.7) will be preceded by two lemmas. Consider the inclusion

$$\mathrm{SO}(2n-1) \times \mathbb{Z}/2 \to \mathrm{O}(2n)$$

of the subgroup consisting of the matrices $\begin{pmatrix} A & 0 \\ 0 & \pm 1 \end{pmatrix}$, $A \in \mathrm{SO}(2n-1)$. We also have the inclusion $\mathrm{SO}(2n) \subsetneq \mathrm{O}(2n)$.

(7.9) Lemma. *The homomorphism*

$$\sigma = (\sigma_1, \sigma_2)\colon \mathrm{RO}(2n) \to \mathrm{RSO}(2n) \times \mathrm{R}(\mathrm{SO}(2n-1) \times \mathbb{Z}/2),$$

which is induced by these two inclusions, is injective.

PROOF. The group O(2n) has two conjugacy classes of Cartan subgroups in the sense of IV, §4, and the two subgroups above contain representatives of these two classes; see (7.11), Ex. 2. □

In (6.4) we have seen that RSO(2n) contains the polynomial ring $\mathbb{Z}[\Lambda^1, \ldots, \Lambda^n]$.

(7.10) Lemma. *The polynomial ring $\mathbb{Z}[\Lambda^1, \ldots, \Lambda^n]$ is the image of the restriction homomorphism* RO(2n) → RSO(2n).

PROOF. Clearly, the image contains this polynomial ring. On the other hand, the automorphism φ from (6.5) is induced via conjugation by an element of O(2n). Hence the image of RO(2n) is contained in the fixed ring of φ. □

PROOF OF THEOREM (7.7). We show that μ is injective by showing that $\sigma \circ \mu$ is injective. So let $x \in S$ be in the kernel of $\sigma \circ \mu$. Then x has a representative of the form

$$x = P_1(\Lambda^1, \ldots, \Lambda^n) + P_2(\Lambda^1, \ldots, \Lambda^{n-1})\Lambda^{2n},$$

where P_1 and P_2 are polynomials in the indicated indeterminates with integral coefficients. Now σ_1 maps Λ^{2n} to 1, and since $\sigma_1\mu(x) = 0$, we conclude from (7.10) that $P_1(\Lambda^1, \ldots, \Lambda^n) + P_2(\Lambda^1, \ldots, \Lambda^{n-1}) = 0$. Thus $P_1 = -P_2$ is independent of Λ^n. As for the other component in (7.9), we have $\mathrm{R}(\mathrm{SO}(2n-1) \times \mathbb{Z}/2) \cong \mathrm{RSO}(2n-1) \otimes \mathrm{R}(\mathbb{Z}/2)$, and

$$\mathrm{R}(\mathbb{Z}/2) = \mathbb{Z}[\omega]/(\omega^2 - 1).$$

Thus $R(SO(2n-1) \times \mathbb{Z}/2)$ is a free $RSO(2n-1)$-module with $1, \omega$ as a basis. Also σ_2 maps Λ^{2n} to ω, and since

$$0 = \sigma_2 \mu(x) = P_1(\Lambda^1, \ldots, \Lambda^{n-1}) - P_1(\Lambda^1, \ldots, \Lambda^{n-1})\omega,$$

we conclude that the polynomial P_1 is zero. Hence $x = 0$. This shows injectivity.

We now show that μ is surjective. Let $x \in RO(2n)$ be given. Then $\sigma_1(x) = P(\Lambda^1, \ldots, \Lambda^n) \in RSO(2n)$ by (7.10). We may consider this polynomial P as an element of S, and then $\sigma_1(x - \mu P) = 0$. Thus, replacing x by $x - \mu P$, we may suppose $\sigma_1(x) = 0$. Next,

$$\sigma_2(x) = Q_1(\Lambda^1, \ldots, \Lambda^{n-1}) + Q_2(\Lambda^1, \ldots, \Lambda^{n-1})\omega,$$

and $Q_2 = -Q_1$ since $\sigma_1(x) = 0$. Thus $\sigma_2(x) = Q_1 - Q_1\omega$. We may consider Q_1 as an element of S. Setting $y = Q_1 - Q_1\Lambda^{2n} \in S$, we have $x = \mu(y)$ since $\sigma(x) = \sigma(\mu y) = (0, Q_1 - Q_1\omega)$, and σ is injective. □

Note that this computation is quite elementary and does not use the discussion on induced representations. It is also valid for $n = 1$, i.e., for the group $O(2)$. In this case the standard representation Λ^1 corresponds to the element $z + z^{-1}$ in the character ring $\mathbb{Z}[z, z^{-1}]$ of $SO(2) \cong S^1$.

(7.11) Exercises

1. The factor group $O(2n)/SO(2n) = P$ acts on $RO(2n)$. Show that there is a well-defined exact sequence

$$0 \to RSO(2n)_P \xrightarrow[\text{ind}]{} RO(2n) \xrightarrow[\cdot(1-\Lambda^{2n})]{} RO(2n) \xrightarrow[\text{res}]{} RSO(2n)^P \to 0.$$

The first term is defined to be the factor group of $RSO(2n)$ by the subgroup of all elements of the form $x - p(x)$, $p \in P$.

2. Find the Cartan subgroups of $O(2n)$ and verify the statement in the proof of (7.9).

3. Give an example of a group extension $1 \to H \to G \to \mathbb{Z}/2 \to 1$, such that there exist irreducible G-modules V of real type which have the form $V = \text{ind}_H U$ with U of complex type.

4. Consider a group extension $1 \to H \to G \to \mathbb{Z}/p \to 1$, where p is an odd prime. We use table II, (6.2). Suppose $V \in \text{Irr}(G, \mathbb{C})_K$ is of type I. Show that $\text{res}_H V \in \text{Irr}(H, \mathbb{C})_K$. Suppose $U \in \text{Irr}(H, \mathbb{C})_K$ is of type II. Show that $\text{ind}_H U \in \text{Irr}(G, \mathbb{C})_K$.

Bibliography

Adams, J. F.

[1] *Lectures on Lie Groups.* New York, Amsterdam: Benjamin, 1969.

Atiyah, M. F.

[1] Characters and cohomology of finite groups. *Inst. Hautes Études Sci. Publ. Math.*, **9**, 23–64 (1961).

Atiyah, M. F., and R. Bott

[1] A Lefschetz fixed point formula for elliptic complexes: II. Applications. *Ann. of Math.*, **88**, 451–491 (1968).

Atiyah, M. F., R. Bott, and A. Shapiro

[1] Clifford modules. *Topology*, **3**, 3–38 (1964).

Atiyah, M. F., and F. Hirzebruch

[1] Vector bundles and homogeneous spaces. *Amer. Math. Soc. Symposium in Pure Math.*, III, 7–38, 1961.

Atiyah, M. F., and G. B. Segal

[1] Equivariant *K*-theory and completion. *J. Differential Geometry*, **3**, 1–18 (1969).

Atiyah, M. F., and D. O. Tall

[1] Group representations, λ-rings, and the *J*-homomorphism. *Topology*, **8**, 253–297 (1969).

Beyl, F. R., and J. Tappe

[1] *Group Extensions, Representations and the Schur Multiplicator.* Lecture Notes in Mathematics, 958. New York, Berlin, Heidelberg: Springer-Verlag, 1982.

Boerner, H.

[1] *Darstellungen von Gruppen.* Berlin, Heidelberg, New York: Springer-Verlag, 1967, 2. Auflage.

Bott, R.

[1] Homogeneous vector bundles. *Ann. of Math.*, **66**, 203–248 (1957).

[2] The index theorem for homogeneous differential operators. In: *Differential and Combinatorial Topology. A Symposium in Honor of Marsten Morse*, pp. 167–186. Princeton: Princeton University Press, 1965.

Bourbaki, N.

[1] *Groupes et Algèbres de Lie.* Chaps. 1–9. Paris: Hermann, 1960–1983.

Bredon, G. E.
[1] *Introduction to Compact Transformation Groups*. New York, London: Academic Press, 1972.
Bröcker, Th.
[1] *Analysis in mehreren Variablen*. Stuttgart: Teubner, 1980.
Bröcker, Th., und K. Jänich
[1] *Introduction to Differential Topology*. Cambridge: Cambridge University Press, 1982; translation of *Einführung in die Differentialtopologie*. Heidelberg: Springer-Verlag, 1973.
Cartan, H., and S. Eilenberg
[1] *Homological Algebra*. Princeton: Princeton University Press, 1956.
Cartier, P.
[1] On H. Weyl's character formula. *Bull. Amer. Math. Soc.*, **67**, 228–230 (1961).
Chevalley, C.
[1] *Theory of Lie Groups, I*. Princeton: Princeton University Press, 1946.
Conner, P. E., and E. E. Floyd
[1] *Differentiable Periodic Maps*. Berlin, Göttingen, Heidelberg: Springer-Verlag, 1964.
Correspondence
[1] *Ann. of Math.*, **69**, 247–251 (1959).
Curtis, C. W., and I. Reiner
[1] *Representation Theory of Finite Groups and Associative Algebras*. New York: Interscience, 1962.
Demazure, M.
[1] *A, B, C, D, E, F*, etc. in: *Séminaire sur les Singularités des Surfaces*. Lecture Notes in Mathematics, 777. New York, Berlin, Heidelberg: Springer-Verlag, 1980.
Dieudonné, J.
[1] *Foundations of Modern Analysis*. New York: Academic Press, 1960.
[2] *Élements d'Analyse*, V. Chapitre XXI. Paris: Gauthier-Villars, 1975.
[3] *Élements d'Analyse*, VII. Chapitre XXIII. Paris: Gauthier-Villars, 1978.
Ebbinghaus, H.-D. *et al.*
[1] *Zahlen* (Grundwissen Mathematik, 1). Berlin, Heidelberg, New York: Springer-Verlag, 1983.
Freudenthal, H.
[1] Zur Berechnung der Charaktere der halbeinfachen Lieschen Gruppen I, II, III.
I: *Indag. Math.*, **16**, 269–376 (1954).
II: *ibid.*, 487–497.
III: *ibid.*, **18**, 511–514 (1956).
Freudenthal, H., and H. de Vries
[1] *Linear Lie Groups*. New York, London: Academic Press, 1969.
Gelfand, I. M., R. A. Minlos, and Z. Ya. Shapiro
[1] *Representations of the Rotation and Lorentz Groups and their Applications*. Oxford, New York: Pergamon Press, 1963.
Gelfand, I. M., and M. A. Naimark
[1] *Unitäre Darstellungen der klassischen Gruppen*. Berlin: Akademie Verlag, 1957.
Guillemin, V., and A. Pollak
[1] *Differential Topology*. Englewood Cliffs, NJ: Prentice-Hall, 1974.
Hewitt, E., and K. A. Ross
[1] *Abstract Harmonic Analysis*, I. Berlin, Heidelberg, New York: Springer-Verlag, 1963.
[2] *Abstract Harmonic Analysis*, II. Berlin, Heidelberg, New York: Springer-Verlag, 1970.

Hochschild, G. P.
[1] *The Structure of Lie Groups.* San Francisco: Holden-Day, 1965.
[2] *Basic Theory of Algebraic Groups and Lie Algebras.* New York, Heidelberg, Berlin: Springer-Verlag, 1981.
Hopf, H.
[1] Über die Abbildungen der dreidimensionalen Sphäre auf die Kugelfläche. *Math. Ann.*, **104**, 639–665 (1931).
[2] Maximale Toroide und singuläre Elemente in geschlossenen Lieschen Gruppen. *Comment. Math. Helv.*, **15**, 59–70 (1943).
Hu, Sze-Tsen
[1] *Homotopy Theory.* New York, London: Academic Press, 1959.
Humphreys, J. E.
[1] *Introduction to Lie Algebras and Representation Theory.* New York, Heidelberg, Berlin: Springer-Verlag, 1972.
[2] *Linear Algebraic Groups.* New York, Heidelberg, Berlin: Springer-Verlag, 1975.
Hunt, G. A.
[1] A theorem of Élie Cartan. *Proc. Amer. Math. Soc.*, **7**, 307–308 (1956).
Iwahori, N.
[1] On real irreducible representations of Lie algebras. *Nagoya Math. J.*, **14**, 59–83 (1959).
Iwahori, N., and Sugiura, M.
[1] A duality theorem for homogeneous manifolds of compact Lie groups. *Osaka J. Math.*, **3**, 139–153 (1966).
Jacobson, N.
[1] *Lie Algebras.* New York: Interscience, 1962.
[2] *Basic Algebra*, I. San Francisco: W. H. Freeman, 1974.
Karoubi, M. *et al.*
[1] *Séminaire Heidelberg–Saarbrücken–Strasbourg sur la K-Theorie.* Lecture Notes in Mathematics, 136, Berlin, Heidelberg, New York: Springer-Verlag, 1970.
Kirillov, A. A.
[1] *Elements of the Theory of Representations.* Berlin, Heidelberg, New York: Springer-Verlag, 1976.
Korn, G. A., and M. Korn
[1] *Mathematical Handbook for Scientists and Engineers*, 2nd ed. New York: McGraw-Hill, 1968.
Kostant, B.
[1] A formula for the multiplicity of a weight. *Trans. Amer. Math. Soc.*, **93**, 53–73 (1959).
[2] Lie algebra cohomology and the generalized Borel–Weil theorem. *Ann. of Math.*, **74**, 329–387 (1961).
Kreĭn, M. G.
[1] A principle of duality for a bicompact group and a square block-algebra. (In Russian). *Dokl. Akad. Nauk SSSR*, (N.S.) **69**, 725–728 (1949).
Lang, S.
[1] *Analysis*, II. Reading, MA.: Addison-Wesley, 1969.
[2] *Linear Algebra*, 2nd ed. Reading, MA.: Addison-Wesley, 1971.
[3] *Differential Manifolds.* Reading, MA.: Addison-Wesley, 1972.
McLeod, J.
[1] The Künneth formula in equivariant *K*-theory. *Algebraic Topology Waterloo 1978, Proceedings*, pp. 316–333. Berlin, Heidelberg, New York: Springer-Verlag, 1979.
Milnor, J.
[1] *Topology from the Differentiable Viewpoint.* Charlottesville: Virginia University Press, 1965.

Montgomery, D., and L. Zippin
[1] *Topological Transformation Groups*. New York: Interscience, 1955.
Mumford, D.
[1] *Algebraic Geometry*, I. *Complex Projective Varieties*. Berlin, Heidelberg, New York: Springer-Verlag, 1976.
Murnaghan, F. D.
[1] *The Theory of Group Representations*. Baltimore: Johns Hopkins Press, 1949.
Naimark, M. A., and A. I. Stern
[1] *Theory of Group Representations*. New York, Heidelberg, Berlin: Springer-Verlag, 1982.
Narasimhan, R.
[1] *Analysis on Real and Complex Manifolds*. Amsterdam: North-Holland, 1968.
Peter, F., and H. Weyl
[1] Die Vollständigkeit der primitiven Darstellungen einer geschlossenen kontinuierlichen Gruppe. *Math. Ann.*, **97**, 737–755 (1927).
Pittie, H. V.
[1] Homogeneous vector bundles on homogeneous spaces. *Topology*, **11**, 199–203 (1972).
Pontrjagin, L. S.
[1] *Topological Groups*. Princeton: Princeton University Press, 1939.
Topologische Gruppen. Leipzig: Teubner, 1957.
Pukansky, L.
[1] *Leçons sur les Représentations des Groupes*. Paris: Dunod, 1967.
Robert, A.
[1] *Introduction to the Representation Theory of Compact and Locally Compact Groups*. Cambridge: Cambridge University Press, 1983.
Segal, G. B.
[1] The representation ring of a compact Lie group. *Publ. Math. IHES*, **34**, 113–128 (1968).
Séminaire "Sophus LIE"
[1] 1e année 1954/1955. École Normale Supérieur. Paris, 1955.
Serre, J.-P.
[1] Représentations linéaires et espaces homogènes Kählériens des groupes de Lie compacts. *Séminaire Bourbaki, Exposé* 100 (1954).
[2] *Lie Algebras and Lie Groups*. New York: Benjamin, 1965.
[3] *Algèbres de Lie Semi-Simples Complexes*. New York: Benjamin, 1966.
[4] *Représentations Linéaires des Groupes finis*. 2nd ed. Paris: Hermann, 1971.
Spivak, M.
[1] *Calculus on Manifolds*. New York: Benjamin, 1965.
Steinberg, R.
[1] A general Clebsch–Gordan theorem. *Bull. Amer. Math. Soc.*, **67**, 406–407 (1961).
[2] On a theorem of Pittie. *Topology*, **14**, 173–177 (1975).
Stiefel, E.
[1] Über eine Beziehung zwischen geschlossenen Lieschen Gruppen und diskontinuierlichen Bewegungsgruppen euklidischer Räume und ihre Anwendung auf die Aufzählung der einfachen Lieschen Gruppen. *Comment. Math. Helv.*, **14**, 350–380 (1941–42).
[2] Kristallographische Bestimmung der Charaktere der geschlossenen Lieschen Gruppen. *Comment. Math. Helv.*, **17**, 165–200 (1944–45).
Sugiura, M.
[1] *Unitary Representations and Harmonic Analysis—An Introduction*. Kodansha Ltd.: Tokyo; Wiley: New York, 1975.
Tannaka, T.
[1] Über den Dualitätssatz der nichtkommutativen Gruppen. *Tôhoku Math. J.*, **45**, 1–12 (1939).

Tits, J.
[1] *Tabellen zu den einfachen Liegruppen und ihren Darstellungen*. Lecture Notes in Mathematics, 40. Berlin, Heidelberg, New York: Springer-Verlag, 1967.
[2] *Liesche Gruppen und Algebren*. Berlin, Heidelberg, New York: Springer-Verlag, 1983.
Vilenkin, N. Ja.
[1] *Special Functions and the Theory of Group Representations*. Trans. of AMS Monographs, 22. Providence, RI: American Mathematical Society, 1968.
van der Waerden, B. L.
[1] Stetigkeitssätze der halbeinfachen Lieschen Gruppen. *Math. Z.*, **36**, 780–786 (1933).
[2] *Gruppen von linearen Transformationen*. Ergebnisse der Math., 4. Berlin: Springer-Verlag, 1935.
Warner, G.
[1] *Harmonic Analysis on Semi-Simple Lie Groups*, I, II. Berlin, Heidelberg, New York: Springer-Verlag, 1972.
Weil, A.
[1] Demonstration topologique d'un théorème fondamental de Cartan. *C.R. Acad. Sci. Paris*, **200**, 518–520 (1935).
[2] *L'Intégration dans les Groupes Topologiques et ses Applications*. Paris: Hermann, 1941.
Weyl, H.
[1] Theorie der Darstellungen kontinuierlicher halbeinfacher Gruppen durch lineare Transformationen.
I. *Math. Zeit.* **23**, 271–309 (1925),
II. **24**, 328–376 (1926),
III. 377–395 (1926),
Nachtrag. 789–791 (1926).
[2] *The Classical Groups*. Princeton: Princeton University Press, 1946.
Whitehead, G. W.
[1] *Elements of Homotopy Theory*. New York, Heidelberg, Berlin: Springer-Verlag, 1978.
Želobenko, D. P.
[1] *Compact Lie Groups and their Representations*. Providence, RI: American Mathematical Society, 1973.
Dynkin, E. B.
[1] *Semisimple subalgebras of semisimple Lie algebras*. Am. Math. Soc. Translations, Ser. 2, **6**, 111–244 (1957).
[2] *Topological characteristics of homomorphisms of compact Lie groups*. Am. Math. Soc. Translations, Ser. 2, **12**, 301–342 (1959).
Malcev, I. I.
[1] *On semisimple subgroups of Lie groups*. Am. Math. Soc. Translations, Ser. 1, **9**, 172–213 (1962).
Mc Kay, W. G., and J. Patera
[1] *Tables of Dimensions, Indices, and Branching Rules for Representations of Simple Lie Algebras*. New York and Basel, Marcel Dekker, 1981.

Symbol Index

Subject Index

Words in italics indicate the relevant main key word with further references.

D

Graduate Texts in Mathematics

continued from page ii

119 ROTMAN. An Introduction to Algebraic Topology.
120 ZIEMER. Weakly Differentiable Functions: Sobolev Spaces and Functions of Bounded Variation.
121 LANG. Cyclotomic Fields I and II. Combined 2nd ed.
122 REMMERT. Theory of Complex Functions. *Readings in Mathematics*
123 EBBINGHAUS/HERMES et al. Numbers. *Readings in Mathematics*
124 DUBROVIN/FOMENKO/NOVIKOV. Modern Geometry—Methods and Applications. Part III.
125 BERENSTEIN/GAY. Complex Variables: An Introduction.
126 BOREL. Linear Algebraic Groups.
127 MASSEY. A Basic Course in Algebraic Topology.
128 RAUCH. Partial Differential Equations.
129 FULTON/HARRIS. Representation Theory: A First Course. *Readings in Mathematics*
130 DODSON/POSTON. Tensor Geometry.
131 LAM. A First Course in Noncommutative Rings.
132 BEARDON. Iteration of Rational Functions.
133 HARRIS. Algebraic Geometry: A First Course.
134 ROMAN. Coding and Information Theory.
135 ROMAN. Advanced Linear Algebra.
136 ADKINS/WEINTRAUB. Algebra: An Approach via Module Theory.
137 AXLER/BOURDON/RAMEY. Harmonic Function Theory.
138 COHEN. A Course in Computational Algebraic Number Theory.
139 BREDON. Topology and Geometry.
140 AUBIN. Optima and Equilibria. An Introduction to Nonlinear Analysis.
141 BECKER/WEISPFENNING/KREDEL. Gröbner Bases. A Computational Approach to Commutative Algebra.
142 LANG. Real and Functional Analysis. 3rd ed.
143 DOOB. Measure Theory.
144 DENNIS/FARB. Noncommutative Algebra.
145 VICK. Homology Theory. An Introduction to Algebraic Topology. 2nd ed.
146 BRIDGES. Computability: A Mathematical Sketchbook.
147 ROSENBERG. Algebraic *K*-Theory and Its Applications.
148 ROTMAN. An Introduction to the Theory of Groups. 4th ed.
149 RATCLIFFE. Foundations of Hyperbolic Manifolds.
150 EISENBUD. Commutative Algebra with a View Toward Algebraic Geometry.
151 SILVERMAN. Advanced Topics in the Arithmetic of Elliptic Curves.
152 ZIEGLER. Lectures on Polytopes.
153 FULTON. Algebraic Topology: A First Course.
154 BROWN/PEARCY. An Introduction to Analysis.
155 KASSEL. Quantum Groups.
156 KECHRIS. Classical Descriptive Set Theory.
157 MALLIAVIN. Integration and Probability.
158 ROMAN. Field Theory.
159 CONWAY. Functions of One Complex Variable II.
160 LANG. Differential and Riemannian Manifolds.
161 BORWEIN/ERDÉLYI. Polynomials and Polynomial Inequalities.
162 ALPERIN/BELL. Groups and Representations.
163 DIXON/MORTIMER. Permutation Groups.